C0-ALO-942

Peter Ax · Multicellular Animals Volume III

Order in Nature – System Made by Man

AKADEMIE DER WISSENSCHAFTEN UND DER LITERATUR · MAINZ

Springer
Berlin
Heidelberg
New York
Hong Kong
London
Milan
Paris
Tokyo

Peter Ax

Multicellular Animals

Order in Nature – System Made by Man

Volume III

With 110 Figures

Springer

Professor Dr. PETER AX
University of Göttingen
Institute for Zoology and Anthropology
Berliner Straße 28
37073 Göttingen, Germany

Translated by
Dr. RICHARD E. DUNMUR
Ditzinger Straße 30
71254 Ditzingen, Germany

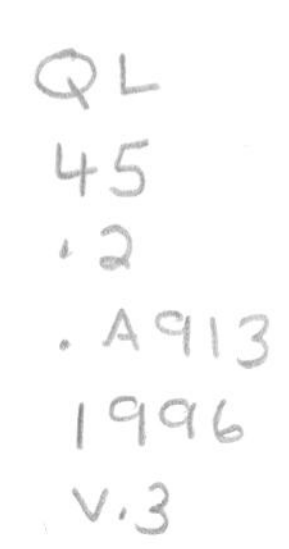

Original edition: Peter Ax. Das System der Metazoa III. Ein Lehrbuch der phylogenetischen Systematik. Spektrum Akademischer Verlag Gustav Fischer, Heidelberg. Akademie der Wissenschaften und der Literatur, Mainz 2001.

ISBN 3-540-00146-8 Springer-Verlag Berlin Heidelberg New York

Library of Congress Cataloging-in-Publication Data

Ax, Peter, 1927 – (System der Metazoa, English)
Multicellular animals: Order in Nature-System Made by Man / Peter Ax.
Includes bibliographical references and index.
ISBN 3-540-00146-8 (hardcover)
1. Metazoa. 2. Phylogeny. I. Title.
QL 45.2A913 1996 591′.012–dc20

Springer-Verlag Berlin Heidelberg New York
a member of BertelsmannSpringer Science+Business Media GmbH

http://www.springer.de

Printed in Germany

Cover design: Erich Kirchner, Heidelberg
Cover illustration. In front: *Craegrus furcatus*. Back: *Gorgonocephalus eucnemis* (Echinodermata).
Behind: *Membranipora membranacea* (Bryozoa).
Typesetting: K+V Fotosatz, Beerfelden
31/3150Wi-5 4 3 2 1 0 – Printed on acid-free paper

Preface

The turtle, toothless, soft bodied, enclosed in boney, shell, is primordial and invincible, slow and changeless. Locked into turtles of time and tradition, inertia and ignorance, darkness and fear, we cajole and pull, struggle and push until we learn that ideas set us free.

Fairbanks, Alaska
University Campus

Sculpture
Liz Biesot 1985

The evolutionary order of organisms as a product of Nature and its representation in a phylogenetic system as a construct of Man are two different things.

The order in Nature consists of relationships between organisms. It is the result of an historic process that we call phylogenesis. When working with this order we must clearly differentiate between its identification and the subsequent description of what has been identified.

When identifying the order we are placing ourselves in the framework of "hypothetical realism" (Vol. 1, p. 11); it is principally impossible to determine how well or how poorly the real world and our cognitive apparatus match. For phylogenetics as an historically directed discipline, however, the following limitation is more relevant. We cannot demonstrate experimentally whether or not that which we have interpreted from the products of phylogenesis corresponds to facts of (hypothetical) reality. Even so, we can separate ourselves from the fog of arbitrary opinions and speculation when our ideas on relationships are disguised as hypotheses that can be confirmed and supplemented at any time as a consequence of empirical experience – but that can just as easily be weakened or rejected. The phylogenetic system of organisms presents its knowledge on the order in Nature in the form of testable hypotheses and hence purports to be a science.

Thus, we come to the description or representation of order in a conceptual system of Man. In the present textbook, higher-ranking, supraspecific taxa from the Porifera to the Mammalia are identified as equivalents of phylogenetic products of Nature on the basis of derived characteristics (autapomorphies). We also present the sister group of every single taxon with demonstration of the commonly derived agreements (synapomorphies). A well-developed methodology is available in phylogenetic systematics for the evaluation of characteristics in the comparison of organisms.

There may be different, competing paths for the form of presentation. Our principle of systematization has led to the abandonment of the categories of traditional classifications. Sister groups (adelphotaxa) of concomitant origin are to be ordered next to each other on one level (coordination). Their combination to a new taxon occurs at the next higher level (superordination), their division at the next lower level (subordination). The result can be presented both as a hierarchical tabulation and as a diagram of phylogenetic relationships.

Let us take the Chordata as an example from this volume. The Tunicata and Vertebrata are coordinated at the level 2 as sister groups. The taxon Chordata is superordinated to them on level 1. Moreover, on level 3, the Acrania and Craniota are subordinated as adelphotaxa of the Vertebrata. A comparable subdivision of the Tunicata is not possible at present.

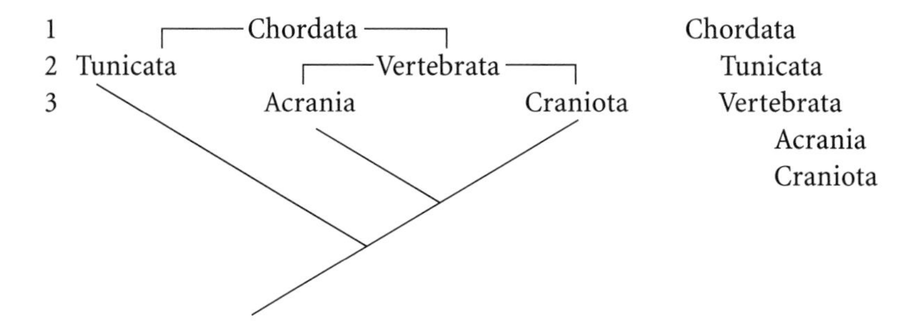

In addition to the diagram of relationships and the hierarchical tabulation I have, in an attempt to achieve optimal clarity, chosen the following measures. Adelphotaxa are always presented again in coordination in the running text when they are to be discussed. Thus, the Tunicata and Vertebrata are placed side by side on p. 125, the Acrania and Craniota on p. 170.

My sincere thanks go to Ms. Renate Grüneberg and Dipl.-Biol. Claudia Wolter who have again provided indispensable help in the preparation of this third volume.

I am extremely grateful to Dr. Carlo Servatius for including this textbook in the publication program of the Academy of Sciences and Literature in Mainz.

Dr. Richard Dunmur, Ditzingen (Germany), provided the translation of the German edition into English. I am obliged to him for the harmonious cooperation.

Peter Ax

Contents

Nemathelminthes and Syndermata

From the viewpoint of phylogenetic systematics, the traditional combination of the Gastrotricha, Nematoda, Nematomorpha, Priapulida, Loricifera, Kinorhyncha, Rotifera, and Acanthocephala to an entity with the name Nemathelminthes (Aschelminthes) is not sufficiently justified. In addition, the mutual relationships between the subtaxa remain unclear because precise questions regarding monophyly and sister group relationships have seldom been posed (LORENZEN 1985).

This situation has changed fundamentally in the last decade – through specific phylogenetic-systematic investigations on the Kinorhyncha (NEBELSICK 1993; NEUHAUS 1994), Seison and the Rotifera (AHLRICHS 1995), the Nematomorpha (SCHMIDT-RHAESA 1996a), and the Priapulida (LEMBERG 1999) as well as, lastly, the Acanthocephala (HERLYN 2000). Now we can split the previous collection of the Nemathelminthes into two higher ranking taxa, both of which can justifiably be established as monophyla – but which have practically nothing else in common. These are firstly the Nemathelminthes with the old name, but with new, stricter borders; this contains as subtaxa the Gastrotricha, Nematoda, Nematomorpha, Priapulida, Loricifera, and Kinorhyncha. The second one is the newly established entity of the Syndermata with the subtaxa Rotifera, Seison, and Acanthocephala.

Fundamental differences occur already peripherally in the construction of the integument. In their ground pattern, the Nemathelminthes possess an extracellular, two-layered cuticle that is secreted from epidermal cells. In contrast, the Syndermata are characterized by intrasyncytial thickening in the skin; here, hard substances are deposited in the epidermis.

Nemathelminthes

Autapomorphies (Fig. 1 → 1)

- Two-layered cuticle.
  Composed of a trilamellar epicuticle and a uniform basal layer (Fig. 2). The epicuticle covers the entire body; it encloses locomotory and sensory cilia of the body surface.

- Locomotory cilia limited to the ventral side.
  The complete coverage of the body with cilia in the ground pattern of the Bilateria (Vol. 1, p. 112) has developed in the stem lineage of the Nemathelminthes into a ventral "ciliary creeping sole." The cilia are reduced dorsally and laterally.

- Terminal oral aperture.
  Displacement of the intestinal opening from subventral to the anterior body pole.

- Cuticle in the pharynx.
  The cuticle of the body surface is taken internally by the pharyngeal epithelium and used to cover the lumen of the pharynx (Fig. 2).

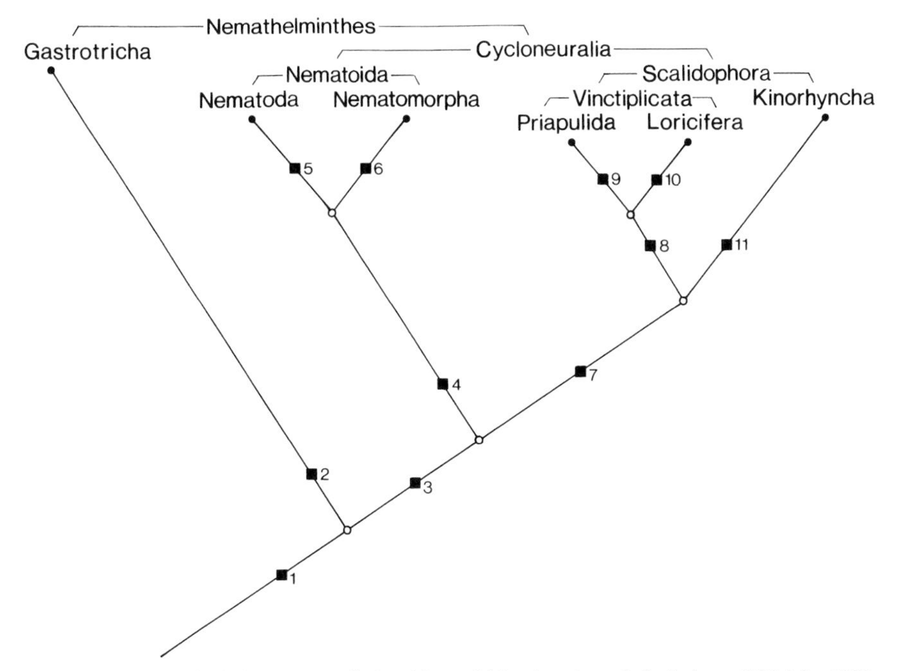

Fig. 1. Diagram of phylogenetic relationships within the Nemathelminthes. (Ahlrichs 1995; Ehlers et al. 1996; Schmidt-Rhaesa 1996a; Lemburg 1999)

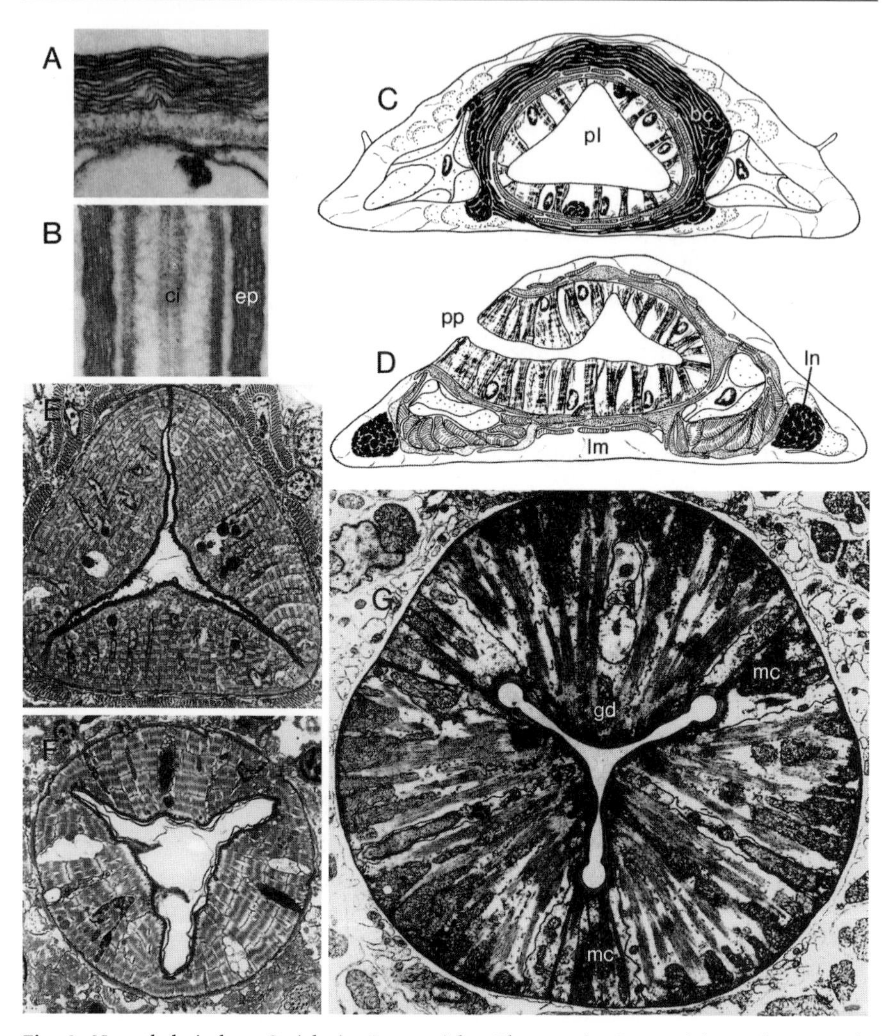

Fig. 2. Nemathelminthes. Cuticle in Gastrotricha. Pharynx in Gastrotricha and Nematoda. **A** Turbanella ocellata (Gastrotricha, Macrodasyida). Section through the dorsal cuticle. Epicuticle consisting of several layers of trilamellar units over a uniform basal layer. **B** *Macrodasys* (Gastrotricha, Macrodasyida). Longitudinal section of locomotory cilium with surrounding epicuticle. **C** and **D** *Turbanella cornuta* (Gastrotricha, Macrodasyida). Body cross sections in the pharynx region. **C** At the level of the brain commissure. Wide, triradiate lumen, with a surface to the ventral side. **D** At the level of the pharyngeal pores. The pharynx lumen with two passages to the outside. Passage and pore in the body wall only to be seen on one side. **E** *Paraturbanella* (Gastrotricha, Macrodasyida). Cross section of the pharynx. Myoepithelial cells with 12 sarcomeres. **F** *Halichaetonotus* (Gastrotricha, Chaetonotida). Cross section of the pharynx. One edge of the triradiate lumen in ventral direction. Myoepithelial cells with four sarcomeres. **G** *Caenorhabditis elegans* (Nematoda). Cross section of the pharynx. Configuration of the lumen as in the Gastrotricha Chaetonotida. Monosarcomeric myoepithelial cells at the surfaces, muscle-free marginal cells at the edges of the lumen. *bc* Brain commissure; *ci* cilium; *ep* epicuticle; *gd* glandular duct; *lm* longitudinal muscle; *ln* longitudinal nerve; *mc* marginal cell; *pl* pharynx lumen; *pp* pharyngeal pore. (**A,B** Rieger & Rieger 1977; **C,D** Teuchert 1977; **E,F** Ruppert 1982; **G** Bird & Bird 1991)

- Cuticularized oesophagus.
  The oesophagus joins the pharynx as a short, straight tube with cuticle. Epithelial cells protruding into the lumen form the boundary to the mid-gut.

With the first item I have taken new ideas about the evolution of the **cuticle** of Nemathelminthes and the value for kinship research into account (LEMBURG 1999). In the ground pattern of the Nemathelminthes, the trilaminar epicuticle consists of two electron-dense laminae with a lighter, less electron-dense lamina in between them. Moving inwards, there is a uniform, protein-containing basal layer. This cuticle probably evolved from a simple glycocalyx (RIEGER 1984).

In the ground pattern of the Gastrotricha, several trilamellar layers are combined to a multilamellar epicuticle. This may be an autapomorphy of the Gastrotricha; however, I will also discuss the possibility of a much earlier evolution in the stem lineage of the Nemathelminthes later (p. 7). In any case, the basal layer at first remains unchanged.

The evolution of chitin is then an important evolutionary step in the stem lineage of the Cycloneuralia (Fig. 1). The cuticle can now by divided into three layers. The trilamellar epicuticle is followed be a protein-containing exocuticle (middle layer) and a fibrillary endocuticle (basal layer) with chitin.

Adult Nematoda and Nematomorpha, however, do not have chitin in their body covering. When they are taken together as Nematoida forming with the Scalidophora, the two sister groups of the Cycloneuralia, the result is surprising: the chitin "just" evolved in the stem lineage of the Cycloneuralia has already been lost again in the Nematoida. We have no other possible interpretation for this. The lack of chitin in the integument of adult Nematoida must be a secondary condition. The reason is simple. The larvae and young stages of the Nematoida have the three-layered cuticle with chitin in the endocuticle from the ground pattern of the Cycloneuralia (SCHMIDT-RHAESA 1996a; NEUHAUS et al. 1995). Furthermore, chitin has recently been detected in the pharyngeal cuticle of adult Nematoda.

"For the Scalidophora no unambiguous autapomorphy concerning the cuticle can be given" (LEMBURG 1998, p. 156). The integument consisting of trilamellar epicuticle, exocuticle and endocuticle with chitin has been taken over by the Scalidophora and continued in the stem lineage of the Vinctiplicata (Priapulida + Loricifera). In contrast, an evolutionary change has occurred in the stem lineage of the Kinorhyncha. Chitin is present in the homogeneous middle layer, whereas the basal layer is free of chitin.

The **pharynx** of the Nemathelminthes with its numerous divergent forms presents further problems. At a first approximation, two states can be distinguished.

1. The Gastrotricha, Nematoda and Loricifera each have a pharynx with epithelial muscle cells in which the myofibrils are arranged radially. The myoepithelium encloses a triradiate lumen (Fig. 2).
2. The Kinorhyncha and Priapulida, on the other hand, possess muscle-free pharyngeal epithelia and a subepithelial pharyngeal musculature. The pharynx has a round lumen in both taxa (Fig. 12).

Myoepithelia have frequently been suggested as a primitive contractile tissue in the Bilateria. Thus, among the Nemathelminthes, the myoepithelially organized pharynx is generally considered as a plesiomorphous state. This assumption is now vigorously rejected (LEMBURG 1999).

When a myoepithelial pharynx is postulated for the stem species of the Nemathelminthes, then the pharyngeal bulbs of the Priapulida and Kinorhyncha with muscle-free epithelia must have arisen secondarily within the Nemathelminthes. This is unlikely. The arrangement and innervation of the circular and longitudinal musculature of the Priapulida are identical to those of the body musculature from which the pharyngeal musculature has developed.

In cases of a myoepithelial pharynx, there are also considerable differences in the structure and distribution of the myofilaments. In the Gastrotricha there are always multisarcomeric myoepithelial cells. In the Nematoda there are monosarcomeric myoepithelial cells on the plains of the pharynx lumen and myofilament-free cells on the apical edges. In the Loricifera (*Nanaloricus mysticus*) there are alternating apical cells with monosarcomeric myofilaments and interapical cells with disarcomeric myofilaments.

For the rest, there is no continuous alternative in the subepithelial pharyngeal musculature. It is absent only in the Nematoda. In the Gastrotricha, there is a network of circular and longitudinal muscles around the myoepithelial cells of the pharynx that continues around the subsequent midgut. The Loricifera also possess circular muscle bundles around the myoepithelial pharynx.

This briefly described situation leads to the following considerations. A pharynx of muscle-free epithelial cells with a round lumen and a weak subepithelial musculature belongs to the ground pattern of the Nemathelminthes (LEMBURG 1999). The evolution of myoepithelial radial muscle cells must have occurred convergently in the Gastrotricha, Nematoda, and Loricifera. The same holds for the triradiate pharynx lumen through which a sucking pharynx with the ability to expand strongly is reached. There are even good arguments to propose a double evolution of the triradiate pharynx within the Gastrotricha (p. 10).

Since the sister group of the Nemathelminthes cannot be determined satisfactorily (see below), the question of the extent to which the outlined features of the ground pattern of the pharynx – muscle-free epithelium, round lumen, subepithelial musculature – are plesiomorphies or autapomorphies of the Nemathelminthes remains open.

There is no body cavity in the ground pattern of the Nemathelminthes. Primarily, millimeter-sized Nemathelminthes such as the Gastrotricha, Kinorhyncha, Loricifera and free-living Nematoda have a **compact organization** with extracellular matrix (ECM) between the tissue and organs of the body (Figs. 2, 11; Vol. I, Fig. 45). When, as a consequence of increases in body size, liquid-filled cavities occur in parasites, these are primary body cavities – defined as "a cavity between ectoderm and endoderm that is surrounded by extracellular matrix" (Vol. I, p. 113). It could be disputed to what extent a cavity created by the dissolution of a compact organization can be included in the term primary body cavity. However, this is merely a question of the definition of a term. In no case is there a particular cavity that characterizes the Nemathelminthes as an entity and, accordingly, there is no justification for the term pseudocoel. This misleading term as well as the group name Pseudocoelomata derived thereof should be abandoned.

Eutely is the term for cell constancy – for the establishment of the individuals of a species by a specific number of cells in the construction of the body. With the absence of cell divisions in the adult, appreciable regeneration processes are hindered. In attempts to characterize the Nemathelminthes, one often speaks of a tendency or trend for eutely. In place of this loose, indefinite formulation we can today make a clear statement for the monophylum Nemathelminthes within the boundaries given here. Eutely most certainly does not belong the ground pattern (SCHMIDT-RHAESA 1996a). The ability for regeneration has been demonstrated for Gastrotricha. In *Pontonema vulgaris* (Nematoda, Enoplida), mitosis starts in cells of the intestinal epithelium when the intestinal tube is interrupted by injury (MALAKHOV 1998). In the Nematomorpha and the entire Scalidophora, there are no indications at all for a possible cell constancy.

The question of the primary **cleavage pattern** in the development of the Nemathelminthes cannot yet be answered satisfactorily.

First of all, the frequent attempts to attribute the development of the Gastrotricha and Nematoda to a common, bilaterally symmetrical cleavage pattern is doomed to failure. To be sure, the few studies on Gastrotricha do support an early determination of the body axes at the four-cell stage (TEUCHERT 1968). For the Nematoda, however, the classical idea of a mosaic development with determination of the fate of each blastomere has undergone a radical change (SCHIERENBERG et al. 1997/98). The early cleavage of certain Enoplida (*Enoplus, Pontonema vulgare*) is characterized by variability in the arrangement of the blastomeres. There is no evidence for a strict determination of blastomeres. Furthermore, the bilateral symmetry only appears later in the embryogenesis at the stage of ca. 500 cells (MALAKHOV 1998). "Our studies suggest that a precise cell lineage is not a necessary attribute of nematode development" (VORONOV & PANCHIN 1998, p. 143).

Secondly, our knowledge of the Scalidophora is very poor. No observations on the Loricifera and Kinorhyncha are available. According to LANG (1953), the cleavage of *Priapulus caudatus* follows a radial pattern. VAN DER LAND summarized the few, sometimes contradictory data on the Priapulida as follows: "Cleavage is total and equal (only occasionally are the cells of a markedly different size), of the bilateral type,..." (1975, p. 62).

Systematization

We now come to a consequent, phylogenetic systematization of the Nemathelminthes and bring the subordinated taxa – nearly all of which have already been mentioned – in the appropriate order (Fig. 1).

Nemathelminthes
  Gastrotricha
  Cycloneuralia
    Nematoida
      Nematoda
      Nematomorpha
    Scalidophora
      Vinctiplicata
        Priapulida
        Loricifera
      Kinorhyncha

The highest placed adelphotaxa of the Nemathelminthes are the **Gastrotricha** with numerous primitive features and the **Cycloneuralia** with a circumpharyngeal nerve ring of uniform density and specific distribution of perikarya and neuropil.

Within the Cycloneuralia, the Nematoida and Scalidophora are placed side by side as sister groups.

At this point, the exclusive existence of longitudinal muscles under the skin and the thus resulting, undulatory movements of the body may be mentioned as an outstanding autapomorphy of the **Nematoida**. As presumed sister groups the Nematoida encompasses the array of free-living and parasitic Nematoda as well as the Nematomorpha with their well-known larval and juvenile parasitism.

In the stem lineage of the **Scalidophora**, a movable proboscis with scalids – long spines for locomotion in soft sediments – has evolved. Clear adelphotaxa relationships have recently been founded for this taxon also (LEMBURG 1999). In the **Vinctiplicata** (Priapulida + Loricifera), a larva having an armored lorica comprising 20 plates exists. As their sister group, the **Kinorhyncha** remain primitive in the direct development; here we can emphasize the serial division of the abdomen into 11 zonites as well as the development with a constant number of six juvenile stages as derived characteristics.

Gastrotricha – Cycloneuralia

Gastrotricha

Today, the Gastrotricha are considered as the adelphotaxon of all other Nemathelminthes that are combined together in the monophylum Cycloneuralia on the basis of several autapomorphies. The Gastrotricha are characterized by an accumulation of primitive features – and, as is often the case with taxa with such a pattern, we are again suddenly faced with the question as to what extent the Gastrotricha may be justified at all as a monophyletic entity. The limitation of locomotory cilia to the ventral side is certainly not suitable for this justification. The name-giving and simultaneously outstanding feature is not an autapomorphy of the Gastrotricha. In comparison with a primary, complete ciliation of the body in the Bilateria – e.g., Plathelminthes or Nemertini – we must rather postulate the limitation to the ventral side with reduction of the dorsal and lateral cilia as an evolutionary event in the stem lineage of the Nemathelminthes. The ventral **ciliary creeping sole** is then only retained by the Gastrotricha as a plesiomorphy – it disappeared completely in the stem lineage of the Cycloneuralia.

The outstanding example of a parallel evolution is provided by the ciliary creeping sole of the Otoplanidae; this is a species-rich taxon of the Proseriata (Plathelminthes) with a dominance in the surf zone of sandy coasts (AX 1956). The creeping sole in this case has without doubt arisen in the interstitial milieu – possibly by intensification of locomotory activity in an unstable substrate. A corresponding evolution of the ventral locomotory system of the Nemathelminthes primarily connected with life in the interstitial spaces of marine sand must be considered as a plausible option.

These considerations shed light on the first candidate for an autapomorphy of the Gastrotricha – the **multilamellar epicuticle** as an accumulation of trilamellar single layers in the periphery of the body. This situation seems to be valid in general for the Macrodasyida and also holds for the taxon *Neodasys* under the Chaetonotida; however, within the Paucitubulata there are cases with a single trilamellar construction (RIEGER & RIEGER 1977). With the question of the functional significance, we again come to the interstitial milieu. BOADEN (1968) argued that the stack of lamella may serve as protection against instability in the sandy bottom – effectively by a type of molting in which lamella are newly formed internally to the same extent that they are peeled or scraped away by sand grains. However, even when this explanation appears to be acceptable, the multilamellar epicuticle need not necessarily be an autapomorphy of the Gastrotricha. The stem species of the Nemathelminthes must have lived somewhere – and this could in the simplest interpretation have been the interstitial spaces of sandy bottoms. In this case, not only the ventral ciliary creeping sole, but also the multilamellar epicuticle as well as the adhesive papilla with a duo-gland adhesive system would be possible ground pattern characteristics of the Nemathelminthes and not just unique features of the Gastrotricha.

An autapomorphy of the Gastrotricha is then obviously a **pharynx** with **multisarcomeric myoepithelial cells** (p. 4). The adoption of a plesiomorphic round pharyngeal lumen in the ground pattern of the Gastrotricha is discussed later.

We now come to **hermaphroditism**, which belongs without doubt in the ground pattern of the Gastrotricha. An evolutionary change to parthenogenetic development took place first within the Chaetonotida. Gonochorism is widespread among the Nemathelminthes. Accordingly, there are good reasons to postulate a dioecious organism for the stem species of the Nemathelminthes and to interpret the hermaphroditism of the Gastrotricha as an autapomorphy. However, the alternative hermaphroditism – gonochorism is not highly valued for the justification of kinship relationships.

For an evaluation of the type of sperm transfer we must jump forward to the situation in the Priapulida. In this case the release of free plesiomorphous sperm in combination with an internal fertilization can belong to the ground pattern (p. 31). A consequence of this hypothesis is the assumption of multiple, independent evolution of a direct sperm transfer in various subtaxa of the Nemathelminthes.

In spite of the hermaphroditism, complex mechanisms of copulation are known for the Macrodasyida (TEUCHERT 1968; RUPPERT 1978a,b). In other words, a direct transfer of sperm and in the same context the evolution of thread-like sperm could constitute autapomorphies of the Gastrotricha. The colonization of the interstitial milieu by the Macrodasyida, however, once more provides a reason for caution. When the cavity system between grains of sand is to be the primary biotope of the Nemathelminthes, the prevailing evolutionary pressure in the dark, underground milieu to combine the sexes may have led to direct sperm transfer in the stem species of the Nemathelminthes. Such a consideration must at least be weighed against the above-mentioned Priapulida hypothesis.

The results of our attempt to validate the monophyly of an entity Gastrotricha are meager. The possible characteristics are summarized in the following survey.

Autapomorphies (Fig. 1 → 2)

- Multilamellar epicuticle in an accumulation of trilamellar individual layers.
- Pharynx with multisarcomeric myoepithelial cells.
- Hermaphroditism.
- Direct transfer of filiform sperm.

Plesiomorphies

Protonephridia were certainly taken up in the stem species of the Gastrotricha from the ground pattern of the Nemathelminthes – in the primitive composition of the single organ comprising a terminal, canal and nephroporus cell, each with one cilium (NEUHAUS 1987; BARTOLOMAEUS & AX 1992).

Thus, we come directly to the number of protonephridia. A pair was postulated for the ground pattern of the Bilateria. However, I consider the question of whether one pair of protonephridia in the apomorphic bottle-shaped Paucitubulata (Chaetonotida) represents the original situation of the Gastrotricha to be unanswered. For *Neodasys* (Chaetonotida), for example, two pairs of protonephridia are known (RUPPERT 1991). Furthermore, among the Macrodasyida several pairs of serially arranged excretory organs are found regularly; *Mesodasys laticaudatus* has 11 pairs of protonephridia. If we want to put one pair in the ground pattern of the Gastrotricha, then a secondary increase must be postulated for the Macrodasyida and for *Neodasys*. However, this is not in harmony with the hypothesis of a stem species of the Gastrotricha with the plesiomorphic, band-shaped habitus of the Macrodasyida that will be discussed later in the text.

The **body wall musculature** shows primitive traits with the existence of circular and longitudinal muscles. However, there is no muscular tube composed of continuous muscle layers. The circular peripheral musculature is arranged in longitudinal rows of barrel hoop-like rings. Progressing inwards the longitudinal musculature follows in the form of isolated muscle cords or muscle bands, each consisting of numerous cells (TEUCHERT 1974; RUPPERT 1991).

We conclude the plesiomorphies with the **nervous system** in which a central alternative between the adelphotaxa Gastrotricha and Cycloneuralia is manifest in the construction of the **brain**. Firstly, the brain in both taxa consists of a nerve ring around the anterior third of the pharynx. In the case of the Gastrotricha, there is a large dorsal commissure and a thin ventral commissure; the dorsal commissure is enclosed above and at the sides by regularly distributed perikarya (WIEDERMANN 1995). This is interpreted as the plesiomorphous state for the Nemathelminthes. In the apomorphous alternative of the Cycloneuralia the brain consists of a uniformly thick ring with a frontocaudal-oriented division in perikarya – neuropil layer – perikarya (p. 14).

The brain sends out several pairs of body-length nerves in a bilaterally symmetrical arrangement; the stronger ventral nerves lie far apart. In *Cephalodasys maximus* there are two ventral, eight lateral and two dorsal nerves (WIEDERMANN 1995).

Systematization

Macrodasyida – Chaetonotida

At the last count by HUMMON (1982), 429 Gastrotricha were known, about 25% of them being Macrodasyida (body length between 120 μm and 2 mm) and the other species being members of the Chaetonotida (lengths of 75–350 μm; *Neodasys* 400–750 μm; d'HONDT 1971; MOCK 1979). With the naming of the two entities Macrodasyida and Chaetonotida, we already encounter the first problem. Since the pioneering works of REMANE starting in 1924 and reaching a climax with the monograph of 1936, the division of the Gastrotricha in the two "orders" Macrodasyida and Chaetonotida is unconditionally accepted in every relevant textbook. However, the justification of the two taxa as monophyla and highest-ranking adelphotaxa of the Gastrotricha is only possible under the following preconditions. Neither the triradiate pharynx of the Macrodasyida with a flattened surface in the ventral direction and lateral pharyngeal pores nor the triradiate pharynx of the Chaetonotida with a ventral edge (as in the Nematoda) belongs in the ground pattern of the Gastrotricha. Instead, we must rather consider a myoepithelial pharynx with round lumen for their stem species (RUPPERT 1982; TRAVIS 1983). The argumentation is certainly unusual because no member of the Gastrotricha with such a situation in the pharynx is known; but it draws its legitimation from two considerations – firstly, from the justification of a pharynx with a round lumen and muscle-free epithelium in the ground pattern of the Nemathelminthes (p. 4) and secondly, from the sheer impossibility to derive one of the two states of the Gastrotricha pharynx from the other. Under these premises, the evolution of a myoepithelium around a round pharyngeal lumen in the stem lineage of the Gastrotricha can be hypothesized. Then, the divergent triradiate adaptations of the pharynx in the Macrodasyida and Chaetonotida evolved from the primary state as alternative autapomorphies (Fig. 2).

Macrodasyida

Autapomorphies. Pharynx with a triradiate lumen with orientation of one flat surface to the ventral side. Paired pharyngeal pores lateral in the posterior region of the pharynx for draining off excess water in the process of taking up particulate food (Fig. 2 C–E).

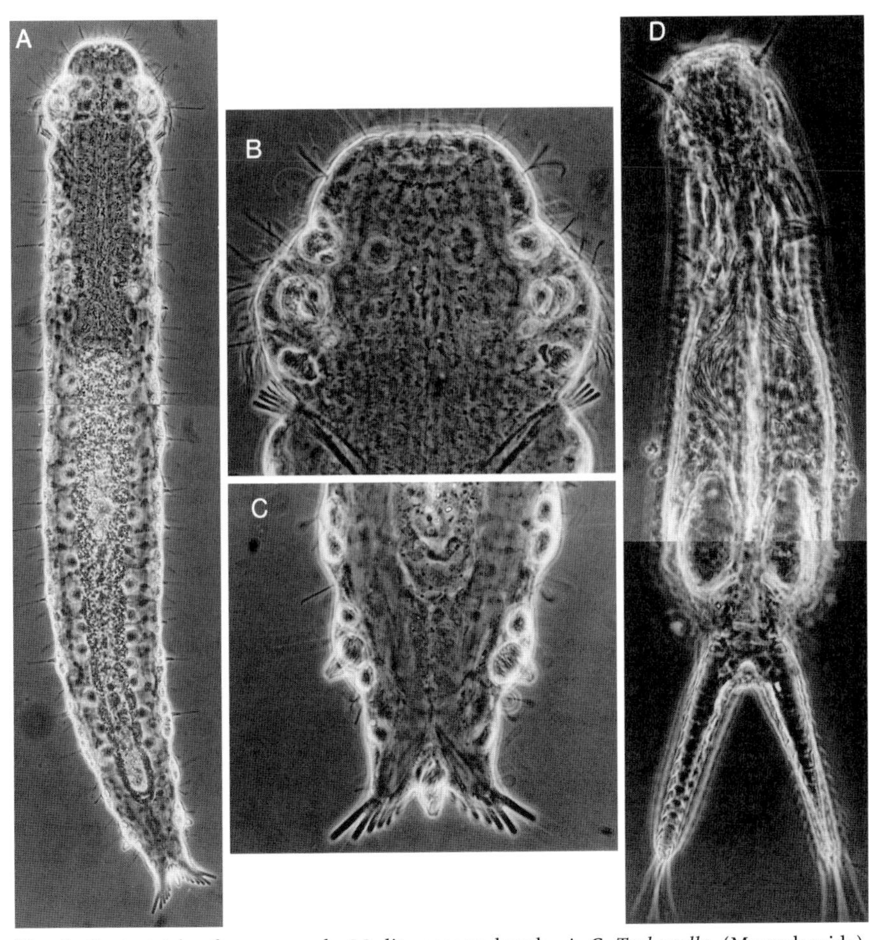

Fig. 3. Gastrotricha from a sandy Mediterranean beach. **A–C** *Turbanella* (Macrodasyida). **A** Habitus. Length 0.3 mm. Lateral two rows of short adhesive tubes (papillae), each with a sensory cilium. **B** Anterior end. Two groups of lateral adhesive tubes behind the head; with insertions of muscle fibrils. **C** Posterior end. Tail flaps with lateral rows of adhesive tubes and a median cone. **D** *Xenotrichula* (Chaetonotida). Length 0.2 mm. With large toes (caudal furca). With hermaphroditism as primitive state within the Chaetonotida. (Originals; Canet Plage, France)

Plesiomorphies. Band-shaped habitus. Rich endowment with adhesive papillae laterally and at the poles of the body.

Macrodasys, Mesodasys, Cephalodasys, Urodasys, Dactylopodola, Turbanella.

Psammobiotic organisms in the cavity systems of gravel sediments to fine sands and from the supralittoral sea coast to water depths of ca. 50 m.

Almost exclusively in the original marine milieu, *Turbanella lutheri* (Fig. 4) is a widespread brackish water species with a tendency to migrate into fresh water (REMANE 1952; KISIELEWSKI 1987b; AX 1993). *Redudasys fornerise* from Brazil is the sole unambiguously fresh water species (KISIELEWSKI 1987a).

Chaetonotida

Autapomorphies. Pharynx with triradiate lumen with orientation of an edge to the ventral side (Fig. 2F).

D'HONDT (1971) classified the Chaetonotida in the two taxa Multitubulata for two *Neodasys* species and Paucitubulata for the extensive remainder. The name Multitubulata is to be deleted because the content is identical with *Neodasys* and, in addition, it has not been justified as a monophylum. The name Paucitubulata covers a valid taxon in the composition of all other Chaetonotida.

Neodasys

The two Atlantic species *Neodasys chaetonotoides* and *Neodasys uchidai* discovered by REMANE jointly colonize a sandy beach of the North Sea island of Sylt (MOCK 1979). The taxon is of particular interest. As psammobiontes with the band-shaped habitus of the Macrodasyida and the expression of weak lateral adhesive papillae, the two *Neodasys* species introduce primitive traits to the ground pattern of the Chaetonotida (Fig. 4).

Paucitubulata

Xenotrichula, Aspidiophorus, Heterolepidoderma, Chaetonotus.

Bottle-shaped habitus (Fig. 4). Two large toes with adhesive papillae at the posterior end. The lack of adhesive devices at the anterior end and on the sides of the body can be interpreted most economically as the product of a single reduction in the stem lineage of the Paucitubulata. In contrast, living in fresh water and the parthenogenesis most certainly do not belong in the ground pattern of the Paucitubulata.

Members of the first mentioned three taxa, further Paucitubulata such as *Halichaetonotus*, and even *Chaetonotus* species live in the original marine milieu. The invasion into diverse fresh water environments with dissolu-

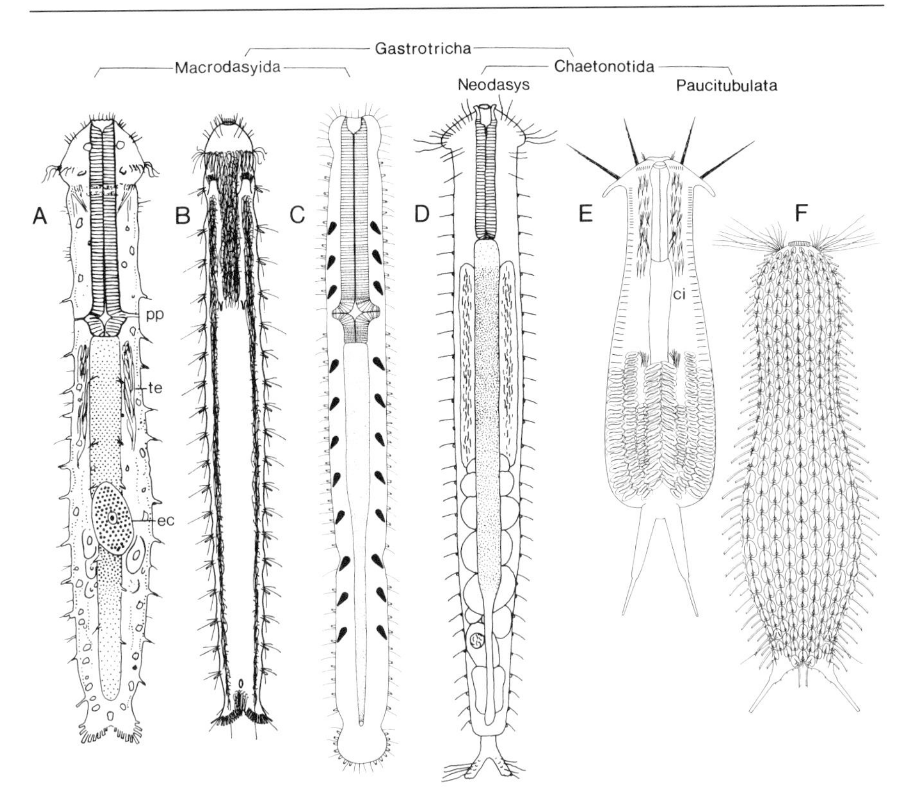

Fig. 4. Gastrotricha. **A** *Turbanella lutheri.* Dorsal view. Demonstration of the lateral pharyngeal pores. Testes and ovaries in a single individual. **B** *Turbanella ocellata.* Ventral view. Locomotory ciliation in the form of two long, lateral ciliary bands and a short median ciliary stripe in the anterior body. Both *Turbanella* species show the long lateral adhesive papillae of the Macrodasyida. **C** *Mesodasys laticaudatus* Distribution of 11 pairs of protonephridia in the longitudinal axis of the body. **D** *Neodasys uchidai.* Representative of the Chaetonotida with primitive features such as band-shaped habitus and rows of lateral (albeit short) adhesive papillae. Pharynx without pharyngeal pores. **E** *Xenotrichula velox.* Ventral view. Cilia on anterior body joined together to locomotory cirri. **F** *Chaetonotus condensus.* Dorsal side with scaly spines. Product of the basal layer of the cuticle. *ci* Cirrus; *ec* egg cell; *pp* pharyngeal pore; *te* testes. (**A** Remane 1952; **B** Hummon 1974; **C** Neuhaus 1984; **D** Remane 1961; **E, F** Mock 1979)

tion of the strong link to the sandy interstitial spaces only occurred within the entity.

The situation is similar with the often apostrophized parthenogenesis under reduction of the male sex organs. Testes are present in *Neodasys* and Xenotrichulidae, sperm has been detected in other Paucitubulata (D'HONDT 1971). A purely parthenogenetic reproduction developed late within the Paucitubulata.

Cycloneuralia

Autapomorphies (Fig. 1 → 3)

- Brain as a circumpharyngeal nerve ring with two apomorphies as opposed to the Gastrotricha.
  1. Course with uniform thickness around the gut.
  2. Frontocaudal division of the ring into three sections: anterior accumulation of perikarya – middle neuropil layer – posterior accumulation of perikarya (Fig. 12 C).
- Three-layered cuticle.
- Chitin in the basal, fine fibrillary layer.
- The evolution of repeated moltings in the development is connected with the expression of chitin.
- Furthermore, a consequence of the chitin cuticle is the complete reduction of locomotory cilia, i.e., the abandonment of the ciliary creeping sole that exists in the adelphotaxon Gastrotricha.
- Close combination of the paired main longitudinal nerve cords on the ventral side.

Nematoida – Scalidophora

Nematoida

Autapomorphies (Fig. 1 → 4)

- Long, thin body.
- Lack of chitin in the basal cuticle layer of the epidermis in adults.
- Lack of circular muscles in the peripheral body musculature.
- Lack of protonephridia.
  In comparison with the existence of nephridial excretory organs in the Gastrotricha and Scalidophora, this must be interpreted as a secondary condition.
- Ventral main nerves fused caudally.
- Evolution of an unpaired dorsal nerve cord.
- Dorsal and ventral inward reaching epidermal cords in which the corresponding nerve cords run.
- Sperm without cilia and without accessory centrioles.
- Undulatory movements of the body.

A **peripheral musculature** consisting solely of **longitudinal muscles** in combination with a rigid cuticle does not allow peristaltic movements or contractions along the body axis. They rather provide the basis for **undu-**

latory movements that are realized in two forms. In the Nematomorpha, the twists occur alternately to the left and to the right (SCHMIDT-RHAESA, pers. comm.), in the Nematoda up and down. The lateral twisting movement is to be included in the ground pattern of the Nematoida. The dorsoventral bending of the body in the Nematoda represents the apomorphous behavior; it is linked to the evolution of lateral epidermal cords that separate the longitudinal muscles into upper and lower groups (see below). For dorsoventral undulations the Nematoda lie with one side on the ground.

Nematoda – Nematomorpha

In the conventional classification roundworms and horsehair worms have long been considered as equal-ranking entities of the Nemathelminthes. Phylogenetic systematics has now justified them as adelphotaxa and accordingly combined them under the name Nematoida (SCHMIDT-RHAESA 1996a).

However, there is a competing hypothesis which orders the Nematomorpha in the Nematoda under the assumption of a close relationship with the subtaxon Mermithida (LORENZEN 1996). Indeed, there are remarkable similarities in the life cycle with larval parasitism and certain correlated features. If we want to identify them as homologous congruences, the specific characteristics of the Nematoda such as head sensilla, lateral organs, lateral epidermal cords, vulva and spicule – that are present in the Mermithida – must have been lost in the stem lineage of the Nematomorpha. There are neither facts nor arguments for this assumption. We must work with the most economical explanation, and that is the interpretation of the similarities between the Mermithida and the Nematomorpha as convergences with considerable differences in detail (SCHMIDT-RHAESA 1997/1998).

Nematoda

The currently described 20,000 species probably represent only a few percent of the actually existing Nematoda. With a high abundance, the roundworms usually are the dominating group of microscopic organisms in marine, limnetic, and terrestrial environments (Fig. 5). Without doubt, the stem species consisted of populations of free-living, millimeter-sized individuals.

In separate stem lineages independent plant pests and animal parasites have evolved repeatedly. They can remain primarily small such as the trichina worm *Trichinella spiralis*, but also grow to meter sizes such as the

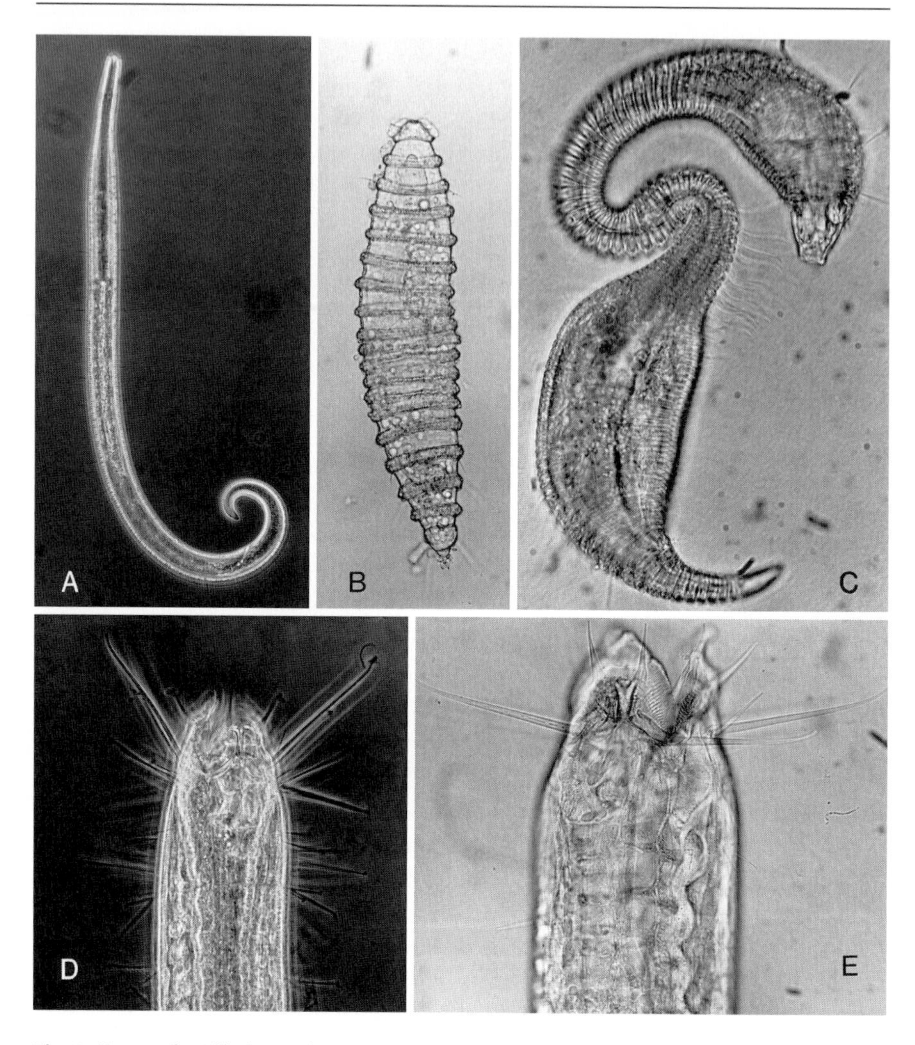

Fig. 5. Nematoda. Life forms from sandy beaches of the Mediterranean Sea. **A** *Theristus heterospiculum*. The small organism of 0.6 mm length moves rapidly between grains of sand by pushing movements of its tail. **B** *Desmoscolex frontalis*. Length 0.2 mm. The organisms stand on subdorsal body setae; the ventral side is directed upwards. **C** *Epsilonema pustulatum*. Length 0.35 mm. Fluke-like or looper-like movements by alternate adhesion of the anterior and posterior ends, supporting stalk bristles ventral in the middle of the body. **D** and **E** *Enoplolaimus subterraneus*. Anterior end with long head setae. Large species of more than 2 mm length. Member of the widespread "stem twister" life form. In the undulating basic movement, the body is pushed along over sediment particles of the ground. (Originals; Sandy beach Canet Plage, France. Analysis of life forms in Gerlach 1954)

Guinea worm *Dracunculus medinensis* – to name but two of the troublesome human parasites.

We will now describe the characteristics of the Nematoda – at first briefly and subsequently with more detailed explanations and justification.

Autapomorphies (Fig. 1 → 5)

- Sensilla at the anterior end arranged in a 6+6+4 pattern.
- Lateral organs (amphids).
- Lateral epidermal cords.
- Dorsoventral undulating movements.
- Gonads opposed to each other.
- Vulva in the middle of the female.
- Cloaca in the male.
- Spicule in the male.
- Amoeboid sperm without acrosome.
- Postembryogenesis with four juvenile stages.
- Pharynx with triradiate, Y-shaped lumen. Apical, muscle-free epithelial cells at the three edges; interapical monosarcomeric myoepithelial cells on the surfaces (p. 4).
- Reduction of subepithelial circular and longitudinal musculature around the pharynx.
- Ventral glands and caudal glands.

At the anterior end, three circlets of sensilla belong to the ground pattern of the Nematoda. A first circlet of six lip sensilla surrounds the opening of the mouth. It is followed by a second circlet of six head sensilla and, somewhat displaced, a third circlet of four head sensilla (Fig. 6C). The papillary to bristle-shaped sensilla are simply organized; they consist primarily of only three cells – a nerve cell with cilium, a jacket cell and a pocket cell.

The **lateral organs** or **amphids** are paired chemoreceptors set deeply under the epidermis. The opening (aperture) leads into a groove (fovea) covered with cuticle. Aperture and fovea have widely varying structures – slit-like to circular, loop-like or spiral (RIEMANN 1972). The fovea is connected via a canal to a subcuticular spindle (fusus). Basally, dendrites with highly modified cilia enter the fusus; they reach far distally into the fovea filled with gel (Fig. 6E).

Pocket-like, insunk caudal sensilla (phasmids) do not belong in the ground pattern, but have rather evolved first within the Nematoda. The existence of phasmids in paired form justifies the Secernentea (Phasmidia) as an extensive monophyletic subtaxon of the Nematoda (Fig. 6H).

Dorsal and ventral epidermal cords were taken over from the ground pattern of the Nematoida. Additional **lateral epidermal cords** within which

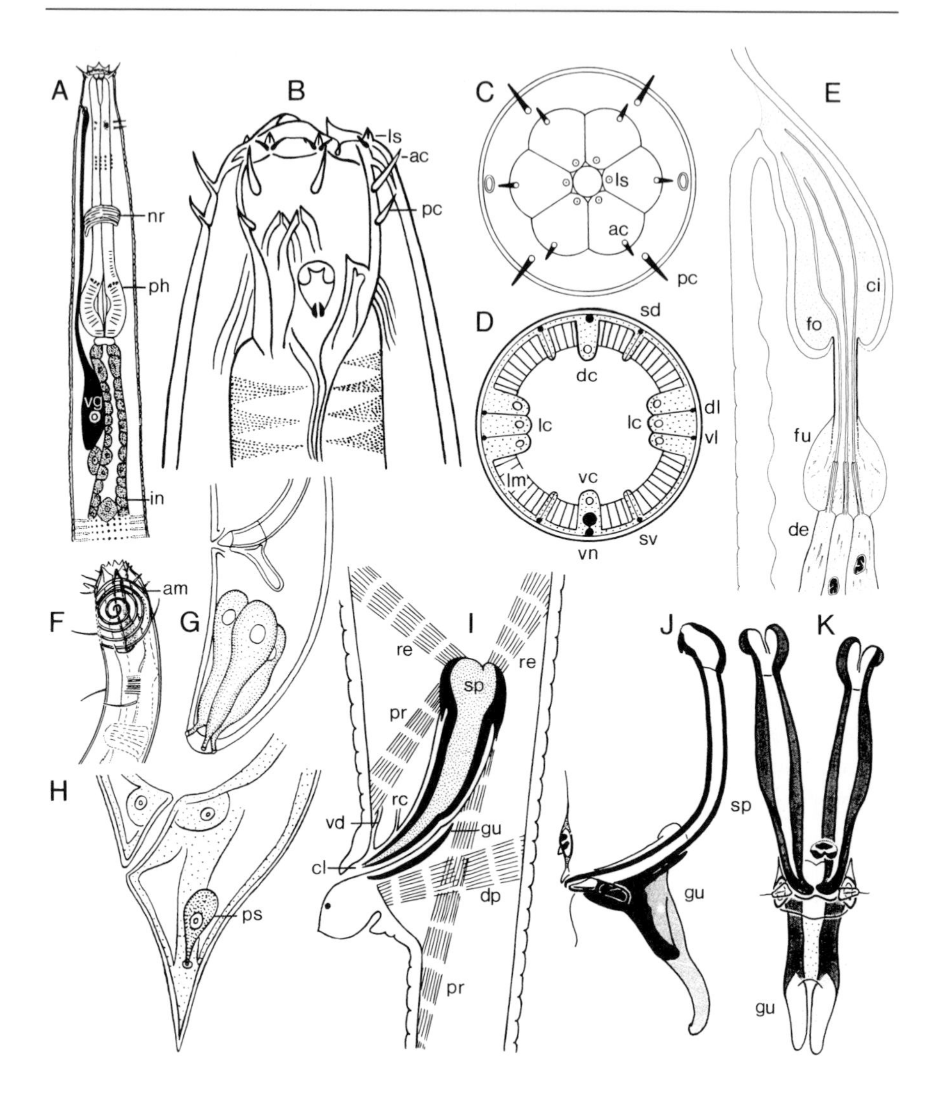

the nuclei of the epidermal cells are concentrated evolved additionally in the stem lineage of the Nematoda; they protrude as wide and high arches into the inside of the body (Fig. 6D).

The lateral epidermal cords divide the body musculature into two groups. A dorsolateral and a ventrolateral group of longitudinal muscles are each innervated by dorsal and ventral nerve cords.

In the **undulating movements,** the bending of the body occurs in **dorsal and ventral directions**. A connection with the development of lateral epidermal cords seems to be obvious. The evolution of two functional groups of longitudinal muscles with alternating contractions must be understood together with the evolution of lateral epidermal cords. However, the dorso-

◀ **Fig. 6.** Nematoda. **A** *Chromadora quadrilinea*. Side view of the anterior end to demonstrate the ventral gland with long cell neck and ventral pore. **B** *Pontonema vulgare*. Anterior end with three circlets of sensilla (six lip sensilla, anterior circlet of six-head sensilla, posterior circlet of four-head sensilla). **C** Ground pattern for the arrangement of sensilla at the head end. Frontal view. **D** *Pontonema vulgare*. Cross section through the body with dorsal and ventral epidermal cords, with paired lateral epidermal cords, with dorsolateral and ventrolateral groups of longitudinal muscles and bilaterally symmetrically arranged longitudinal nerves. **E** *Tobrilus aberrans*. Longitudinal section of a lateral organ. **F** *Onyx macramphis*. Large lateral organ (amphid) with spirally rolled-up aperture. **G** *Diplopeltula*. Posterior end with caudal glands. **H** *Neoaplectana*. Caudal end with a lateral phasmid behind the hindgut/anus. **I** *Pratylenchus penetrans*. Longitudinal section of the spicular apparatus. Spicule with a pair of protractor and retractor muscles. The dorsally positioned gubernacula are accessory hard parts in the spicular apparatus. **J** and **K** *Odontobius ceti*. **J** Lateral view of spiculum and gubernaculum. **K** Top view of paired spicules and gubernacula. *ac* Anterior head sensilla; *am* amphid; *ci* cilium; *cl* cloaca; *dc* dorsal epidermal cord with dorsal nerve; *de* dendrite; *dl* dorsolateral nerve; *dp* depressor muscle; *fo* fovea; *fu* fusus; *gu* gubernaculum; *lc* lateral epidermal cord; *lm* longitudinal muscle; *ls* lip sensilla; *nr* nerve ring; *pc* posterior head sensilla; *ph* pharynx; *pr* protractor muscle; *ps* phasmid; *rc* rectum; *re* retractor muscle; *sd* subdorsal nerve; *sp* spicule; *sv* subventral nerve; *vc* ventral epidermal cord; *vd* vas deferens; *vg* ventral gland; *vl* ventrolateral nerve. (**A, C, H** Bird & Bird 1991; **B, D, G** Malakhov 1994; **E** Storch & Riemann 1973; **F** Blome & Riemann 1994; **J, K** Lorenzen 1986)

ventral undulations never occur vertically in space. Nematodes lie with an arbitrary side of the body on the substrate and thus move in a horizontal plane. Thereby, they usually lift themselves over bumps in the ground. I suggest that the widespread **lifting undulation** ("Stemmschlängeln") is the primary mode of movement in the ground pattern of the Nematoda. On smooth surfaces the undulatory movements give the picture of helpless weaving to and fro without covering any appreciable distance.

A parallel alignment of paired gonads at the same level is standard for the Bilateria. It is realized in the Gastrotricha and can be included in the ground pattern of the Nemathelminthes as a plesiomorphy. The **opposing orientation of the gonads** to the back and front must accordingly represent an autapomorphy of the Nematoda (LORENZEN 1981).

Sexual openings are usually at the posterior end. The development of the female **vulva in the mid-body region** is the apomorphous result of a forward displacement.

Males of the Nematoda have a **cloaca**. Vas deferens, rectum, and spicular apparatus are united to a common pore. The combination of the three elements is a singular phenomenon of the Nematoda. A homology with the identically named cloaca of the Gordiida within the Nematomorpha seems to be unlikely.

Paired **spicula** in the form of a scimitar lie in separate spicule pockets as dorsal protrusions of the cloaca (Fig. 6 I–K). The spicules of the Nematoda do not serve to transfer sperm to the sexual partner as is usually the case with copulatory organs. The spicules are instead pure anchoring organs that are introduced into the vulva of the female for a firm union of the sexes. After this, sperm emerge directly from the vas deferens into the female genital system.

The amoeboid, motile **sperms** possess neither acrosome nor cilia. The lack of cilia is shared by the Nematoda and the Nematomorpha; it was taken over from the ground pattern of the Nematoida.

In their development, Nematoda constantly pass through **four moltings.** This is independent of lifestyle and body size (LORENZEN 1996).

We conclude the survey with two different glands, the evolution of which is most economically interpreted in the stem lineage of the Nematoda and which accordingly should be included in the ground pattern of the Nematoda as derived characteristics. The ventral gland and the caudal gland, however, occur only in a wider distribution in the paraphyletic "Adenophorea".

The **ventral gland** is primarily a large cell lying in the anterior body which, via a long duct, opens shortly behind the mouth (Fig. 6 A). Only rarely is there a multiplication to two or three glands. There is no evidence for the often proposed interpretation of ventral glands as an excretory system (BIRD & BIRD 1991). Furthermore, there is no justification for an evolutionary derivation from protonephridia – they are already reduced in the stem lineage of the Nematoida.

Three unicellular **caudal glands** are present in marine and limnetic species of the "Adenophorea" (Fig. 6 G). Reductions in the terrestrial Nematoda have led to their complete absence. In general, caudal glands are then absent in the Secernentea. The caudal glands produce a sticky substance. This serves for adhesion to the substrate, coating of living tubes in soft sediments (CULLEN 1973) and in fibrous structures even for trapping detritus, bacteria and macromolecules for nourishment (RIEMANN & SCHRAGE 1978).

The "provisional system" of the Nematoda (LORENZEN 1996) with the "Adenophorea (Aphasmidia)" and Secernentea (Phasmidia) as highest-ranking taxa belongs to the chain of traditional classifications in which, as in an identification key, a paraphylum with primitive features and a monophylum with autapomorphies are simply placed side by side.

Time has run out for these classifications. In order to demonstrate the mistake once more, I will mention for comparison the out-of-date division of the insects into Apterygota and Pterygota. In this context, the names Aphasmidia and Phasmidia are welcome. With their phasmids, the Phasmidia have specific sensory equipment in the ground pattern just as the Pterygota have two pairs of wings; both taxa are identified as monophyla through autapomorphies. The Aphasmidia have no phasmids, the Apterygota no wings – a primary deficiency in each case. However, neither for the Apterygota nor for the Aphasmidia are apomorphous characteristics for the justification as valid system entities known. Thus, just as the "Apterygota" has been abandoned for the insects, so must the "Aphasmidia" be eliminated from the system of the Nematoda.

In a molecular evolutionary framework, BLAXTER et al. (1995) identified five major clades of the Nematoda. Various subtaxa were, however, omitted in this process. Furthermore, there are no details about morphological autapomorphies of the presumed monophyla to control the compatibility with molecular data.

A consequent phylogenetic system for the Nematoda has not yet been written.

Nematomorpha

Life cycles with parasitic larvae and free-living adults remain an exception among the animals – and just for this reason the Nematomorpha have attracted much attention from zoologists in the past. The impressive body length with a tiny diameter is an additional factor. *Gordius fulgur* from the Fiji Islands (Fig. 7A) exceeds 2 m in length. The long, thin, and at the same time elastic body has led to the name horsehair worms. Comprehensive studies have been performed recently (SCHMIDT-RHAESA 1996–1999).

In the taxon *Nectonema* four marine species stand opposite about 300 limnetic Gordiida as sister group (p. 25).

Autapomorphies (Fig. 1 → 6)

- Life cycle with three phases.
  Free-living, aquatic larvae – parasitic larvae and parasitic juvenile stage in Arthropoda – free-living adults in aquatic milieu.
- Brain mainly ventral under the pharynx. Only a weak dorsal commissure.
- Brain with internal neuropil and perikarya distributed uniformly in the periphery.
- Cuticle of adults with fibrils in a crossed arrangement.
- Longitudinal musculature of numerous cells (polymyarian) with peripheral arrangement of the myofibrils in the flat, basal parts (cyclomyarian).
- Gut with subterminal opening of the mouth as opposed to terminal mouth in the ground pattern of the Nemathelminthes.
- Pharynx without musculature.
- Larvae with two (or three) circlets of hooks.

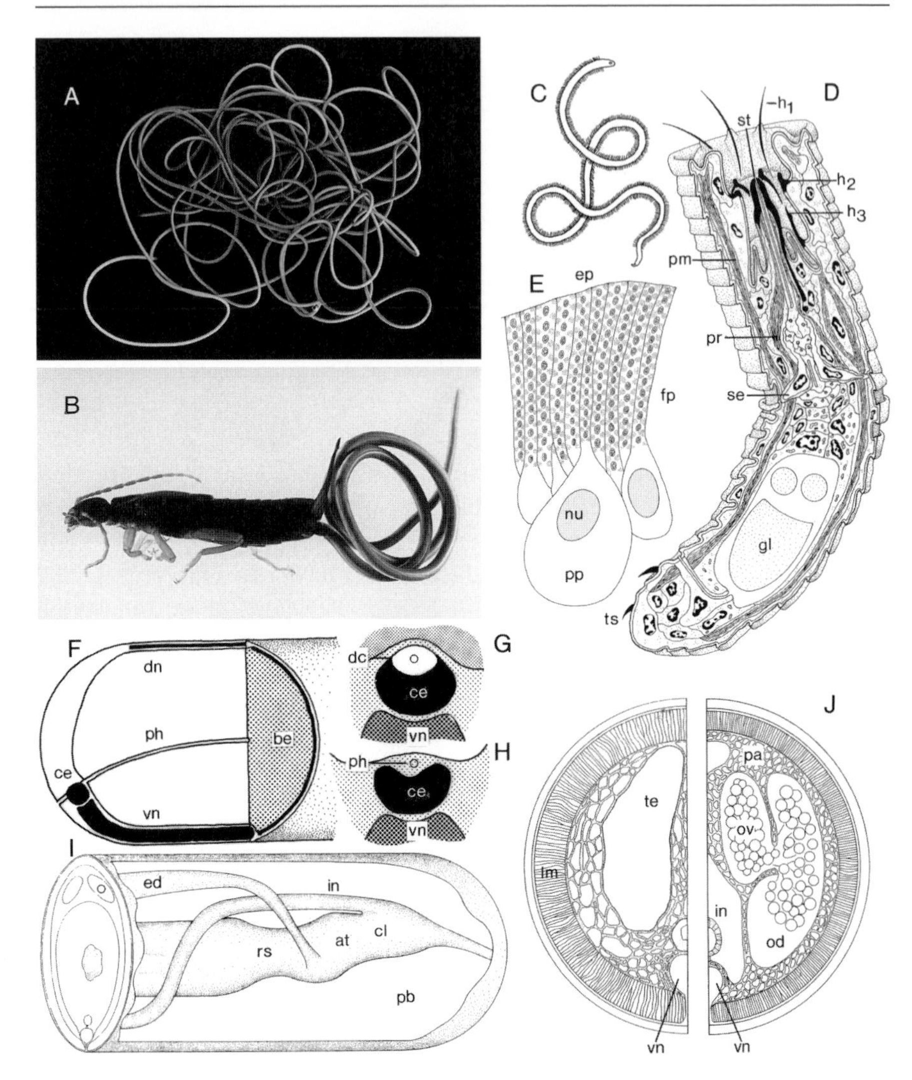

Fig. 7. Nematomorpha. **A** *Gordius fulgur* (Southeast Asia). Female, length 2.24 m. **B** Member of the Gordiida on emergence from an earwig *Forficula auricularia* (Dermaptera). **C** *Nectonema agile* with swim bristles. **D** Larva of *Paragordius varius*, longitudinal section. **E** *Nectonema munidae.* Muscle cells from longitudinal musculature of the body wall. **F** *Nectonema munidae.* Diagram of the nerve system in the anterior end. Longitudinal section. **G** *Nectonema munidae.* Brain of a juvenile stage with weak dorsal commissure. Cross section. **H** *Nectonema munidae.* Brain of an adult. Dorsal commissure reduced. **I** *Gordius aquaticus.* Diagram of the female sexual organs in the posterior end. **J** Cross sections through developmental stages of Gordiida. *Left* Male with testes in parenchyma. *Right* Female with ovary and oviduct. *at* Atrium; *be* basiepidermal nerve system; *ce* cerebrum; *cl* cloaca; *dc* dorsal commissure; *dn* dorsal nerve; *ep* epidermis; *fp* fibrillary part of longitudinal muscle cell; *gl* gland; *h 1–3* hooks on three circlets; *in* interstitial space; *lm* longitudinal muscle; *nu* nucleus; *od* oviduct; *ov* ovary; *pa* parenchyma; *pb* primary body cavity; *ph* pharynx; *pm* parietal muscle; *pp* protoplasmic part of longitudinal muscle cell; *pr* proboscis muscle; *rs* seminal receptacle; *se* septum; *st* stylet; *te* testis; *ts* terminal spine; *vn* ventral nerve. (**A, B** Originals; Schmidt-Rhaesa; **C** Schmidt-Rhaesa 1996d; **D, I, J** Schmidt-Rhaesa 1997c; **E** Schmidt-Rhaesa 1998; **F, G, H** Schmidt-Rhaesa 1996c)

Ground Pattern

In a presentation of the major features of the ground pattern, we will precisely describe the autapomorphies of the Nematomorpha. In addition, the characteristics of the adelphotaxa *Nectonema* and the Gordiida will be laid down in this section.

In comparison with the Nematoda, plesiomorphies are lack of head sensilla and lateral organs, of lateral epidermal cords and sexual apparatus (vulva, spicule).

The **three phases of the life cycle** are characterized as follows (SCHMIDT-RHAESA 1997c).

1. Free-living larvae (Fig. 7D). Hatching in water. Length 50–150 μm. Division of the body by a cellular septum into two regions. The preseptum carries circlets of cuticular retractable hooks. It is possible that the two circlets of hooks of *Nectonema* belong to the ground pattern. The Gordiida possess three circlets and additionally three cuticular stylets. Numerous glands. No digestive tract. For the Gordiida existence in water is limited to a few days. Encystment of the larvae is possible.
2. Larvae and juvenile stage as parasites. The routes of infection of Arthropoda by *Nectonema* are unknown and those by Gordiida have not yet been completely analyzed. Suspected usual mode: oral uptake as larva by the host from water or as cyst from terrestrial environment - penetration of the host's gut by hooks and stylets - residence in body cavities. The postseptum of the larva grows to a juvenile stage of several centimeters in length (Fig. 7B). Only one molting occurs. The larval cuticle with hooks and stylets is discarded shortly before emergence from the host.
   Food is only taken up in the host - and then in dissolved form. Absorption of amino acids by the epidermis and gut cells has been demonstrated (SKALING et al. 1988).
3. Free-living adult (Fig. 7A). Development in water from the released juvenile stage. Gonochorism, copulation, internal fertilization, and deposition of eggs in water can be traced back to the stem species of the Nematoida; they constitute plesiomorphies in the life cycle of the Nematomorpha.

The adult Gordiida live in the benthic zone of various fresh waters from lakes to ponds. The colonization of the bottom may be original, but that of fresh water almost certainly is not primitive. The marine *Nectonema* species probably live in the original milieu while existence in the pelagial as an organism with swim bristles is possibly an apomorphous behavior.

We now come to the **nerve system**. In contrast to the uniformly thick ring of the Cycloneuralia, the **brain** of the Nematomorpha lies almost exclusively subpharyngeally on the ventral side. A weak dorsal commissure has disappeared in the development of *Nectonema* and is absent in some species of Gordiida. In comparison with the primitive three-part sequence perikarya - neuropil - perikarya in the brain ring of the Cycloneuralia, perikarya are uniformly distributed in the brain of Nematomorpha.

A ventral and a dorsal **nerve cord** in a thickened epidermal cord belong in the ground pattern of the Nematomorpha as a plesiomorphy, but are only realized in *Nectonema*. In the Gordiida, the ventral cord is displaced subepidermally into the body cavity and a dorsal cord is absent (Fig. 7 F–J).

The **cuticle** of the adult consists of an external epicuticle and an internal fibril layer. The fibrils are arranged in a crossed fashion.

Two characteristics from the **longitudinal musculature** below the body wall have to be mentioned. The muscle layer consists of numerous, basally flattened cells; the perikarya are oriented towards the body cavity. In the flattened cell parts the myofibrils are arranged around the central cytoplasm with mitochondria (Fig. 7 E).

The Nematomorpha do not take up any solid food. This fact has led to extensive **simplifications** in the **gut** but, however, does not explain the subterminal displacement of the opening of the mouth. It belongs in the ground pattern of the Nematomorpha but may, together with the pharynx, be completely absent - as in *Gordius aquaticus* (SCHMIDT-RHAESA 1997c). When present, the pharynx is a cuticularized tube with round lumen and without any musculature. In cross section the thin gut consists of the few cells with microvilli to the lumen. In *Nectonema* it has a blind end, but in the Gordiida joins together with the sexual organs to a cloaca.

The famous **parenchyma** of the Nematomorpha is of little value for the determination of kinship relations because it is not present in *Nectonema*, but only occurs in the Gordiida - and, thus, hardly constitutes a feature of the ground pattern of the horsehair worms. Cellular parenchymatous tissue generally fills the space between body wall and organs except for insignificant clefts of the primary body cavity. Only in mature females is the parenchyma limited to the periphery of a large central cavity with egg cells (Fig. 7 J).

There are large differences between the sexual organs of *Nectonema* and those of the Gordiida; they will thus be discussed separately.

Nectonema. The females possess an extensive gonoparenchyma as presumed site for the formation of eggs; a homology with the parenchyma of the Gordiida has been denied. In ripe females a large central cavity is then filled with eggs; now neither gonoparenchyma nor an epithelial gonad wall are present. The male is characterized by an unpaired dorsal sperm sac. In both female (♀) and male (♂) the sexual openings are terminal.

Gordiida. Females with two dorsolateral ovarian tubes. The eggs originate from the epithelium of lateral protrusions which form the ovaries.

The oviducts join together to form an atrium from the front of which a seminal receptacle emerges and into which the gut enters behind. The thus formed cloaca has its terminal end at the posterior end of the body. The male possesses two tube-shaped testes. The two vas deferens also join with the gut here to give a cloaca with a subterminal exit.

It seems reasonable to discuss the cloaca of the Gordiida with regard to a possible homology with the cloaca of the Nematoda. However, I consider this to be problematic because a cloaca only exists in the male sex of the Nematoda and is lacking completely in the taxon *Nectonema* of the Nematomorpha.

Systematization

Nectonema – Gordiida

Nectonema

Autapomorphies: ? Swim bristles = bundle of hair-like structures of the epicuticle (Fig. 7C). Blind closure of gut. ♀ with gonoparenchyma, ♂ with unpaired sperm sac.

Of the four described species only *N. agile* (Atlantic, Mediterranean) and *N. munidae* (Norway) have been repeatedly detected. Host organisms are always Decapoda (Crustacea). SCHMIDT-RHAESA (1996a) analyzed a population of the crab *Munida tenuimana* (Kors Fjord, near Bergen, 670 m depth). In 951 individuals 15 ♀ and 2 ♂ specimens of *Nectonema munidae* were found.

Gordiida

Autapomorphies: Life in fresh water. Nerve system: reduction of the dorsal nerve cord; subepidermal displacement of the ventral cord into the body cavity. Parenchyma. Ovarian tubes with ovaries as lateral protrusions. ? Cloaca in both sexes.

Gordius (with *G. aquaticus* as the first species of the Nematomorpha to be described), *Paragordius, Parachordodes, Gordionus.*

Although the infection by larvae of a broad spectrum of invertebrates and even some vertebrates living in water has been detected, the development to juvenile stages ripe for emergence only takes place in arthropods. Coleoptera are the most frequent hosts in Europe (SCHMIDT-RHAESA 1997c).

Scalidophora

The Scalidophora form the sister group of the Nematoida with good justification as a monophylum (LEMBURG 1995a). They encompass the Priapulida, Loricifera, and Kinorhyncha. The former two entities have been recently recognized as adelphotaxa and accordingly combined to a new taxon under the name Vinctiplicata (LEMBURG 1999).

As locomotory organ, the Scalidophora have an eversible proboscis with long spines in a radially symmetrical arrangement. Inside the hollow **scalids** there are epidermis cells and sensory cells. This is the major difference to the larval hooks of pure cuticle in the Nematomorpha.

Autapomorphies (Fig. 1 → 7)

- Division of the body into three sections: introvert (proboscis) - neck - abdomen.
- Introvert.
  Motile locomotory organ that can be everted and retracted.
- Scalids.
  Regular circlets of cuticular spines arranged on the introvert in a radially symmetrical pattern. They probably occur in 30 longitudinal rows (p. 29).
  The cavity under the cuticle of the scalids is filled with epidermis and receptor cells.
- Mouth cone.
  Extendable structure with scalid-like sensilla (oral styles, buccal papillae).
- Frontal displacement of the circumpharyngeal nerve ring.
  Only in the three taxa of the Scalidophora does the nerve ring lie far forward at the boundary between mouth cone and introvert. This position is considered as an apomorphy in comparison with the Nematoda in which the nerve ring courses around the anterior third of the pharynx.
- Flosculi.
  Tiny sensory "spots" in the skin (Fig. 10B). At least one receptor cell runs in a canal through the cuticle to a pore at the point. A ring of cuticular papillae surrounds the pore (LEMBERG 1995b).

Vinctiplicata – Kinorhyncha

The development through larvae with longitudinally folded lorica and subadult postlarvae represents the most prominent synapomorphy between the Priapulida and the Loricifera (LEMBURG 1999). In contrast, the direct development of the Kinorhyncha must be considered as a plesiomorphy.

Vinctiplicata

Autapomorphies (Fig. 1 → 8)

- Life cycle with indirect development.
  Larvae with specific larval characteristics are followed by postlarval stages with the principal organization of the adult. Metamorphosis occurs in the cuticle of the last larva. The postlarvae molt until they reach sexual maturity.
- Larva with tessellation of the neck region.
  Formation of a closure apparatus of rectangular cuticular fields. Primary arrangement of the fields in 20 longitudinal rows and about 5–8 transverse rows. After retraction of the introvert the cuticular fields close together in a dome-like fashion over the frontal opening.
- Abdomen of the larvae with longitudinal folded lorica.
  Subdivision of the abdominal cuticle into 20 (or 22) small plates displaced from each other by longitudinal ridges. Primarily, the lorica has a round cross section. (Decrease in the number of plates within the Priapulida, increase within the Loricifera.)
- Adult with urogenital system.
  In the Priapulida and the Loricifera the paired protonephridia and gonads are each combined to a urogenital system that is fixed to the body wall by a ligament and its ECM.
  In the Priapulida, the protonephridia opens into the posterior part of the gonoduct. In the Loricifera, the exact nature of the connection between the two organs still requires clarification (LEMBURG 1999).

Priapulida – Loricifera

Priapulida

A species-poor taxon of the sea bottom; with *Halicryptus spinulosus* reaching to the low-salt, brackish water of the Bay of Finland (PURASJOKI 1944). The currently described 18 species represent 3 completely different life forms (LEMBURG 1999; Fig. 8).

A

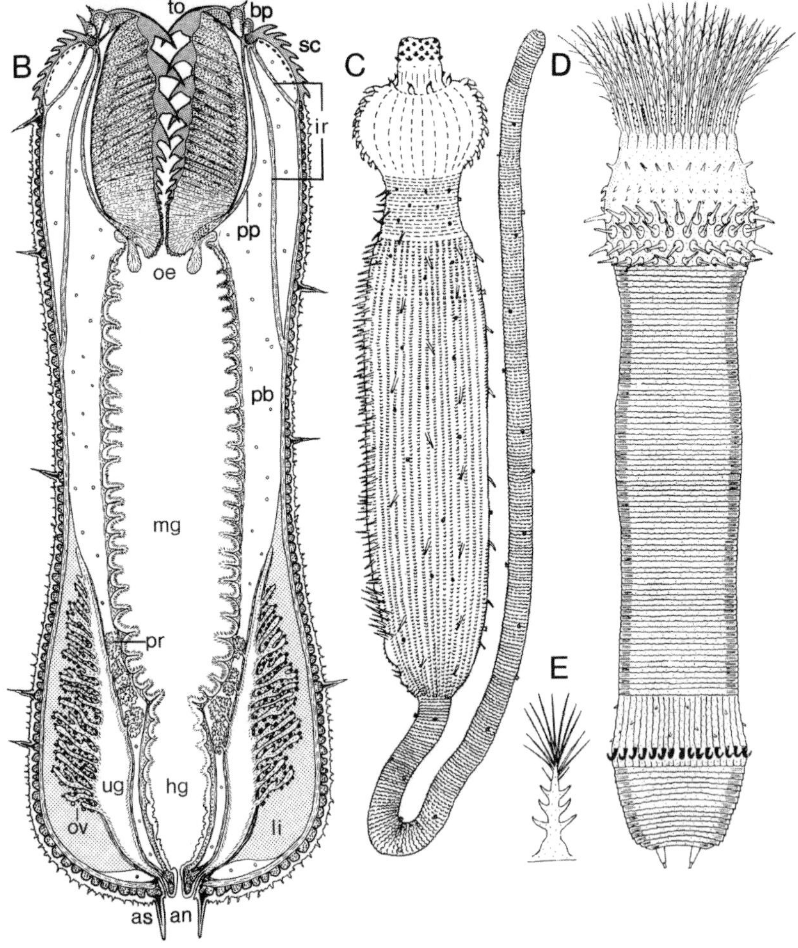
B
C
D
E
to
bp
sc
ir
pp
oe
pb
mg
pr
ug
hg
ov
li
as
an

◀ **Fig. 8.** Priapulida. **A** *Priapulus caudatus* with tuft-like caudal appendage. **B** *Halicryptus spinulosus.* Scheme of a horizontal section. Female. **C** *Tubiluchus corallicola.* Male with simple, long tail. **D** *Maccabeus tentaculatus* with a large crown of double spines or bristles. **E** *Maccabeus tentaculatus.* Circumoral scalid from a shaft with hair bristles. *an* Anus; *as* anal setae; *bp* buccal papillae; *hg* hindgut; *ir* introvert retractor; *li* ligament; *mg* midgut; *oe* oesophagus; *ov* ovary; *pb* primary body cavity; *pp* pharynx protractor; *pr* protonephridium; *sc* scalid; *to* tooth of the pharynx; *ug* urogenital tract. (**A** Original; Puget Sound, Washington, USA; **B** Lemburg 1999; **C, D, E** Adrianov & Malakhov 1996)

1. The supraspecific taxa *Halicryptus, Priapulus,* and *Priapulopsis* encompass macroscopic organisms of centimeter length. On foraging through surface layers of soft sediment by alternately protruding and retracting the introvert, they catch larger prey with the teeth of the pharynx.
2. Microscopic 2–3 mm-long species of the taxa *Tubiluchus* and *Meiopriapulus* are also motile organisms. As inhabitants of the interstitial system between the grains of sand, they are microphagic organisms.
3. *Maccabeus* (*Chaetostephanus*) species with a corresponding body size of around 3 mm are, in contrast, hemisessile "trappers"; they inhabit frail tubes of their own production. Twenty-five branched double spines or bristles form a large ring around the mouth cone on which eight long shafts with hair bristles stand. When small prey organisms get caught in this "tentacle apparatus", the bristle trap closes and the prey is drawn into the pharynx (POR & BROMLEY 1974; SALVINI-PLAWEN 1974). Morphologically, these specialized bristles correspond to the scalids of the introvert.

Autapomorphies (Fig. 1 → 9)

- Scalids in 25 longitudinal rows.
  With the exception of 8 scalids in the first circlet, the remaining scalids occur in 25 rows over the introvert. As compared with the 30 longitudinal rows in the Loricifera and Kinorhyncha (? ground pattern of the Scalidophora) this number is interpreted as an apomorphous condition.
- Scalids with apical and lateral receptor tubules.
- Completely dentate pharynx with cuspidate teeth.
- Simultaneous formation of new teeth in a pocket-like diverticulum at the end of the dentated pharynx.
- Cuticularized oesophagus following the pharynx extends as a U-shaped turned over tissue fold into the midgut. With a strong, multicellular sphincter.
- Completely unpaired state of the ventral nerve system in introvert and abdomen. In comparison, in the Loricifera and Kinorhyncha there are, at least in the introvert, two separate longitudinal nerves; in the ground pattern of the Scalidophora this also possibly holds for the ventral longitudinal nerves in the abdomen.

- Primary body cavity highly expanded.
- Closed body wall musculature in the abdomen in combination with the evolution of a novel mode of locomotion by means of peristaltic contraction waves.
- Protonephridia. Terminal complexes composed of several terminal cells combine to highly branched "solenocyte trees"; these run by means of a common canal in the main canal of the system.

From the above list of autapomorphies, the complete occupation of the inner wall of the pharynx with **cuticular teeth** as well as the zone of their new formation at the posterior end of the pharynx are the most conspicuous features. This complex is unique within the Nemathelminthes - and can be interpreted as an evolutionary novelty that has evolved in the stem lineage of the Priapulida.

The construction of the **pharynx musculature** is, in contrast, considered as a plesiomorphy (LEMBURG 1999). It is not a myoepithelial musculature, but rather a subepithelial combination of circular muscle bundles interspersed with right-angled tooth abductors which join at the periphery to a longitudinal muscle sheath. This construction can be derived from the subepithelial body musculature in which the longitudinal muscles (comparable with the tooth abductors) insert at intervals between the circular muscle bundles of the epidermis.

From now on, I will concentrate on considerations about three interpretatively related circumstances.

Body Division. Three sections - introvert, neck, abdomen - are placed in the ground pattern of the Scalidophora because this division is realized throughout the Kinorhyncha (that do not have larvae), in larvae and adults of the Loricifera and finally also in larvae of the Priapulida. We must, however, differentiate among the adults of the Priapulida. In the microscopically small, interstitial *Tubiluchus* species, the introvert and abdomen are separated by a distinct neck region. In the macroscopic species, the neck is indicated either solely by a constriction between introvert and abdomen (*Priapulus*) or it is completely lacking (*Halicryptus*).

On comparison of the three taxa, Kinorhyncha, Loricifera and Priapulida, the absence of a neck region in the large Priapulida must be considered as an apomorphous state (LEMBURG 1999). I consider this statement to be important since it casts light on the problem of body size in the ground pattern of the Priapulida.

Body Size. Taking two species of already mentioned taxa, *Halicryptus spinulosus* reaches a length of maximal 4 cm and *Priapulus caudatus* grows to as much as 20 cm in length. In contrast, *Tubiluchus, Meiopriapulus* and *Maccabeus* remain in the order of 2–3 mm.

What belongs in the ground pattern of the Priapulida? This question has been a subject of controversial discussion even in the most recent literature (AHLRICHS 1995; LORENZEN 1996; ADRIANOV & MALAKHOV 1996; LEMBURG 1999). The hypothesis of an original size in the centimeter range may appear plausible in the light of the widespread "miniaturization" of marine organisms upon conquest of the interstitial milieu. However, this is not only not compatible with the mentioned apomorphous body division of larger Priapulida. With the exception of some of the Priapulida, all free-living Nemathelminthes are small organisms of millimeter length. The Gastrotricha, Nematoda, Kinorhyncha, Loricifera as well as the interstitial Priapulida must have all independently shrunk to millimeter size if the macroscopic state within the Priapulida was an original feature that must inevitably be included in the ground pattern of the Nemathelminthes. The improbability of this extremely laborious assumption is obvious. In other words, the centimeter dimensions must be interpreted as the result of a secondary increase of body size within the Priapulida.

Union of Egg and Sperm Cells. However, this argumentation is not free of conflicts with the evaluation of certain elements of reproductive biology.

Priapulus caudatus has the widespread, primitive sperm pattern of the Acrosomata with uniform acrosome as well as four large mitochondria and two centrioles arranged at right-angles behind a round nucleus (Vol. I, Fig. 14). Linked to this is the free emission of large amounts of gametes into the surrounding water and external fertilization. *Tubiluchus* species represent, so to speak, the antipode. There, there are stretched sperms with a division of the acrosome into two intertwined halves and two also spirally rolled up outgrowths of the nucleus (Fig. 9B). In addition, the existence of sperm in the urogenital tract of female animals provides unambiguous evidence for internal fertilization (STORCH & HIGGINS 1989; ALBERTI & STORCH 1988).

The interstitial species *Meiopriapulus fijiensis* occupies the interesting middle position. In spite of its millimeter size, it has primitive sperm in which merely the accessory centriole is absent (Fig. 9E). The sperm pattern supports the assumption of free release; however, the sperm must apparently find its way to the female since vivipary has been demonstrated in this case (STORCH, HIGGINS & MORSE 1989; HIGGINS & STORCH 1991; LEMBURG & SCHMIDT-RHAESA 1999).

Taken together, a microscopic, interstitial stem species with a free release of primitive sperm, but an internal fertilization, could belong in the ground pattern. The release of large numbers of male and female gametes into the water of the biotope would, in this case, be a derived behavior, evolved in connection with the high production of egg and sperm cells upon the secondary increase in size and volume of the body.

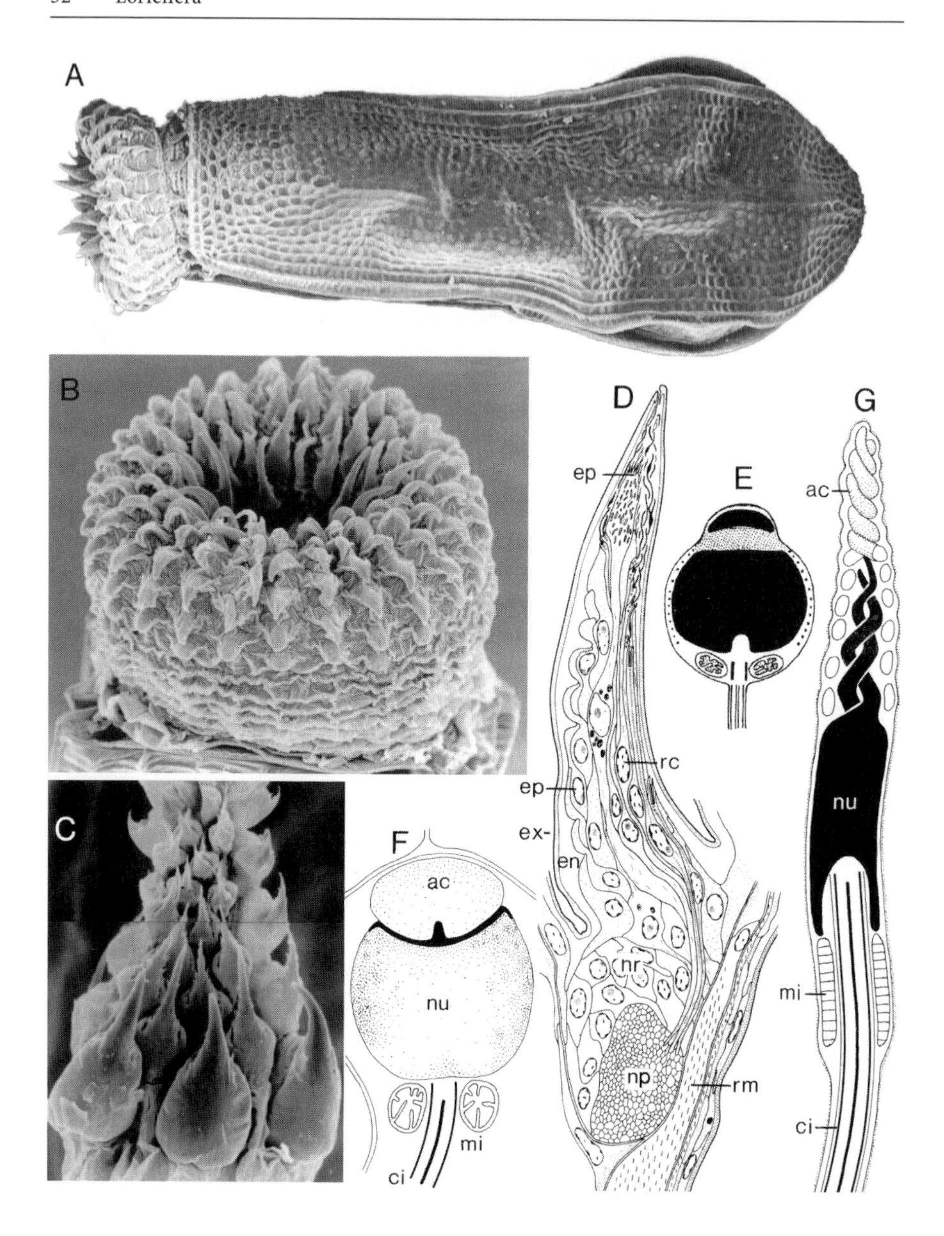

Loricifera

As the smallest of the multicellular animals the Loricifera were discovered very late. To date, about 80 species have been detected worldwide in marine milieu – from the tidal zone to a depth of 8260 m. They live in the interstitial of coarse sand of the littoral zone as well as on red clay in deep sea trenches of the hadal zone (KRISTENSEN 1983, 1991 a, b; HIGGINS & KRISTENSEN 1986, 1988).

◀ **Fig. 9.** Priapulida. **A** *Halicryptus spinulosus.* Larva. Introvert with scalids – short neck region – long abdomen with lorica. Surface of the lorica divided into fields. **B** *Priapulus caudatus.* Larva. Oblique view of the introvert. **C** *Halicryptus spinulosus.* Larva. Protruded pharynx with teeth. **D** *Halicryptus spinulosus.* Larva. Longitudinal section through a scalid of the first circlet that is filled with epidermis and receptor cells. At the apex a terminal pore with two receptor cells. The cilium of one cell passes through the pore and reaches the surface; the other cilium ends in the pore canal. **E** *Meiopriapulus fijiensis.* Primitive sperm, but accessory centriole is lacking. **F** *Tubiluchus corallicola.* Young spermatid with plesiomorphous attributes: acrosomal vesicle, round nucleus, roundish mitochondria. **G** *Tubiluchus corallicola.* Fully differentiated, apomorphous sperm. Acrosome and anterior part of the nucleus spirally rolled up. *ac* Acrosome; *ci* cilium; *en* endocuticle; *ep* epidermis cell; *ex* exocuticle; *mi* mitochondrion; *np* neuropil of the nerve ring; *nr* nerve ring; *nu* nucleus; *rc* receptor cell; *rm* retractor muscle of the introvert. (**A, B, C** Originals; Lemburg; **D** Lemburg 1995a; **E** Adrianov and Malakhov 1996 after Storch et al. 1989; **F, G** Storch & Higgins 1989)

With body lengths between 80 and 385 μm, they are midgets of unbelievable complexity. Adult Loricifera with a length of 100 μm are in the same order of size as small unicellular organisms, but have over 200 multicellular scalids with muscles on the introvert.

Nanaloricus, Pliciloricus, Rugiloricus.

Autapomorphies (Fig. 1 → 10)

- Adult with lorica.
  In the interpretation as persistence of a larval feature, the condition becomes an autapomorphy in the ground pattern of the Loricifera.
- Mouth cone at the anterior end.
  Differentiation of the introvert to an apex, retractable like a telescope. Small terminal opening of the mouth.
- Cuticularized buccal tube.
  Connection between the mouth and the pharynx.
- Pharynx with myoepithelium.
  Construction of muscle epithelial cells with radially radiating myofibrils. Y-like pharynx lumen with three edges, one of which is directed to the ventral side.
- Spinoscalids on the introvert.
- Mobile spine with basal joint and musculature.
- Neck region with trichoscalids.
  Clearly distinguishable from the other scalids by long cuticular hairs. Convergency to homonymous formations in the Kinorhyncha.
- Larvae with toes.
  Two structures at the posterior end. It has not been clarified whether propeller-like locomotory devices or rod-like structures with anchor function belong in the ground pattern of the Loricifera.

Larvae and **adults** of the Loricifera all show a morphological division into the three parts introvert (head), neck region, and abdomen with lorica (Fig. 10). In comparison to the adelphotaxon Priapulida, therefore, the existence of a lorica in the adult organism should be emphasized as being a unique character. The hypothesis of taking over or continuing a primary larval characteristic in the organization of the adult offers the simplest explanation with which the lorica of the adult is taken as an autapomorphy of the Loricifera.

The slim **mouth cone** tapering to a point at the anterior end of the introvert is also unique (Fig. 10 A, B). It cannot be inverted, but is comprised rather of three telescope-like retractable parts. The terminal opening of the mouth is followed by a cuticular **buccal tube**; this is short in *Pliciloricus enigmaticus* and, still in the mouth cone, opens into the muscular pharynx - in *Nanaloricus mysticus* it runs as a long duct to the pharynx in the region of the introvert-abdomen boundary.

The **pharynx wall** consists of myoepithelial cells with radially arranged muscle fibrils. The myoepithelium encloses a triradiate, Y-shaped lumen. When the pharynx in the adelphotaxon Priapulida with a round lumen, muscle-free epithelium and subepithelial musculature represents plesiomorphous features (p. 30), the triradiate, myoepithelial construction of the pharynx in Loricifera must have arisen as an evolutionary novelty in their stem lineage.

We now come to the provision of the introvert with nine rows of impressive **scalids** that possess different names according to their different structures. The first row of club-like clavoscalids is followed by eight rows of variable, mostly spine-shaped spinoscalids. In the neck region 15 trichoscalids are placed around the body.

The structure of the above discussed **lorica** varies within the taxon. The strongly sclerotized lorica of *Nanaloricus mysticus* consist of 6 plates with 15 spines at the anterior edges. In contrast, the lorica of *Pliciloricus* and *Rugiloricus* are only weakly cuticularized and are divided into small fields by numerous longitudinal folds.

Finally, the larvae of the Loricifera with their two prominent **toes** also represent a marked autapomorphy of the Loricifera (Fig. 10 C, D). The leaf-shaped, broadened toes of *Nanaloricus* are each equipped with 8–12 pairs of cross-striated muscles; this can probably only be the propeller for swimming locomotion. In contrast, the toes of *Pliciloricus* and *Rugiloricus* appear to have differentiated to adhesive devices; they consist of hollow spines with glandular tissue.

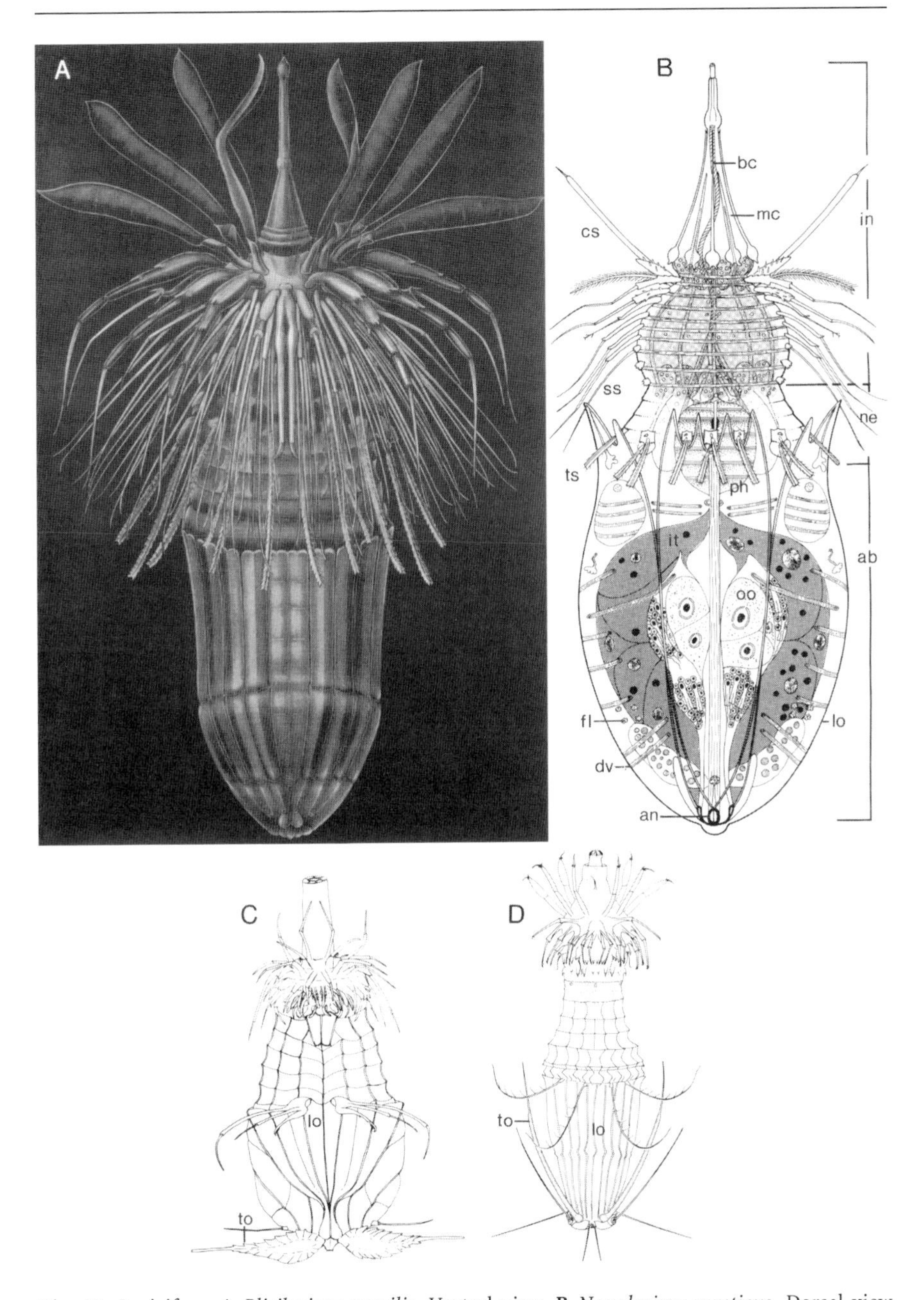

Fig. 10. Loricifera. **A** *Pliciloricus gracilis*. Ventral view. **B** *Nanaloricus mysticus*. Dorsal view of a female. **C** *Nanaloricus mysticus*. Larva with leaf-shaped toes. Ventral view. **D** *Pliciloricus profundus*. Larva with spine-shaped toes. Ventral view. *ab* Abdomen; *an* anus; *bc* buccal tube; *cs* clavoscalid; *dv* dorsoventral muscle; *fl* flosculum; *in* introvert; *it* intestine; *mc* mouth cone; *lo* lorica; *ne* neck region; *oo* oocyte; *ph* pharynx; *ss* spinoscalid; *to* toe; *ts* trichoscalid. (**A**, **D** Higgins & Kristensen 1986; **B** Kristensen 1991; **C** Kristensen 1983)

Kinorhyncha

By way of direct development, the Kinorhyncha remain more primitive than the Priapulida and the Loricifera. Due to the lack of a larva they do not have the characteristic lorica of the Vinctiplicata. On the other hand, the Kinorhyncha are already identified as a monophylum solely by the apomorphous division of the abdomen into 11 zonites.

At present, around 120 species are recognized (ADRIANOV & MALAKHOV 1994). Without exception, they are microscopic organisms of the sea bottom with body lengths between 180 µm and 1 mm. Mud in slack-water regions from the littoral to the deep sea should be emphasized as colonization areas. However, Kinorhyncha have also conquered the surf zones of sandy shores (*Cateria styx, C. gerlachi*) as well as the interstitial spaces of sublittoral coarse sand (*Zelinkaderes submersus*) (GERLACH 1956, 1969; HIGGINS 1968, 1990).

Systematization

As yet, there is no phylogenetic system for the Kinorhyncha. I will name some taxa in their distribution to the common high-ranking entities Cyclorhagida and Homalorhagida.

"Cyclorhagida"

Only the head (introvert) can be inverted. The neck is equipped with 14–16 cuticular plates; after retraction of the introvert the placids form a dome-like closure at the anterior end.

Zelinkaderes (Fig. 11 A–D), *Cateria, Semnoderes, Echinoderes.*

Homalorhagida

Introvert and neck are invaginated upon retraction of the anterior end. The first zonite of the abdomen forms a closure apparatus.

Paracentrophyes, Kinorhynchus, Pycnophyes (Fig. 11 E, F).

The mode of invagination and closure in the "Cyclorhagida" is a primitive characteristic of the ground pattern; similarly, further diagnostic features of the traditional classification are plesiomorphies. The "Cyclorhagida" are a paraphylum. In contrast, the Homalorhagida with their ability to retract the first two body sections appear to be a well-founded monophylum of the Kinorhyncha.

ADRIANOV & MALAKHOV (1994) retain the division into Cyclorhagida and Homalorhagida, but do order the taxa one after the other according to the increasing number of apomorphous characteristics. If we translate their "scheme of phylogenetic relationships within Kinorhyncha" into the language of phylogenetic systematics, then the taxon *Zelinkaderes* with a

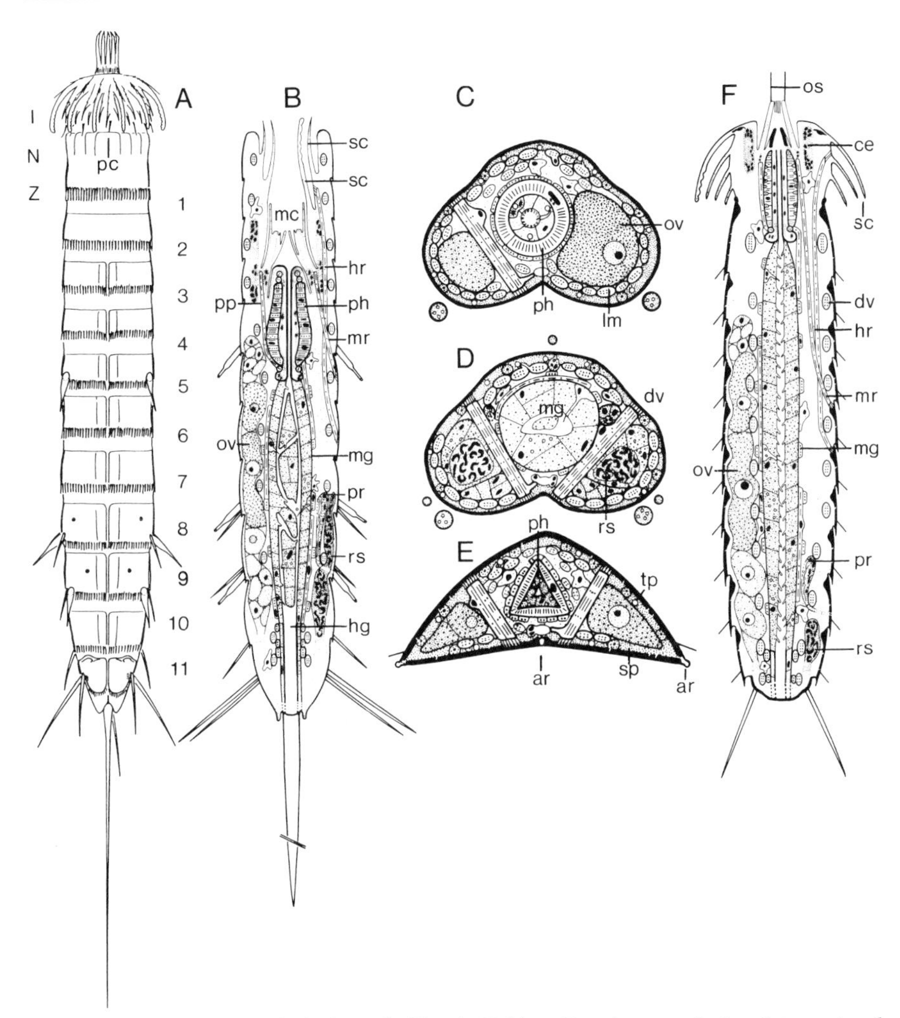

Fig. 11. Kinorhyncha. **A** *Zelinkaderes floridensis.* Habitus. Female, ventral view. Introvert and mouth cone extended. *I* Introvert (head); *N* neck region with placids; *Z* zonite of the abdomen. **B** *Zelinkaderes floridensis.* Longitudinal section. Mouth cone and introvert retracted. **C** and **D** *Zelinkaderes floridensis.* Cross section through zonites 5 and 9. **E** *Pycnophyes dentatus.* Cross section through zonite 4 (pharynx retracted). **F** *Pycnophyes dentatus.* Longitudinal section. Introvert and mouth cone extended. *ar* Articulation between cuticular plates of zonites; *ce* brain; *dv* dorsoventral muscle; *hg* hindgut; *hr* head retractor; *lm* longitudinal muscle; *mc* mouth cone; *mg* midgut; *mr* mouth cone retractor; *os* oral style; *ov* ovary; *pc* placids; *ph* pharynx; *pp* pharynx protractor; *pr* protonephridium; *rs* seminal receptacle; *sc* scalids; *sp* sternal plate; *tp* tergal plate. (**A** Higgins 1990; **B–F** Neuhaus 1994)

series of primitive characteristics forms the sister group to all other Kinorhyncha. This is of interest because NEUHAUS (1994) placed ultrastructural studies on *Zelinkaderes floridensis* in the center of an elaboration of the ground pattern of the Kinorhyncha. We will discuss this after a short summary of the autapomorphies.

Autapomorphies (Fig. 1 → 11)

- Division of the abdomen into 11 zonites. Combined with this is the evolution of paired dorsoventral muscles and peripheral longitudinal muscles as well as paired ganglia in the zonites.
- Cuticle of zonites 3–11 ventral with longitudinal indentation.
- Neck region with closure apparatus of at least 14 placids.
- Zonites 1–10 with paired lateral spines and an unpaired dorsal spine. Zonite 11 with two lateral spines on each side and a spine at the posterior end.
- Eversible mouth cone with external and internal oral styles.
- Pharynx with multilamellar epicuticle.
- In the pharynx musculature cells with circular muscles and radial muscles join together in a regular, alternating pattern.
- Midgut with monociliary sensomotory cells.
- Hindgut with slit-like, transverse lumen.
- Internal fertilization.
- Six juvenile stages in the development.

Ground Pattern

At the beginning of a discussion of major characteristics of the ground pattern, we must insist upon a logical and consistent **terminology for the division of the body**. From the ground pattern of the Scalidophora, the Kinorhyncha have taken over a division into three regions. When these are designated for the Priapulida and Loricifera as introvert, neck region, and abdomen, then the homologous sections of the Kinorhyncha should also have these names (even when the term trunk instead of abdomen may appear to be more common).

On this basis, we can take the division of the abdomen into 11 parts as the central autapomorphy of the Kinorhyncha. These are frequently called zonites. Unfortunately, at the same time, the introvert and neck are also included as zonites. However, introvert and neck have nothing to do with the subordinate evolution of the abdominal zonites. Introvert and neck are not serial equivalents of parts of the abdomen; the term zonite is not suitable here. Finally, the word "segment" has also been inappropriately used for the Kinorhyncha. It results in a body of 13 segments, whereby the head and neck are also counted as segments. However, the subdivision of the abdomen into zonites can be interpreted with desirable certainty as an evolutionary novelty within the Scalidophora; it is not homologous with the segmentation of the body of the Articulata. The homonymous designation can lead to misunderstanding; it has to be rejected for the Kinorhyncha.

The result of this discussion is the following unimpeachable terminology for the body division:

introvert – neck region – abdomen with zonites 1–11 (Fig. 11 A).

We have already mentioned the original microscopic dimensions of the **habitus.** A roundish cross section only slightly flattened on the ventral side of the abdomen constitutes a further plesiomorphous characteristic of the ground pattern. This arises from the primary state of the **integument.** A weakly developed cuticle covers the abdomen in uniform thickness without the differentiation of plates. Complete cuticular rings are present in zonites 1 and 2. There are ventromedial longitudinal indentations in zonites 3–11. The evolution to one strong, semicircular dorsal plate and two flattened ventral plates occurred first within the Kinorhyncha (Fig. 11 E).

The **introvert** carries a minimum of three circles of spinoscalids and one circle of trichoscalids. The spinoscalids surround the head in a radially symmetrical pattern.

The **neck** possesses a closure apparatus consisting of at least 14 weakly differentiated placids.

Motile **integument spines** belong in the ground pattern in the following arrangement: zonites 1 to 10 – paired lateral spines, an unpaired median dorsal spine. Zonite 11 – two lateral spines on both sides and a terminal spine.

Primary body cavity. No enclosure by epithelial or myoepithelial cells.

An eversible **mouth cone** exists as a ring-shaped epidermal fold around the anterior section of the pharynx. It is equipped with external and internal oral styles as well as a basal nerve ring.

With regard to the construction of the **pharynx** (Fig. 12 C–E) with a multilamellar epicuticle, the two-layered pharynx wall should be emphasized. An inner, muscle-free epithelium is followed by an outer monolayer of muscle cells. The arrangement of these muscle cells is unique within the Nemathelminthes. About 30 cells are joined together in the longitudinal direction with a regular alternating pattern of circular and radial muscles. Thereby the individual cells each completely encircle the pharynx epithelium. In other words, in any cross section one will only find one muscle cell with nucleus, whereas for other Nemathelminthes several muscle cells will always be seen. The described ordering pattern of the pharynx musculature is apparently an autapomorphy of the Kinorhyncha. In contrast, the round pharynx lumen as in *Zelinkaderes* is a primitive characteristic of the ground pattern. Cross sections showing a triradiate lumen as well as a dorsally directed edge are often found in the literature. However, this apomorphous state only evolved within the Kinorhyncha and is realized solely in species of the Homalorhagida. This leads to the conclusion of a convergent evolution to the pharynx of the Loricifera with a triradiate lumen; in addi-

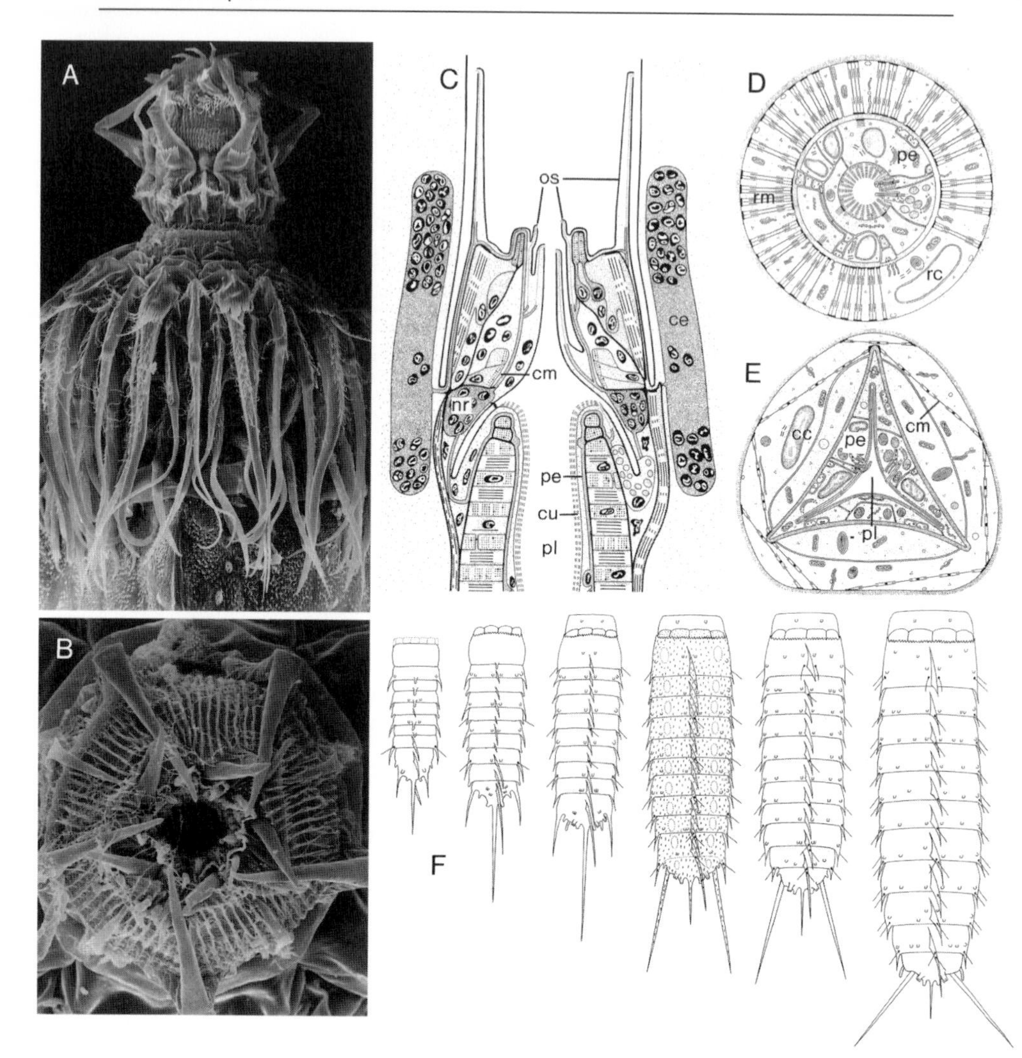

Fig. 12. Kinorhyncha. **A** and **B** *Paracentrophyes praedictus.* **A** Ventral view of the anterior end. Mouth cone with oral styles. Introvert with scalids. **B** Top view of the mouth cone. The five outermost oral styles bend over the trap of the mouth cone towards the central lumen. **C** *Zelinkaderes floridensis.* Longitudinal section through the brain, retracted mouth cone and anterior section of the pharynx. **D** *Zelinkaderes floridensis.* Cross section through the pharynx. Epithelial cells surround a round lumen. In the periphery a closed muscle cell with radial myofibrils. **E** *Pycnophyes dentatus.* Cross section through the pharynx with triradiate lumen. Peripheral muscle cells with ring-shaped myofibrils. **F** *Paracentrophyes praedictus.* Direct development via juvenile stages. Distribution of cuticular hairs and attachment of dorsoventral muscles are shown in stage 4. *cc* Circular muscle cells; *ce* brain; *cm* circular myofibrils; *cu* cuticle of the pharynx; *nr* nerve ring in the mouth cone; *or* oral style; *pe* pharynx epithelium; *pl* pharynx lumen; *rc* radial muscle cell; *rm* radial myofibrils. (**A, B** REM photographs. Originals; Neuhaus; **C–E** Neuhaus 1994; **F** Neuhaus 1999)

tion, one surface thereof is directed to the dorsal side (KRISTENSEN 1991b, Fig. 41).

In the **midgut**, monociliary sensomotory cells are found between nonciliated intestinal cells. The **hindgut** is transversely flattened, slit-shaped and equipped with a complicated system of dilators.

The intraepidermal **nerve system** has taken over the widely forwards shifted brain ring from the ground pattern of the Scalidophora. In the abdomen, there are a pair of dorsal longitudinal nerves, a pair of ventrolateral longitudinal nerves and a strong ventral nerve cord. The origin of the latter as a product of the fusion of paired ventral nerves is documented by distinctly separated, paired ganglia in the zonites of the abdomen (NEBELSICK 1993); its evolution is connected with the division of the abdomen into zonites and is accordingly interpreted as an autapomorphy of the Kinorhyncha.

Just this division has also led to the major apomorphous elements in the **musculature**. A pair of dorsoventral muscles per zonite as well as uniformly distributed longitudinal muscles in the periphery that extend from one zonite to the next effect the extension and contraction of the abdomen upon locomotion.

Paired **protonephridia** are positioned dorsolaterally in zonites 8 and 9 (Fig. 11B,F). In the ground pattern a protonephridium consists of several (? three) biciliated terminal cells, a canal cell, and a nephropore cell (KRISTENSEN & HIGGINS 1991; NEUHAUS 1994).

The Kinorhyncha are **gonochorists**. Paired gonads extend over several zonites of the abdomen. Genital pores are found at the boundary to the last zonite. The seminal receptacle of the female for taking in foreign sperm and the penile spine of the male provide clear arguments for internal fertilization in the stem species of the Kinorhyncha; however, copulation has only been observed once (NEUHAUS 1999).

We conclude with the **direct development** which, as such, represents a plesiomorphy in the ground pattern of the Kinorhyncha. However, the mode of ontogenesis with constantly six juvenile stages which follow each either with molting constitutes a specific characteristic (Fig. 12F). Stage 1 hatches with nine zonites. The two missing zonites arise in a subcaudal growth zone (NEUHAUS 1995, 1999).

On the Position of the Nemathelminthes in the Phylogenetic System of the Bilateria

Finally, we must return to the question of the relationship of the Nemathelminthes with the Spiralia or the Radialia (Vol. I, p. 130). When the justification for these two extensive entities as monophyla is maintained, there are four competing answers. The Nemathelminthes are (1) the sister group of a subtaxon of the Spiralia or (2) of the entire Spiralia – the Nemathelminthes are (3) the sister group of a subtaxon of the Radialia or (4) of all Radialia.

Let us start at the end. In the monophylum Radialia as a union of the Phoronida, Bryozoa, Brachiopoda and Deuterostomia (p. 62), there is no subgroup that could be a potential adelphotaxon of the Nemathelminthes. A remarkable agreement between the Nemathelminthes and the entire Radialia is, however, a continuous gut with anus that most certainly does not belong in the ground pattern of the Bilateria or the Spiralia. An economic explanation is the unique evolution of an anus in the common stem lineage of an entity from the Nemathelminthes + Radialia (BARTOLOMAEUS 1993). As far as I can see, however, there are no further agreements that might be interpreted as being in harmony with the evaluation of the anus as synapomorphies between the Nemathelminthes and the Radialia. The question of a possible sister group relationship between these two taxa must for now remain unanswered.

We turn to the Spiralia, but will set aside the newly posed question of its monophyletic character for a moment. The difficulties in proposing a specific cleavage pattern for the stem species of the Nemathelminthes are exposed above. For the present discussion, it can remain undecided whether it is a radial or a bilaterally symmetrical cleavage. In my opinion, there is no qualified justification for the derivation of the early ontogenesis of the Nemathelminthes from the spiral cleavage – either from the duet cleavage of the Acoela or from the widespread quartet cleavage. This situation would have to be reconsidered if a certain subtaxon of the Spiralia should convincingly offer itself for other reasons as the sister group of the Nemathelminthes, but this is not the case.

In contrast, the derivation of the cleavage in the Nemathelminthes from a stem species with the plesiomorphous radial cleavage of the Bilateria does not present any problem. Consequently, a sister relationship between the Nemathelminthes and the entire Spiralia was hypothesized (AHLRICHS 1995; EHLERS et al. 1996) – in other words, a separation of the stem lineages of the Nemathelminthes and the Spiralia before the evolution of spiral cleavage. One argument for this hypothesis is the construction of the nerve system out of a dorsofrontal cerebral ganglion with regularly distributed perikarya and several pairs of longitudinal nerves in the body. Such a state has been realized in the Gastrotricha and is postulated for the ground pattern of the Spiralia.

However, the Spiralia and also the Plathelminthes as monophyla are being increasingly discussed controversially (RIUTORT et al. 1993; KATAYAMA et al. 1993; HASZPRUNAR 1996a, b; CARRANZA et al. 1997; LITTLEWOOD et al. 1999). This involves the separation of the Acoelomorpha (Acoela + Nemertodermatida) from the Plathelminthes as well as their interpretation as basal Bilateria with a primary lack of cerebral ganglion and longitudinal body nerves. This will hardly be the last word. However, if the mentioned deficits should be a plesiomorphy, then the comparison of the Nemathelminthes by means of the nerve system must be extended to Spiralia entities with ganglioneural organization - and these have the spiral quartet cleavage. With this we return to the less promising search for a sister group of the Nemathelminthes with just this cleavage pattern.

It seems to me that the time is not yet ripe for a convincing solution of the phylogenetic relation of the monophylum Nemathelminthes.

Syndermata

The Rotifera, the taxon *Seison* and the parasitic Acanthocephala (Fig. 13) are interpreted as members of a monophylum and combined under the name Syndermata (AHLRICHS 1995). I present in the first step a set of de-

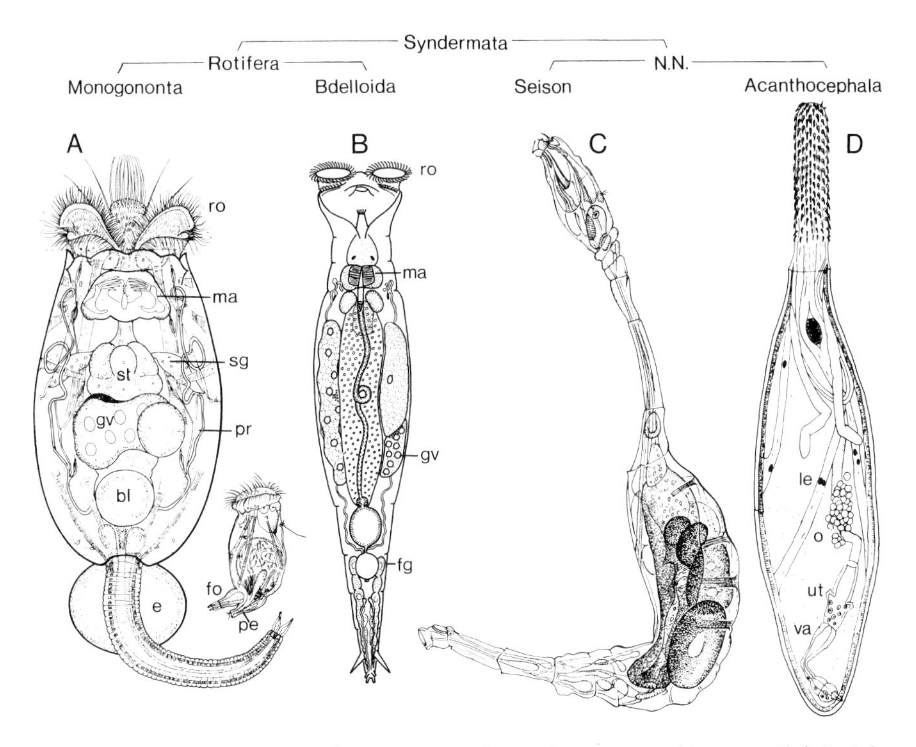

Fig. 13. Syndermata. Members of the highest-ranking subtaxa. **A** *Brachionus* sp. (*left* ♀, *right* dwarf male). **B** *Philodina citrina* (♀). **C** *Seison annulatus* (♀). **D** *Plagiorhynchus cylindraceus* (♀). *bl* Urinary bladder; *e* egg; *fg* foot gland; *fo* foot; *gv* germovitellarium; *le* lemniscus; *ma* mastax; *o* ovarian balls; *pe* penis; *pr* protonephridium; *ro* wheel organ; *st* stomach; *sg* stomach gland; *ut* uterus; *va* vagina. (**A** Nogrady et al. 1993; **B** Bunke & Schmidt 1976; **C** de Beauchamp 1965; **D** Schmidt 1985)

rived characteristics that are realized in all three taxa to justify the monophylum. Only then will I discuss the pharyngeal masticatory system with trophi and the ciliary corona or wheel organ at the anterior end as two outstanding autapomorphies that also probably arose in the stem lineage of the Syndermata, but then got lost again within the entity. The masticatory system is absent in the parasitic Acanthocephala; the corona was probably reduced already in the common stem lineage of *Seison* + Acanthocephala.

Autapomorphies (Fig. 14 → 1)

- Syncytial epidermis. Fusion of epidermis cells with abandonment of the membranes in the development.
- Intrasyncytial thickenings in the periphery of the epidermis (Fig. 15A). The firm body covering is not a cuticle as an outward directed exudate, but an "internal layer".
- Canals as infoldings of the outer cell membrane penetrate the intrasyncytial thickenings and run in the underlying cytoplasm; they end here in sack-like swellings (AHLRICHS 1997; Fig. 15B).
- Insertion of the cilium at the anterior end of sperm.
  The cilium runs backwards - either along the plasma membrane of the filiform sperm (*Seison*, Acanthocephala) or in the cytoplasm (Rotifera).
- Primary body cavity in the form of cavities between skin and gut in which the organs lie. All boundaries covered with extracellular matrix (ECM).

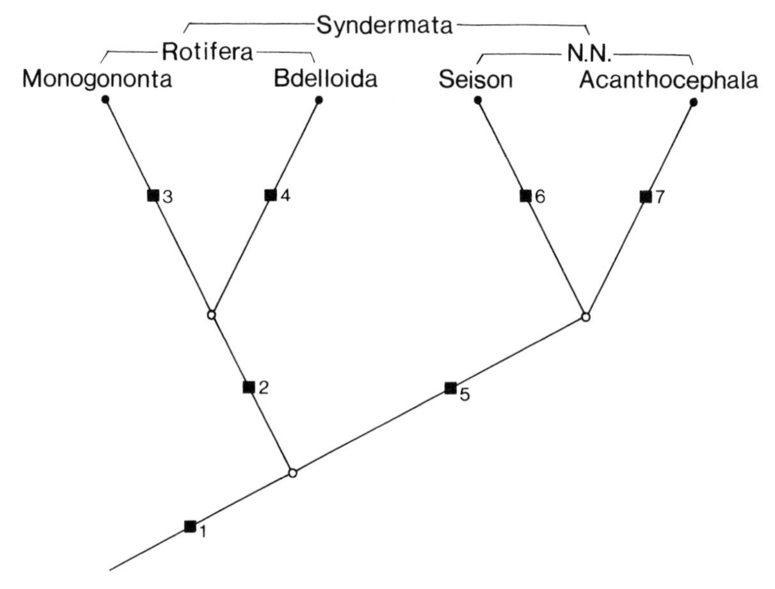

Fig. 14. Diagram of the phylogenetic relationships within the Syndermata

In comparison with the compact organization of the Plathelminthes, Gnathostomulida, Nemertini and in the ground pattern of the Nemathelminthes, an evolution of spaces between organs must be postulated in the stem lineage of the Syndermata.

The pharynx of the Rotifera and the taxon *Seison* is modified as a **mastax** with **seven extracellular hard parts** that are joined by ligaments. They are designated as **trophi**. In the Rotifera, these are the paired unci, manubria, and rami as well as the unpaired fulcrum (Fig. 15). An evaluation of the manifold evolutionary changes under the rules of phylogenetic systematics is still lacking (MARKEVICH & KUTIKOVA 1989).

In *Seison*, only the central elements fulcrum and rami can be unequivocally homologized with the same-named hard parts of the Rotifera (AHL-

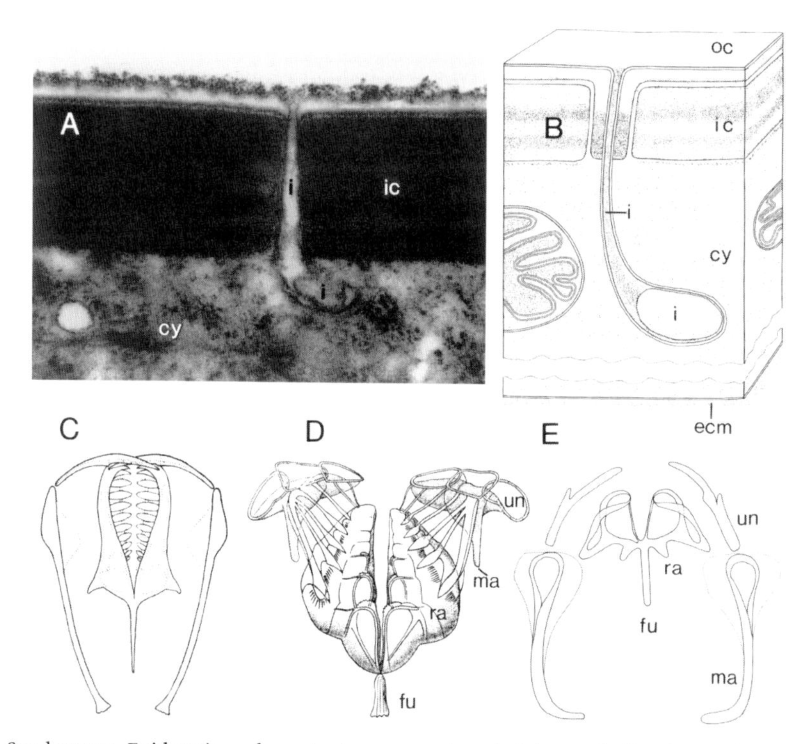

Fig. 15. Syndermata. Epidermis and masticatory apparatus. **A** *Seison nebaliae*. TEM photograph. Section through the external part of the epidermis with glycocalyx on the cell membrane and intrasyncytial thickening. **B** *Seison annulatus*. Block diagram of the epidermis from electron microscope investigations. **C–E** Various states of the masticatory apparatus of the Rotifera Monogononta with seven trophi. **C** *Dicranophorus tegillus*. Ventral view. **D** *Epiphanes senta*. Top view. **E** *Proales germanica*. Ventral view. *cy* Cytoplasm; *ecm* extracellular matrix; *fu* fulcrum; *i* ingrowth of the outer membrane; *ic* intrasyncytial thickening (internal layer) of the epidermis; *ma* manubrium; *oc* outer cell membrane; *ra* ramus; *un* uncus. (**A, B** Ahlrichs 1997; **C** Remane 1929–1933; **D** de Beauchamp 1965; **E** Tzschaschel 1979)

RICHS 1995). The dorsal reinforcement of the rami in *Seison* is designated as a pseudoepipharynx to be on the safe side, and the lateral rods as pseudomanubria (Fig. 18 A, B). However, the existence of an identical number of seven hard parts in the masticatory apparatus of the entities Rotifera and *Seison* is hardly a coincidence – and the characteristics of *Seison* are explicable in connection with feeding as epibionts (see below). I have no reservations about setting a masticatory system with seven trophi in the ground pattern of the Syndermata.

The evolution of a masticatory system or chewing apparatus for grasping food, for its possible diminution and further transport to the gut must be considered together with the evolution of a system for the collection of nutrient particles. This system is recruited from the **head ciliation** which, with a ventral buccal field and a ring-shaped circumapical band, forms the basal elements of the **corona.** Probably, only the buccal field for brushing up particles during a slow sliding over the substrate was present in the stem species of the Syndermata (LORENZEN 1996). In any case, a circumapical band is absent in the Bdelloida; the two renowned ciliary disks are interpreted as the sole derivative of the buccal field (REMANE 1950). The circumapical band probably first arose in the Monogononta.

In any case, the **retrocerebral organ** – an extensive, multinuclear mucous gland behind the brain with apical opening – is inseparably linked with the wheel organ. If the secretion is a lubricant for the cilia, then the retrocerebral organ must have evolved together with the corona in the stem lineage of the Syndermata.

The stem species of the Syndermata was certainly a **millimeter-sized organism** with division of the body into a head, a tapered neck, broadened trunk and foot with two toes.

Equally certain is that the stem species of the Syndermata was a **dioecious organism** with males and females having about the same size. This is realized in the taxon *Seison*.

For the Acanthocephala, **sexual dimorphism** with markedly smaller males can be set in the ground pattern of the entity (see below) on account of its broad distribution. The interpretation is not problematic; the sexual dimorphism evolved in the stem lineage of the Acanthocephala separately from the other Syndermata.

However, what do we know about the evolution of sexual relations among the Rotifera with dwarf males and heterogony in the Monogononta as well as pure parthenogenesis in the Bdelloida? It is precisely the question of whether miniaturization of the male already occurred in the stem lineage of the Rotifera and pushed the Monogononta and Bdelloida on a common pathway. If yes, then not much can have happened in the stem lineage of the Rotifera. In the taxa *Proales* and *Rhinoglena* (Monogononta), the males are only slightly smaller than the females and possess a fully developed gut. The further evolutionary size reduction and a corresponding increase in the activity to find a sexual partner, as well as shorter life span with the cessation of feeding (EPP & LEWIS 1979) have certainly taken place first within the Monogononta. All this has nothing to do with the reduction of the male in the stem lineage of the Bdelloida. Here, we are only concerned with the question if the Bdelloida took over first steps of evolutionary change in their stem lineage from the ground pattern of the Rotifera, or if the reduction of the male occurred totally independently. I think that this question is still unanswered.

Systematization

In the phylogenetic systematization of the Syndermata we come to a reduction in the traditional extent of the Rotifera. The taxon *Seison* is interpreted as the sister group of the Acanthocephala and must, therefore, be removed from the entity Rotifera. The Rotifera with the remaining Monogononta and Bdelloida as well as an entity of *Seison* + Acanthocephala form the highest-ranking adelphotaxa of the Syndermata (Figs. 13, 14).

Syndermata
- **Rotifera**
  - **Monogononta**
  - **Bdelloida**
- **N.N.**
  - **Seison**
  - **Acanthocephala**

Rotifera – N.N. (Seison + Acanthocephala)

Rotifera

Following the interpretation of *Seison* as the sister group of the Acanthocephala (see below), we must now consider the justification of the remaining Rotifera (Rotatoria)[1] – Monogononta + Bdelloida – as a monophylum. The state of the female gonads provides a convincing autapomorphy.

[1] The name Rotifera is common in the English literature, in German, however, the rotifers are generally called Rotatoria. In a textbook that is published in English and German, one and the same taxon cannot have two different names. I have chosen the name Rotifera because the name Rotifères from CUVIER (1798) has been in use longer than the name Rotatoria from EHRENBERG in the year 1832 (RICCI 1983).

Autapomorphies (Fig. 14 → 2)

- Differentiation of the ovary into germarium and vitellarium.
- Filtration structure of the protonephridium with longitudinal slits (Fig. 16B).
  The plesiomorphous condition with rows of pores in the wall of the terminal hollow cylinder represents *Seison* (Fig. 16A).

Very similar to the Neoophora within the Plathelminthes (Vol. I, p. 167), the Rotifera within the Syndermata possess female gonads with a division into two sections. Now, solely the **germarium** forms oocytes capable of de-

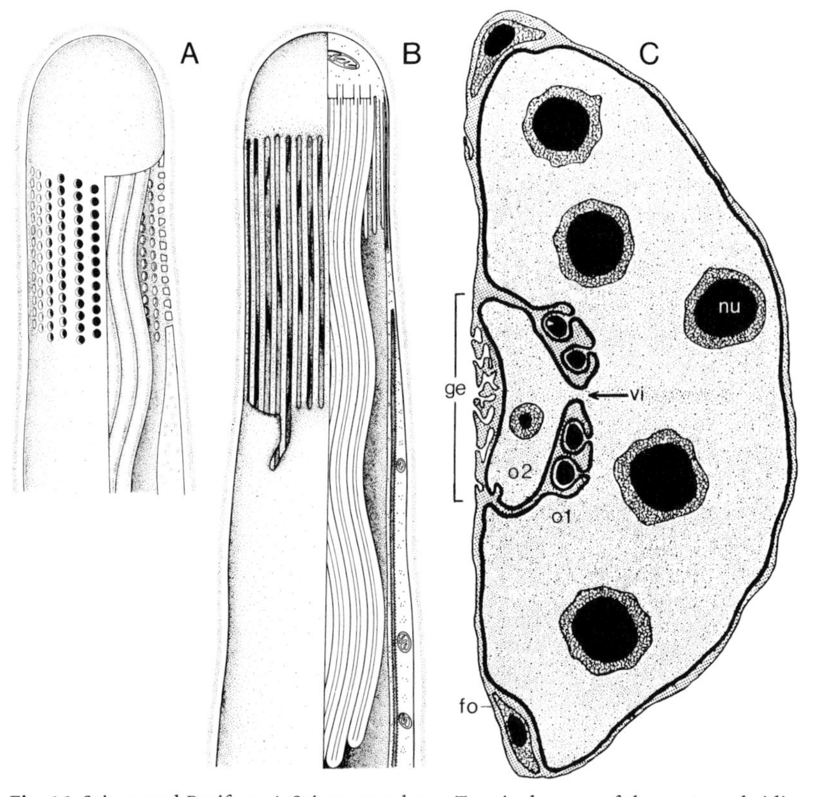

Fig. 16. Seison and Rotifera. **A** *Seison annulatus*. Terminal organ of the protonephridium with rows of pores in the wall of the hollow cylinder. **B** *Proales reinhardti* (Rotifera). Filtration structure of the protonephridium with longitudinal slits. **C** *Philodina rosea* (Rotifera). Longitudinal section through the syncytial germovitellarium, surrounded by a syncytial follicle layer (*fo*). The germarium lies on the inner edge of the vitellarium; it consist of a row of small cells, the plasma-poor, unripe oocytes. The large vitellarium has eight nuclei. Each oocyte is connected with the vitellarium via a cytoplasmatic bridge. *ge* Germarium; *nu* nucleus of the vitellarium; *o1* young oocyte; *o2* developing oocyte; *vi* vitellarium with flow of nutrient substances into an oocyte (*arrow*). (**A** Ahlrichs 1993a; **B** Ahlrichs 1993b; **C** Amsellem & Ricci 1982)

velopment. The **vitellarium** becomes a producer and supplier of nutrient substances. In comparison to the normal, uniform ovary in which nutrients for the embryo are formed in the developing oocyte itself, this is an apomorphous condition.

As compared with the original ovaries of *Seison*, a germovitellarium composed of a small germarium with plasma-poor germocytes (oocytes) and a large syncytial vitellarium with few polyploid nuclei (Fig. 16C) has evolved in the stem lineage of the Rotifera. Cytoplasmic bridges link the oocytes with the nutrient-rich vitellarium section. The germovitellarium forms a single coherent syncytium (CLEMENT & WURDAK 1991). Mitochondria, ribosomes, and endoplasmic reticulum migrate with fat and yolk granules over the cytoplasmic bridges into the oocytes.

A fundamental difference to the germovitellarium of the Plathelminthes must be strongly emphasized. The latter consist of germocytes and cellular yolk producers. In the Neoophora germocytes and vitellocytes are combined to form an ectolecithal "egg"; the embryo receives its nutrition from the decomposing yolk cells.

Monogononta – Bdelloida

The two traditional taxa, Monogononta and Bdelloida, are presumably monophyla and additionally adelphotaxa of the Rotifera. I follow this opinion with reservation because an unpaired germovitellarium can only weakly justify the monophyly of the Monogononta.

Monogononta

Autapomorphies (Fig. 14 → 3)

- Unpaired germovitellarium (Fig. 13A).
  The ovaries of *Seison* are paired organs, just like the germovitellaria of the Bdelloida. Under these circumstances, the unpaired germovitellarium must be identified as an apomorphous state – irrespective of whether it is the result of a fusion of primarily paired gonads or the reduction of one germovitellarium. Without doubt, the hypothesis of a single evolution of the unpaired gonad in the stem lineage of an entity of Monogononta offers the most economical explanation.

Brachionus, Epiphanes, Notholca, Proales, Trichocerca, Asplanchna, Floscularia, Collotheca. 1600 species.

Bdelloida

Autapomorphies (Fig. 14 → 4)

- Wheel organ of two disks of cilia that represent differentiation products of the ventral buccal field (Fig. 13 B).
- Purely parthenogenetic reproduction with complete reduction of the male.
- Fluke-like locomotion. Alternating anchoring to the substrate with the proboscis at the anterior end and the toes at the posterior end.
  Philodina, Rotaria, Habrotrocha. 360 species.

N.N.: Seison + Acanthocephala

Autapomorphies (Fig. 14 → 5)

- Sperm with "dense bodies".
- Filament bundles in the epidermal syncytium.
- Epibiotic way of life with the following components.
- (a) Reduction of head ciliation.
- (b) Feeding by sucking with uptake of bacteria and (?) haemolymph from the body of the host organism.
- (c) Modeling of the rami of the masticatory apparatus to a suction tube. The rami arch upwards frontally, almost lay together dorsally, and form a half-cylinder that is open at the bottom (Fig. 18 A, B).

The Acanthocephala have a characteristic lacunar system in the epidermis (p. 57). It remains unclear whether cavities and possible canals in the skin of *Seison* can be homologized with this system (AHLRICHS 1997).

The justification of a monophylum from the two *Seison* species as epibionts and the Acanthocephala as endoparasites begins with two characteristics that scarcely have anything to do with the discussed way of life.

Firstly, the identical expression of two staggered rows of "**dense bodies**" in the posterior half of the sperm (Fig. 17 B) can be discussed without difficulty as a synapomorphy between *Seison* and the Acanthocephala. In contrast, the homologization of the filament bundles in the epidermis of *Seison* and Acanthocephala appears problematic. Within the taxon *Seison*, filament bundles are only fully formed in *Seison nebaliae* (AHLRICHS 1997). Furthermore, there are considerable differences between the loose arrangement of the filaments in *Seison nebaliae* and their consolidation to a "felt layer" in the Acanthocephala (DUNAGAN & MILLER 1991).

The way of life of *Seison* as an **epibiont** on crustaceans may represent a first step on the way to endoparasitism of the Acanthocephala – and thus have arisen in a common stem lineage of these two entities. In this perspective, the haustellate feeding with modeling of the rami of the masticatory apparatus to a suction tube as well as the reduction of the no longer necessary head ciliation may be interpreted as possible autapomorphies of a stem species of *Seison* + Acanthocephala; but they can also be assessed differently (see below).

In any case, the functional system of an epibiotic way of life may represent an explanation for the unusual development of the pseudopharynges and pseudomanubria in the pharyngeal hard parts of *Seison*.

Seison – Acanthocephala

Seison

There are only two species – *Seison annulatus* and *S. nebaliae* – that live as epibionts on crustaceans of the taxon *Nebalia* (Leptostraca, Malacostraca). As microscopic organisms of 1–2 mm in length their appearance is unmistakable, with a weakly swollen head, extremely long neck with sections that retract like a telescope, a thickened trunk, and the foot with a terminal adhesive disk (Fig. 17A). Equally unique are the slow waving movements over the host while being anchored by the foot – and occasional change of location in the manner of a spanworm.

In their sympatric existence on one host, there seems to be a differentiated mode of settlement between the two species. *Seison annulatus* preferentially lives under the carapace on the epipodits of the thoracopods, while *Seison nebaliae* resides more on the surface of the body (AHLRICHS 1995). *Seison annulatus* probably sucks haemolymph from *Nebalia* and *S. nebaliae* eats particulate matter (SEGERS & MELONE 1998).

For the **analysis of the pattern of features** of *Seison*, we can take the Rotifera for comparison.

Are there **features** whose manifestation in *Seison* is **more primitive** than that in the Rotifera? This is without doubt the case. I would name the identical size of male and female as well as the uniform ovary that is not separated into germarium and vitellarium (Fig. 17C). In addition, *Seison* has paired gonads. These states belong in the ground pattern of the Syndermata and are thus plesiomorphies for the taxon *Seison*.

On the other hand, there are unambiguous **apomorphous characteristics** or feature states for which, nevertheless, a strict interpretation is difficult. We have treated habitus and way of life on *Nebalia*, lack of head ciliation, chewing apparatus with suction tube and haustellate uptake of food above as a coherent complex of characteristics and presented arguments for its inclusion in the ground pattern of an entity consisting of *Seison* +

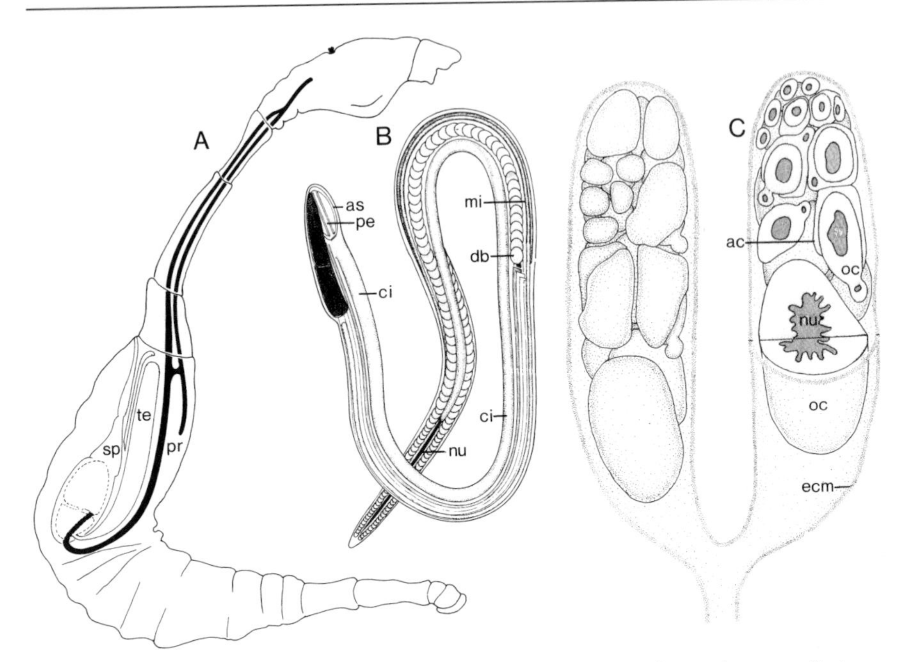

Fig. 17. *Seison.* **A** *Seison annulatus* (♂). Protonephridial system and sexual organs. Protonephridium opens into a spermatophore-forming organ. **B** *Seison nebaliae.* Sperm. Anterior section with acrosome, perforatorium and two electron-dense vesicles. Posterior section with a mitochondrion and two rows of dense bodies. With long, thread-like nucleus. One cilium inserts at the front and runs along the outside of the sperm to the back. **C** *Seison.* Paired ovary. Surrounded by extracellular matrix. Early oocytes are partially or completely surrounded by accessory cells. *ac* Accessory cell; *as* acrosome; *ci* cilium; *de* dense body; *ecm* extracellular matrix; *mi* mitochondrion; *nu* nucleus; *oc* oocyte; *pe* perforatorium; *pr* protonephridium; *sp* organ for forming spermatophores; *te* testis. (Ahlrichs 1995)

Acanthocephala. However, we cannot exclude that these features evolved only in the independent pathway of *Seison* after the separation from the stem lineage of the Acanthocephala.

The possibility remains that the following characteristics in the protonephridial system and male genital organs, which are independent from this feature complex, represent genuine characters of *Seison.*

Autapomorphies (Fig. 14 → 6)

- Protonephridia with extensive supporting structures in the canal region. This consists of two long rods that lie in the neck region parallel to the lumina of the canals in the cytoplasm.
- Formation of spermatophores in a glandular organ of the male genital apparatus.

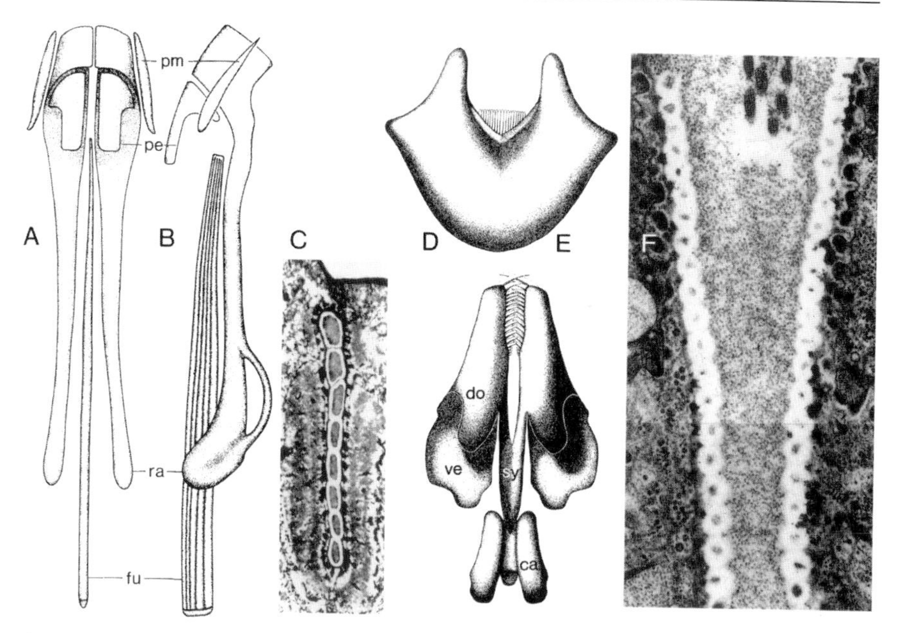

Fig. 18. **A–C** *Seison* (cont.). *Seison nebaliae* with masticatory apparatus of seven trophi elements. **A** Dorsal view. **B** Side view. **C** Electron microscope cross section through the fulcrum. Constructed from tubes with an electron-dense axis. **D–F** *Gnathostomula paradoxa* (Gnathostomulida). **D, E** Jaw apparatus with an unpaired basal plate (**D**) and a jaw (**E**) with two halves that are linked in a caudal symphysis. Reconstruction from serial sections. Dorsal view. **F** Cross section through the internal wall of the jaw half. Again, construction of tubes with electron-dense axis and pale ring. *ca* Cauda; *do* dorsal lamella; *pe* pseudoepipharynx; *pm* pseudomanubrium; *ra* ramus; *sy* symphysis; *ve* ventral lamella. (A–C Ahlrichs 1995; D–F Herlyn & Ehlers 1997)

Acanthocephala

After the invasion of a first host organism by the stem species, over 1000 species of parasites evolved in the phylogenesis of the Acanthocephala. The stem species was a small organism – perhaps only a few millimeters long like the species in the adelphotaxon *Seison* or about a centimeter long like many fish parasites. A considerable increase in the body size followed first in the terrestrial Tetrapoda. The length of females of the porcine parasite *Macracanthorhynchus hirudinaceus* exceeds 50 cm.

Since the *Seison* species live as epibionts on marine Malacostraca, it can be assumed that aquatic Crustacea were the first hosts of the endoparasitic thorny-headed worms. With the uptake of infected crustaceans as food by aquatic Vertebrata, the colonization of Acanthocephala in the intestine of fish occurred. The Arthropoda then became intermediary hosts, and aquatic Gnathostomata the final hosts. As a consequence of the invasion of the

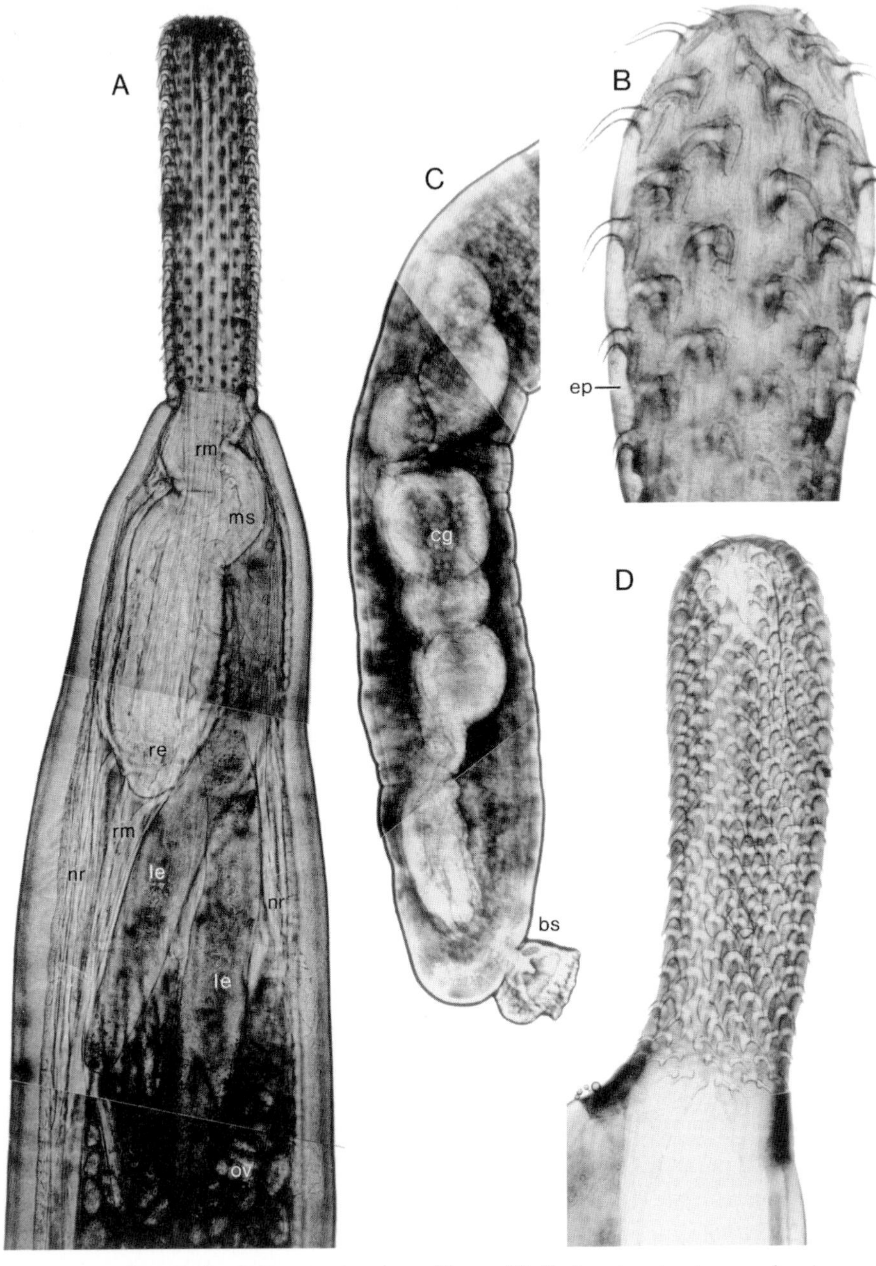

Fig. 19. Acanthocephala. **A** *Paratenuisentis ambiguus* (♀). Proboscis retractors, neck retractors, receptaculum and the two lemnisci are visible in the presoma, a few ovarian balls in the adjacent trunk. **B** *Acanthocephalus anguillae* (♀). Demonstration of the position of the hook roots under the epidermis. **C** and **D** *Echinorhynchus truttae*. **C** Posterior end (♂) with row of cement glands and extended bursa. **D** Proboscis (♀). *bs* Bursa; *cg* cement glands; *ep* epidermis; *le* lemniscus; *ms* muscle sac; *nr* neck retractor; *ov* ovarian balls; *re* receptaculum; *rm* retractor muscle of the proboscis. (Originals: Herlyn)

terrestrial realm, the spectrum of intermediary hosts expanded to include Isopoda, Myriapoda and Insecta, that of the final hosts to terrestrial Tetrapoda (HERLYN 2000).

Now all Acanthocephala pass through an obligate change of hosts. The larvae, designated as Acanthor and Acanthella, live in intermediary hosts; here, development to infection-ripe Cystacanthus occurs. The adult takes up liquid nourishment in the intestine of the final host through its skin; it own gut has been lost in the evolution of the Acanthocephala. The Acanthor larvae develop in the "eggs" within the body of the female.

Autapomorphies (Fig. 14 → 7)

- Closed endoparasitic life cycle with a member of the Arthropoda as intermediary host and a species of the Vertebrata as final host.
- Division of the body into two sections.
  The presoma consists of a proboscis with hooks and the neck region. This is followed by a uniform trunk (Fig. 20).
- The proboscis hooks are derivatives of the ECM under the epidermis.
- An extensive lacunar system runs through the epidermis.
- Lemnisci developing as two long ingrowths of the presomal epidermis.
- Syncytial organization of the musculature from circomyarian muscle cords.
- Central musculature in the anterior end with unpaired retractor and sack-shaped receptaculum as antagonist on movement of the proboscis. Receptaculum surrounded by constrictor or protrusor, respectively. Neck retractor.
- Retinacula. Two muscles that cover the lateral longitudinal nerves between receptaculum and body wall.
- Cerebral ganglion in a ventral position.
- Sexual dimorphism. Male smaller than female.
- Ligament sacs. Probably primarily unpaired formation in both sexes.
- Male with two testes, cement glands, penis and bursa.
- In the female the ovaries decompose into single balls. With uterine bell and vagina.
- Acanthor develops in ovarian balls; it is encased in a egg shell of four egg membranes with keratin.
- Acanthor with complete syncytial organization. Hooks at the anterior end are intraepidermal differentiations.

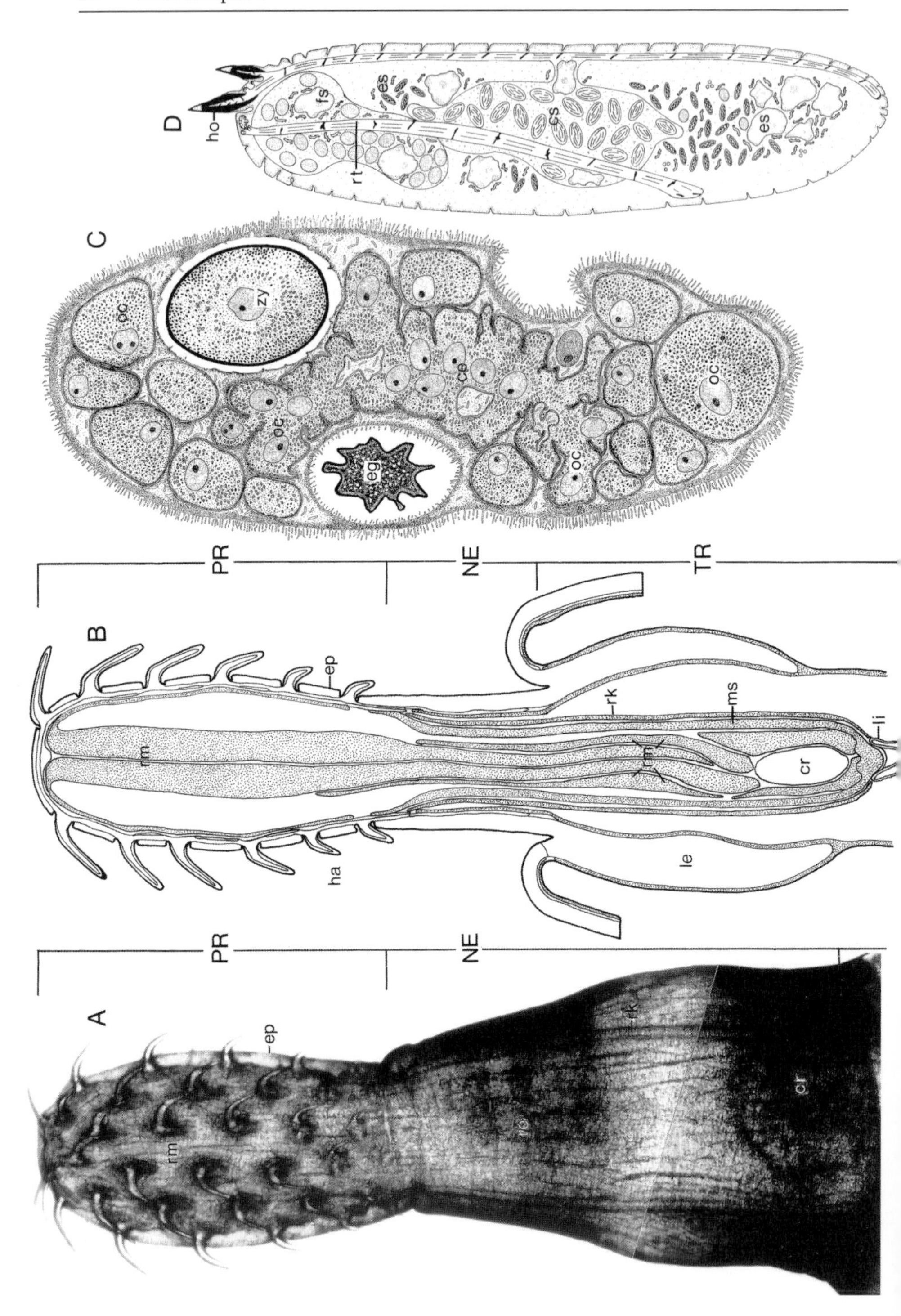
A
B
C
D
PR
NE
TR
ep
rm
rk
cr
le
ha
ms
li
oc
zy
ce
eg
ho
fs
es
cs
rt

◀ **Fig. 20.** Acanthocephala. **A** *Acanthocephalus anguillae*. Anterior end (♀). **B** *Acanthocephalus anguillae* (♀). Reconstruction of a horizontal section through the anterior end. *PR* Proboscis; *NE* neck; *TR* trunk. **C** *Moniliformis moniliformis*. Organization of a freely floating ovarian ball. **D** *Polymorphis minutus*. Longitudinal section of an acanthor with intraepidermal hooks. Construction of three syncytia. *ce* Central syncytium in ovarian ball; *cr* cerebral ganglion; *cs* central syncytium of the acanthor; *eg* egg; *ep* epidermis; *es* epidermal syncytium of the acanthor; *fs* frontal syncytium; *ha* hooks on the proboscis of an adult; *ho* hooks on the acanthor; *le* lemniscus; *li* ligament sac; *ms* muscle sac of the receptaculum; *oc* oocyte; *re* receptaculum; *rk* receptaculum constrictor; *rm* retractor muscle of the adult; *rt* retractor muscle of the acanthor; *zy* zygote. (**A, B** Herlyn 2000; **C** Crompton 1985; **D** Albrecht et al. 1997)

Ground Pattern

The autapomorphies will be discussed in more detail in the context of the ground pattern of the Acanthocephala (HERLYN 2000).

The division of the body into **presoma** and **trunk** with a **proboscis equipped with hooks** at the anterior end has certainly arisen as a characteristic in the stem lineage of the Acanthocephala. In view of the briefly described, possible relation with the way of life of *Seison* its origin upon the transition to endoparasitism is more probable than the derivation of the organ from a locomotory device in an antecedent living free in sediment (VON HAFFNER 1950). One way or the other, the hooks are characterized by a unique construction. The hooks of the Acanthocephala are not cuticular exudates of the epidermis; they also do not belong in the complex of intrasyncytial thickenings in the epidermis; the hooks are rather formed in the extracellular matrix under the epidermis (Fig. 19B). As derivatives of the ECM, they are usually completely covered by the epidermis. Finally, the widespread division of the individual hooks into shaft and root should be mentioned as an apomorphous feature of the ground pattern.

We now come to the **epidermis**. Its development as syncytium and the intrasyncytial thickenings constitute plesiomorphies. They reach back to the ground pattern of the Syndermata; this is also true for the covering of the epidermis that merely consists of a glycocalyx. In contrast, the lacunar system in the skin must, according to the stand of examination, be considered as an autapomorphy of the Acanthocephala (p. 50); a net-like condition may possibly belong in the ground pattern (see below).

The **lemnisci** are impressive structures of the skin. These are primarily two long, sausage-shaped ingrowths that extend from the neck region to the trunk (Figs. 19A, 20B). Presomal epidermis and lemnisci form a coherent syncytium. A homology of the lateral lemnisci of the neck with the apical epidermis bulges of the wheel organ of the Rotifera (LORENZEN 1985) is unlikely (RICCI 1998; HERLYN 2000).

The syncytial structure of the Acanthocephala continues from the epidermis into the musculature. Circomyarian muscle cords arrange themselves to tubes in which the central cytoplasm is surrounded by contractile filaments in a ring-like pattern.

The first outstanding feature in the **musculature** of the **presoma** is the central, unpaired proboscis retractor (Fig. 20 A, B). The muscle consists of several anastomizing, circomyarian muscle cords; it has four nuclei. The retractor runs from the apex of the proboscis through the receptaculum in the anterior trunk; the single muscle cords end here in the longitudinal musculature of the body wall.

The receptaculum serves as the antagonist of the retractor. The central component is a muscle sac with six nuclei that inserts in the anterior body wall. A contraction of the receptaculum increases the hydrostatic pressure in the presoma and thus triggers the evagination of the proboscis. The muscle sac is probably a fusion product of longitudinal muscle cords (HERLYN 2000). A muscle around the receptaculum with differing orientation of the contractile filaments among the species probably promotes the increase of the internal pressure in the prosoma as a constrictor or protrusor. In the system of movements at the anterior end, a neck retractor can withdraw the neck into the trunk when the proboscis is extended.

The determination of the position of the **cerebral ganglion** (Fig. 20 A, B) of the Acanthocephala on the ventral side is oriented to the position of the two main longitudinal nerves. They stem from the ground pattern of the Syndermata as two nerve cords in ventrolateral position. The cerebral ganglion lies on the same side as these nerves. Its ventral position should be interpreted as an apomorphy in comparison with the dorsofrontal position of the cerebral ganglion in *Seison* and the Rotifera.

Sexual dimorphism with a small male and large female belongs in the ground pattern of the Acanthocephala.

Unique features that are not comparable with other Bilateria are seen in the construction of the **sexual organs**. They begin already with the peculiar ligament sacs in the trunk, the wall of which is presumably primarily syncytially organized, and secondarily may consist of pure ECM (HERLYN 2000). For the male sex an unpaired ligament sac that is retained throughout the entire life span may be postulated for the ground pattern; it encloses the male organs. Similarly, an unpaired ligament sac has been discussed for the female in the ground pattern, even when two median ligament sacs occur frequently within the Acanthocephala. The ligament sacs can persist, but are also torn during the development process. The tissue between two ligament sacs in contact with each other is called the ligament strand.

Central parts of the male genital apparatus are two testes, up to eight cement glands and a penis that opens into a terminal bursa (Fig. 19 C). During copulation, the male grasps the posterior end of the female with the bursa extended. After transfer of sperm, the female sexual opening is sealed with secretions from the cement glands. In the female, the ovaries decompose to ovarian balls (Fig. 20C) that float freely – either in the ligament sac or, in the case of their detachment, in the body cavity. Inside the balls the oocytes differentiate from a central syncytium; they are fertilized here by invading sperm. Development occurs in the ovarian balls to the stage of the acanthor with an egg shell of four egg membranes. The "eggs" that are now ready for laying are cast out in a uterine bell and pass through the vagina to the intestine of the host.

The **acanthor** is organized from three syncytia (Fig. 20D). The frontal and central syncytia are enclosed by the epidermal syncytium. The characteristic hooks at the anterior end of the larvae are special differentiations in the epidermal thickenings of the syncytium (ALBRECHT et al. 1997); they thus have nothing to do with the proboscis hooks of ECM in the adult.

Systematization

The usual division of the Acanthocephala into three equal-ranking orders is currently in the process of transformation into a phylogenetic system (HERLYN 2000). Once again, one traditional taxon has proved to be a paraphylum on the basis of primitive features; these are the Palaeacanthocephala. On the other hand, the Eoacanthocephala and the Archiacanthocephala are each justified as monophyla by means of autapomorphies and can also be interpreted as adelphotaxa.

"Palaeacanthocephala"

The only possible autapomorphy constitutes the spindle-shaped eggs in comparison with the ovoid eggs in the other Acanthocephala; they could, however, belong as a primary state in the ground pattern of the Acanthocephala. The lacunar system can be built up either as a network (ground pattern) or from longitudinal and circular vessels with lateral main longitudinal vessels.

Acanthocephalus, Echinorhynchus, Pomphorhynchus, Polymorphus.

Monophylum Eoacanthocephala + Archiacanthocephala

Autapomorphies. Lacunar system of the trunk epidermis from regularly arranged circular and longitudinal vessels with two median main longitudinal vessels – a dorsal and a ventral vessel. Overlapping of the circular musculature of the trunk with the presomal circular musculature.

Eoacanthocephala

Autapomorphies. Sac-like ingrowth of the epidermis at the apex of the proboscis with two or three nuclei. Unpaired cement gland with eight nuclei; interpretable as a fusion product of eight mononuclear cement glands.

Paratenuisentis, Neoechinorhynchus.

Archiacanthocephala

Autapomorphies. Two apical organs in the proboscis linked with the cerebral ganglion by two nerves; with tetranuclear supporting cell. Terrestrial life cycle with thick-shelled eggs; members of the Tracheata as intermediary host, Aves and Mammalia as final hosts.

Plesiomorphous in comparison with the sister group are eight mononuclear cement glands in the male.

Gigantorhynchus, Oligacanthocephalus, Macracanthorhynchus.

On the Relationships of the Syndermata

As with the Nemathelminthes we will, in conclusion, address the question of the adelphotaxon of the Syndermata. At present, the Gnathostomulida (Vol. I, p. 134) are considered as the favorite. The argumentation is based on a comparison of the jaw apparatus of the Gnathostomulida with the chewing apparatus in the ground pattern of the Syndermata (REISINGER 1961; RIEGER & TYLER 1995; AHLRICHS 1995; HERLYN & EHLERS 1997).

We will start with the ultrastructural agreements in the construction of the hard parts. In both taxa, they consist of single cuticular tubes with an electron-dark axis and a surrounding electron-light ring (Fig. 18). It could be assumed that the axis of the tube is a continuation of the cytoplasm that secretes the surrounding hard substance (AHLRICHS 1995) – somewhat like the formation of Annelida chaetae from tubes in which numerous microvilli of a chaetoblast secrete a fibrillary hard substance around themselves (Vol. II, p. 44). According to recent studies on *Gnathostomula paradoxa*, however, the electron-dense axis of the rods also consists of cuticular material (HERLYN 1996; HERLYN & EHLERS 1997); the cuticle rods lack a basal contact with the pharynx epithelium, and there is no evidence for a cellular membrane about the electron-dense filling. The mode of formation of the cuticular rods in the hard parts of Syndermata and Gnathostomulida is unknown. In addition, I can refer to the example of the chaetae in Annelida, Brachiopoda and Cephalopoda. These are the products of identical, but independent solutions of evolution at the ultrastructural level. In spite of concordant fine structure and genesis of the chaetae, prior hypotheses of relationships force the conclusion on convergency (Vol. II, p. 44).

Thus, we come to a comparison of the pharyngeal hard parts at the light microscopic level. In discussion is the hypothesis of a homology of the jaw of Gnathostomulida with the rami of Syndermata as well as the symphysis that joins the two jaws in Gnathostomulida with the fulcrum (AHRLICHS 1995). According to this hypothesis, a system with three elements (jaws + symphysis = rami + fulcrum) must be isolated from the given facts in the Gnathostomulida and Syndermata and postulated for a common stem species of the two entities. It remains, however, questionable to separate the functional unit between the basal plate and jaws in the Gnathostomulida and to hypothesize two subsequent steps for their evolution. The basal plate lifts diatoms, bacteria and detritus from sand grains which are then grasped by the jaws and transported into the oesophagus (HERLYN & EHLERS 1997). The derivation of the functional unit of the mastax in the Syndermata with seven trophi from an organism with three hard parts seems equally questionable.

In my opinion, there are no good reasons for the assumption of a homology between specific parts of the jaw apparatus of the Gnathostomulida and the chewing apparatus of the Syndermata. The Gnathostomulida are not a convincing candidate for the sister group of the Syndermata.

Note Added in Proof:
The above arguments are equally valid for *Limnognathia maerski* (KRISTENSEN & FUNCH 2000). Tubes with electron-dense axes and a light ring in the jaw cannot per se be taken as autapomorphies of an entity composed of Gnathostomulida, *Limnognathia maerski* and Syndermata - just as the identical ultrastructure and genesis of the chaetae or setae cannot justify a monophylum made up of Annelida, Cephalopoda and Brachiopoda (see above). For the example under discussion, convincing hypotheses about the homology of parts of the jaw apparatus in the three taxa would be required; these do not exist. In the remaining organization of *Limnognathia maerski* there are no agreements with the Gnathostomulida or the Rotifera that can be interpreted with good justification as synapomorphies with one of the two taxa.

Superordinated taxa with identical diagnoses for one species (family Limnognathiidae, order Limnognathia, class Micrognathozoa) are rejected for the phylogenetic system of organisms as empty formalisms (Vol. II, p. XI).

Radialia

The "Tentaculata" entities Bryozoa, Phoronida and Brachiopoda can be united with the Deuterostomia[2] in a taxon that was given the name Radialia by JEFFERIES (1986; Vol. I, p. 130). The major justifications for the monophyly of an entity Radialia arise from the construction and function of the tentacle apparatus at the anterior end as well as the secondary body cavity, which must be demonstrated in detail. On the other hand, the name-giving radial cleavage is an ancient plesiomorphous feature of the Bilateria.

There are, however, still difficulties with the "Tentaculata" ("Lophophorata") of the traditional classification. Why do we question them as an entity when the tentacle apparatus in the form of a horseshoe-shaped lophophore appears to be such a prominent agreement between the Bryozoa, Phoronida and Brachiopoda? The tentacle apparatus of the Deuterostomia *Rhabdopleura* and *Cephalodiscus* can be homologized with the lophophore of these taxa (p. 69). Thus, the lophophore logically becomes a plesiomorphy of the Bryozoa, Phoronida and Brachiopoda – taken over from the ground pattern of the Radialia. Otherwise, I can see no other feature for a possible justification of the Tentaculata as a monophylum.

On the other hand, I hesitate to recognize agreements as autapomorphies with which the Brachiopoda could be identified as the sister group of the Deuterostomia. The Brachiopoda are a possible candidate on account of the concordant enterocoely in the formation of the coelom by proliferation of mesodermal cells from the archenteral epithelium (LÜTER 1998a, 2000b).

In any case, at the present state of knowledge, the "Tentaculata" are an uncertain entity without demonstrated autapomorphies. We must set them in quotation marks and leave unconnected the stem lineages of the three monophyla Bryozoa, Phoronida and Brachiopoda.

Autapomorphies (Fig. 21 → 1)

- Biphasic life cycle.
  Solitary adult (plesiomorphy) in the benthic zone – planktotrophic larvae in the pelagic zone.
- Tentacle apparatus and filtration mechanism.
  Form: Horseshoe-shaped tentacle carrier (lophophore) in the anterior body consisting of two arms; edges of the arms equipped with tentacles (Fig. 22A).
  Ultrastructure: Tentacles with three groups of cilia. Monociliated epithelial cells of the tentacle arranged to five longitudinal rows with the fol-

[2] The Chaetognatha are not included in the Deuterostomia, but are presented as a monophylum with unclarified adelphotaxa relationships at the end of this volume.

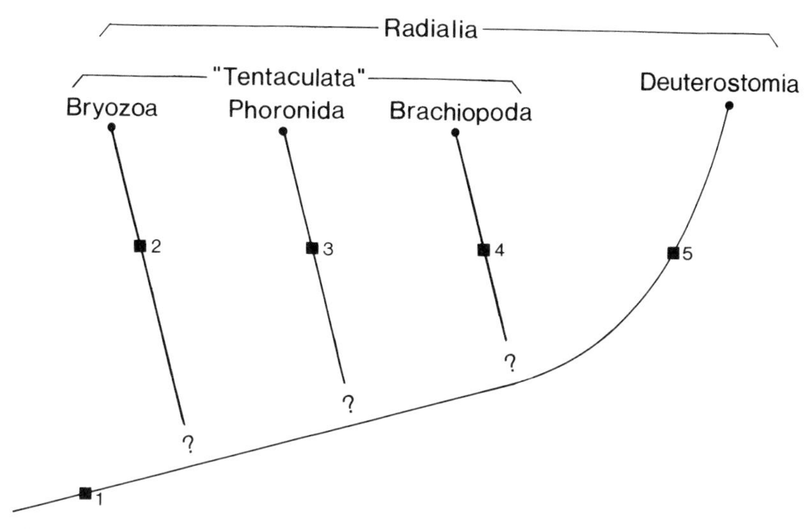

Fig. 21. Extent of the monophylum Radialia with the "Tentaculata" and Deuterostomia. The kinship relationships of the three Tentaculata entities Bryozoa, Phoronida and Brachiopoda among themselves and to the Deuterostomia are still the subject of controversy

lowing differentiation. (1) Inner side with a frontal band of several rows of cilia. (2) On each side of these a laterofrontal row of single cilia. (3) On both sides a lateral cilia-rich band (Fig. 23 A).

Water flow: Entry at the top in the tentacle apparatus. Flow directed towards the mouth. Lateral deviation and exit to the sides between the individual tentacles. Motor for the water flow is the lateral bands of cilia.

Filtration and particle transport: The two rows of laterofrontal cilia form a mechanical filter. Trapped fine particles are forced back into the central flow by beats of the tentacles and flickered towards the oral opening by the frontal row of cilia (Fig. 22 D, F, G).

- Body division and coelom.

  Division into two parts: an anterior section with tentacle apparatus and a trunk without appendages. Coelom of entodermal origin with two consecutive cavities. A mesocoel is attached to the tentacle apparatus; it supplies the individual tentacles through canal-like ducts. The second coelom section lines the interior of the trunk; it is designated as metacoel.

- Blood vessel system.

  With differentiation of the blastocoel to a circular vessel in the anterior body.

- Metanephridia.

  Nephridial organs composed of an ectodermal nephridial canal (taken over from protonephridia) and a mesodermal ciliated funnel. A pair of metanephridia in the metacoelom.

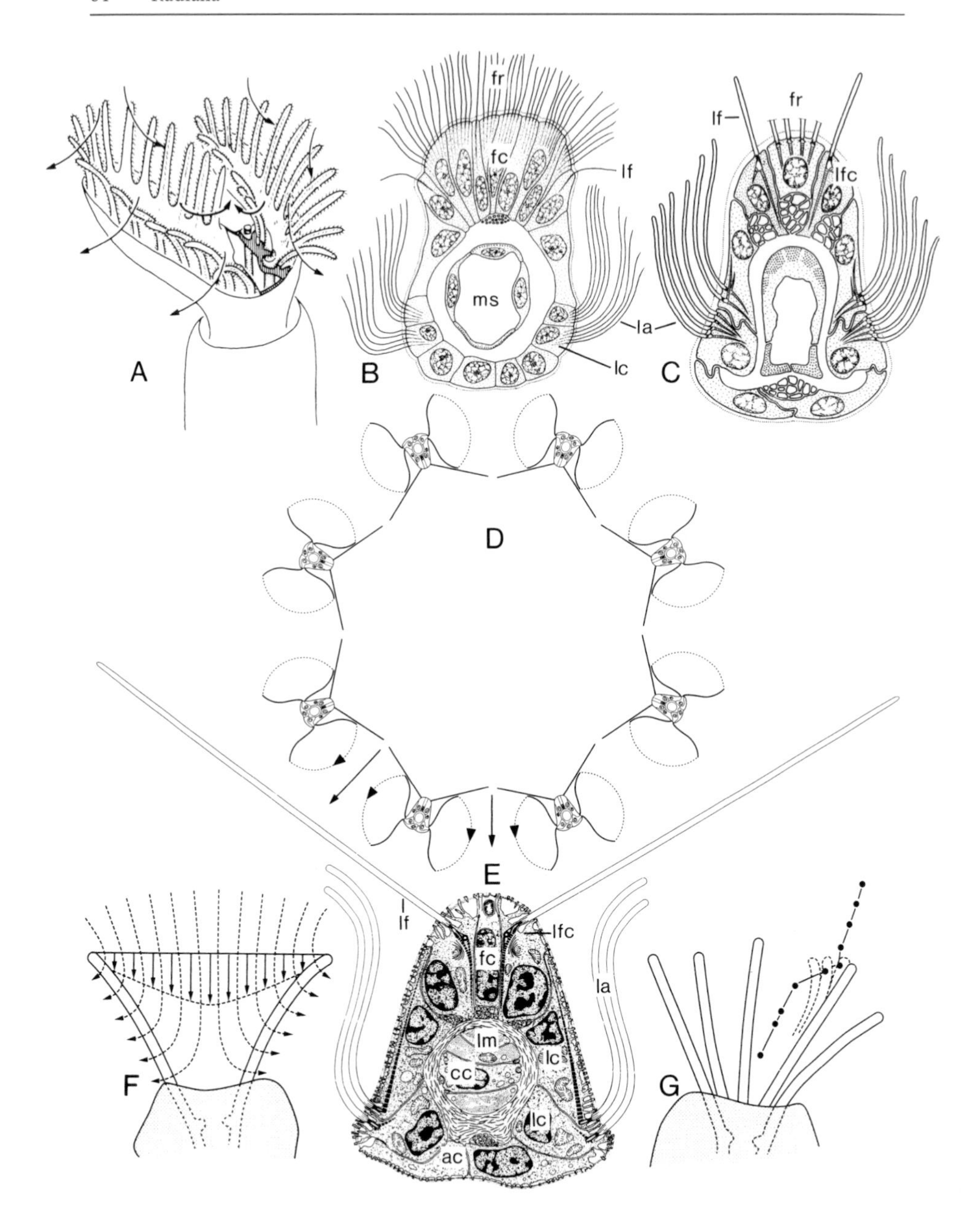
A
B
C
D
E
F
G
fr
fc
lf
ms
la
lc
lfc
lm
cc
ac
l

◀ **Fig. 22.** Radialia: Bryozoa. Structure and function of the tentacle apparatus. **A** and **B** Phylactolaemata. Primitive development of the filter device in the form of a lophophore. **A** *Plumatella*. Two arms arranged to a horseshoe; they are equipped with tentacles at the edges. The water flows from above into the spaces surrounded by the tentacles and emerges laterally between the individual tentacles (*arrows*). **B** *Plumatella fungosa*. Cross section of a tentacle with the following ciliation: (1) frontal cilia emerge from a broad band of multiciliated cells; (2) there is a single row of laterofrontal cilia produced by monociliated sensory cells; (3) finally, several lateral cilia are formed from two rows of multiciliated cells. **C–G** Gymnolaemata. Apomorphous development of a filter apparatus in the form of a crown of tentacles arranged in circles. In the ciliation of the tentacles there are in the ground pattern three groups of cilia (agreement with the Phylactolaemata). **C** *Electra pilosa*. Cross section of a tentacle. A single row of multiciliated cells on the frontal side as an apomorphy in the ground pattern of the Gymnolaemata. **D–G** *Crisia eburnea*. More strongly derived tentacle apparatus within the Gymnolaemata: crown of eight tentacles. Frontal cell row without cilia. The two lateral rows of two cilia originate from two rows of monociliated cells. Central part of the tentacle without lumen, filled with longitudinal muscles and central cells. **D** Cross section through the middle part of a tentacle crown. The effective beat of the lateral cilia runs from the center outwards (*arrowheads*) and creates the force for the flow of water through the crown. The stiffly placed, laterofrontal cilia form the filter. **E** Tentacle (cross section) with the mentioned apomorphies. **F** Water flow through the tentacle crown (*broken lines*) and distribution of particle velocity in the entrance region. **G** Tentacle flicking. The beat of a tentacle brings trapped particles back to the central flow directed towards the mouth. *ac* Abfrontal cell; *cc* central cell; *fc* frontal cell; *fr* frontal cilia; *la* lateral cilia; *lc* lateral cell; *lf* laterofrontal cilium; *lfc* laterofrontal cell; *lm* longitudinal muscle; *ms* mesocoel. (**A** Gilmour 1978; **B,C** Nielsen 1995; **D–G** Nielsen & Riisgard 1998)

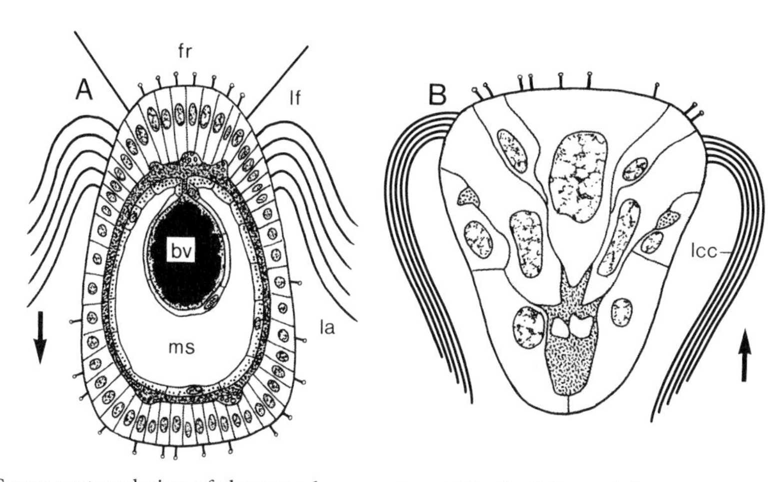

Fig. 23. Convergent evolution of the tentacle apparatus of the Radialia and the Kamptozoa (Spiralia) in the case of a principally concordant function as filter device. Cross sections of tentacles. **A** *Phoronis* (Phoronida, Radialia). Tentacle with coelom. Three rows of cilia from monociliated cells. The lateral cilia draw a stream of water from the center of the tentacle apparatus outwards to the sides. **B** *Loxosomella* (Kamptozoa). Solid tentacle with five rows of cilia from multiciliated cells. The lateral, compound cilia transport water from the sides into the inside of the tentacle apparatus. *bv* Blood vessel; *fr* frontal cilia; *la* lateral cilia; *lcc* lateral compound cilia; *lf* laterofrontal cilium; *ms* mesocoel. (Nielsen 1998)

Life Form

Was the stem species of the Radialia a solitary or a colonial organism? Irrespective of whether an entity of the Spiralia+Nemathelminthes or solely the Nemathelminthes is the sister group of the Radialia (p. 45), the thus realized solitary life form is to be set in the ground pattern of the Radialia (outgroup comparison). In other words, the Phoronida and Brachiopoda are to be interpreted as entities with original solitarily living individuals while the colony-forming of the Bryozoa represents an autapomorphy. Similarly, the colonial state of *Rhabdopleura* and *Cephalodiscus* must be derived – originated as an evolutionary novelty in the stem lineage of the Stomochordata (p. 136).

Body Division and Coelom

The traditional concept of the trimerous division of the body (protosome – mesosome – metasome) and a correspondingly organized body cavity with three sections (protocoel – mesocoel – metacoel) in **all Radialia** is retained even in the newest textbooks (HERRMANN 1996). This interpretation is certainly not valid, at least for the "Tentaculata" entities Phoronida and Brachiopoda. Ultrastructural investigations have not been able to detect a mesodermal covering with an independent, epithelial cell layer either in the epistome of Phoronida (as well as in the episphere of actinotrocha larva) or in the unpaired median tentacle and epistome of planktotrophic juvenile stages of *Lingula* (Brachiopoda). In the Phoronida and Brachiopoda, there is no coelom section that can be interpreted as a protocoel (BARTOLOMAEUS 1993, 2001; LÜTER 1996, 1998 a, 2000 b). According to this perspective, the existence of a protocoel in the epistome of the Phylactolaemata (Bryozoa) is doubtful. Here, also, there is no closed coelom; instead, there is rather a wide, open connection between the space in the epistome and that in the mesocoel (MUKAI et al. 1997).

With the absence of a coelom, the main basis for an evaluation of the epistome as a section of a trimerically divided body is lacking. In addition, the preordered question about a homology of the flap with the homonymous designation epistome is a subject of controversy. It is answered negatively in a comparison of the Phoronida and the Brachiopoda (LÜTER 1998 a).

Thus, the problem of the body division reduces to the interpretation of the division into a section with tentacle apparatus and a trunk without appendages. In other words, the problem is the homology of these two parts of the "Tentaculata" with the mesosome and metasome of the Deuterostomia.

The situation in the Phoronida is unambiguous. According to studies on *Phoronis muelleri* there is a ring-shaped mesocoel with a myoepithelial covering at the base of the lophophore from which branches into the individual tentacles originate. A diaphragm with a myoepithelial double layer separates the mesocoel from the metacoel of the trunk (or the hyposphere). Muscle fibers pass through the coelom.

In the Brachiopoda, an undivided coelom anlage originates from the archenteron. The tentacles are supplied from frontal outgrowths of the, at first, uniform coelom. A division into mesocoel and metacoel by extracellular matrix follow later (LÜTER 1998a, 2000b). Comparable analyses of the Bryozoa are lacking; apparently, mesocoel and metacoel are also completely separated in this case.

The presented findings lead to the following economic interpretations. Two coelom spaces lying directly one after the other belong in the ground pattern of the Radialia – the mesocoel and the metacoel (BARTOLOMAEUS 1993, 2001). This is the primary condition in the entity Radialia – disregarding the difficulties of a more detailed specification. Mesocoel and metacoel can be homologized with the same-named coelom cavities of the Deuterostomia. The trimeric body with a separated protosome and the accompanying, independent protocoel then developed first in the stem lineage of the Deuterostomia.

We will now attempt to explain the evolution of the two-part coelom of the Radialia. In light of the justifiable requirement for a solitary organism in the ground pattern of the Radialia (p. 66), we must orient ourselves on the Phoronida. Widespread and probably also primitive is the colonization of self-made tubes in soft sediments (sand, mud). In such a sedentary organism, the coelom has two elementary functions: (1) stabilization of the individual tentacles in the filter device (mesocoel) and (2) anchoring and freeing the body in the tube by changing the pressure in the coelom liquid (metacoel).

These are fundamentally different requirements in comparison to the function of a uniform, undivided coelom as liquid-filled cushion serving for a directed locomotion in sediment – as has been postulated for the ground pattern of the Pulvinifera (Vol. II, p. 38). According to functional aspects, the development of a secondary body cavity in the Pulvinifera and in the Radialia are in no way connected. In other words, there are good reasons for the hypothesis of the independent, unique evolution of a coelom divided into two consecutive sections in the stem lineage of the Radialia.

Blood Vessel System

The Radialia have a blood vessel system in the primary body cavity. The Bryozoa are a negative exception.

In the ontogenesis of the Phoronida, the blastocoel of the actinotrocha larva transforms to a ring canal of the vascular system (BARTOLOMAEUS

1993). I emphasize this entity in order to interpret the situation in the Bryozoa. We must assume that the evolution of a blood vessel system was linked with the above-mentioned evolution of the coelom in the primarily solitary Radialia. Why? Because the following picture would arise when one, in reverse, would identify the condition of the Bryozoa as the primary one: solitary stem species of the Radialia without blood vessels – taking over this state in the stem lineage of the colonial Bryozoa – and only then the evolution of blood vessels in solitary Radialia. In my opinion, this is a highly unlikely mental exercise. However, the secondary lack must not necessarily be the result of a miniaturization of the Bryozoa. *Rhabdopleura* and *Cephalodiscus* are millimeter-sized Deuterostomia that have a blood vessel system.

Nephridial Organs

Among the Radialia, protonephridia exist only in the actinotrocha larva of the Phoronida. The paired organs consist here of complexes of blindly closed, monociliated terminal cells in the blastocoel, a canal of numerous monociliated wall cells and a pore in the epidermis (Vol. I, p. 122). Since we have presented reasons for the assumption of a homology in all Bilateria (loc. cit. p. 126), protonephridia belong in the ground pattern of the Radialia; they must, therefore, have been lost in the Bryozoa, Brachiopoda and Deuterostomia.

Metanephridia exist in the Phoronida and Brachiopoda. We will discuss them in succession.

In the Phoronida, the protonephridial canal is included directly in the construction of the composite metanephridium. In contrast, the terminal cells are discarded during metamorphosis to the adult; a funnel made up of monociliated cells of the metacoel wall takes their place (Vol. I, Fig. 50).

The Brachiopoda also have composite nephridia consisting of an ectodermal nephridial canal and a mesodermal ciliated funnel. At first, the canal is formed ventrolaterally as an invagination of the ectoderm; connection to the coelom epithelium occurs later. The following situation must be emphasized. The metanephridia of the Brachiopoda arise without a larval protonephridium as precursor (LÜTER 1998a). This is an apomorphy in comparison to the situation in the Phoronida.

In their function, metanephridia are strictly linked to the existence of a secondary body cavity, but can develop without participation of the coelom. This is true for the uniform nephridial anlage of the Annelida. On the basis of the fundamental difference, the like-named metanephridia of the Annelida and the Radialia are not taken as homologous organs (Vol. I, p. 123). This emphasizes the following statement. Composite metanephridia of ectoderm and mesoderm have arisen in connection with the evolution of the coelom in the stem lineage of the Radialia. Together with the two-part coelom, they form an autapomorphy of the Radialia.

We are jumping ahead when, at this point, we mention the axial complex as a novel excretory organ in the ground pattern of the Deuterostomia (p. 93). On the basis of the composition from two tissues of differing origins, a homology with the mentioned metanephridia has been discussed (BARTOLOMAEUS 1993). There are, however, serious differences in the incorporation of the axial complex in the protosome of a trimerous body and the function of the protocoel as a nephridial store (NIELSEN 1995). I can leave this problem open since its solution does not appear to be significant for the systematization of the Radialia.

Tentacle Apparatus as Food Filtration Organ

The tentacle apparatus and the mechanism for sorting organic particles from water together form the central characteristic of the Radialia. The elements of this autapomorphy must, therefore, be discussed in more detail.

We will start with the form. A lophophore (tentacle carrier) of two arms arranged in the shape of a horseshoe belongs in the ground pattern of the Radialia. The out- and insides of the edges are equipped with tentacles. There is no comparable tentacle apparatus outside the Radialia. The principle of the most economical explanation thus dictates at this point the hypothesis of a unique evolution of the lophophore.

What is the justification for the interpretation of the horseshoe as plesiomorphous state of the lophophore? Quite simply its realization in the Phylactolaemata as a subtaxon of the monophyletic Bryozoa, in the Phoronida, and in the Brachiopoda. On the basis of these facts, the closed, ring-shaped tentacle crown of the Gymnolaemata must represent a derived state within the Bryozoa. In contrast, a tentacle apparatus with two arms is continued in the stem lineage of the Deuterostomia. It characterizes *Rhabdopleura* – albeit less in the form of the strict horseshoe, but rather as freely mobile arms with tentacles. In comparison with this, the evolution of several pairs of arms in *Cephalodiscus* and their arrangement to a complex filter basket is a clear apomorphy.

Equally unique as its form as a lophophore is the function of the tentacle apparatus as a food filter – which again supports the hypothesis of a unique evolution. In the "upstream collecting system" (NIELSEN & ROSTGAARD 1976) of the tentacle-bearing Radialia, a stream of water is passed from above into the tentacle apparatus. The cleaned water exits laterally between the individual tentacles of the apparatus (Fig. 22).

The ciliary covering of the tentacles provides the basis for this mechanism. Three groups of cilia belong in the ground pattern: (1) a frontal, longitudinal band on the inner side of the tentacle with several cilia in the cross section, (2) on each a laterofrontal row of single, longer and stiffer cilia as well as (3) two lateral, cilia-rich longitudinal bands. All cilia originate from monociliated cells (Fig. 23 A).

The outlined ground pattern with three differing groups of cilia is realized in the Bryozoa (Phylactolaemata and Gymnolaemata), the Phoronida and within the Brachiopoda (*Lingula, Glottidia*). However, the frontal and lateral bands in the ground pattern of the Bryozoa are produced by multiciliated cells (Fig. 22 B, C); this is an apomorphous state. There are contradictory reports for *Rhabdopleura* and *Cephalodiscus*. Frontal and laterofrontal cilia (GILMOUR 1979) are disputed; only lateral bands of two longitudinal rows with cilia are supposed to exist (NIELSEN & RIISGARD 1998).

Now we come to the function. First of all, the following situation for the tentacle-bearing Radialia is in general clarified. The lateral cilia form the motor for the water stream that flows through the tentacle apparatus. In contrast, the mechanisms of filtration and transport of fine particles have only been studied exactly in the Gymnolaemata with an apomorphous circular arrangement of the tentacles (RIISGARD & MANRIQUES 1997; NIELSEN & RIISGARD 1998; NIELSEN 1998b; Fig. 22 D–G).

Particles that enter the center of the tentacle crown as the location with the highest water flow velocity can be transported directly to the mouth without any contact; with an amount of about 5% this is insignificant for feeding.

Most of the particles pass through a mechanical filter made up of the two rows of laterofrontal cilia. There are two routes for the further transport of the particles trapped here. (1) Fine particles trigger inward-directed beats of the tentacles; the tentacle flicking pushes them back into the central water stream and then further towards the mouth. (2) Particles are transported towards the mouth by the action of the frontal cilia on the surface of the tentacles[3]. It is still disputed just how the particles arrive at the frontal band of cilia. Widespread opinions of a reversed beat of the cilia in the lateral bands (e.g., STRATHMANN 1973) could not be confirmed by investigations of the Bryozoa (NIELSEN & RIISGARD 1998).

At this point, I have to discuss the hypotheses about a possible double evolution of the described tentacle apparatus with the "upstream-collecting filter-feeding system". Two variants are mentioned in the literature.

In comparison with the Bryozoa, our knowledge about the structure and functional analysis in the Phoronida, Brachiopoda, and Deuterostomia is certainly not complete. We must wait until this lack has been remedied.

[3] This route is not applicable in the Cyclostomata, a subtaxon of the Gymnolaemata with a secondary lack of the frontal band of cilia.

For the time being, I, at least, cannot detect any arguments that would allow the hypothesis of a convergent evolution of the filter apparatus in the Bryozoa and the other three mentioned entities (NIELSEN & RIISGARD 1998).

In the second case, a separating line is drawn between the Lophophorata" and the Deuterostomia. The statement, "My findings are consistent with the hypothesis that the tentaculated arms of pterobranchs are homologous to the lophophore of brachiopods, phoronids and bryozoans" (HALANYCH 1993), experiences a complete reversal following the results of sequence analysis of 18S rDNA (HALANYCH 1996b). Molecules are now considered as undisputed censors of morphology. "The amazing similarities" between the feeding apparatus of the "Lophophorata" (Bryozoa, Phoronida, Brachiopoda) and the "Pterobranchia" (*Rhabdopleura, Cephalodiscus*) are now interpreted as the products of convergent evolution. As is usual in such cases, the explanation is readily at hand. "The high degree of morphological convergence presumably results from similar selective regimes" or "comparable selective pressures".

This unreflected misuse of selection as an explanatory principle must be vehemently contradicted. Certainly, the tentacle apparatus under discussion arose under the influence of selection. When there are, however, extensive agreements down to details in the ultrastructure between different organisms, then the hypothesis of a unique evolution is the better, more logical conclusion. In such a case, the supporters of convergent evolution must provide the evidence. We require more than an unprovable dogma of selection processes pointing in the same direction. What selection in the convergent evolution of a tentacular feeding apparatus can demonstrably produce is illustrated by the example of the Kamptozoa. The assumption of the convergent evolution of the tentacle crown is forced a posteriori here because the Kamptozoa are not related to the Radialia, but rather form a subtaxon of the Spiralia – probably the adelphotaxon of the Mollusca (Vol. II, p. 22). In the Kamptozoa the individual tentacle has five rows of multiciliated cells directed towards the atrium (Fig. 23B); the outermost "lateral" cells carry long, compound cilia that beat against the frontal side of the tentacle (NIELSEN & JESPERSEN 1997). Accordingly, the Kamptozoa operate with a "downstream collecting system" (NIELSEN & ROSTGAARD 1976) in which the lateral cilia bundles draw the water from the sides between the tentacles into the atrium. Even with a principally concordant function, the differences between the feeding apparatus of the Radialia with that of the Kamptozoa could hardly be greater.

However, we have not yet reached the end of this argumentation. HALANYCH (1996b) postulated convergence because the molecules suggested a relationship of the "Lophophorata" with the Mollusca and the Annelida and this led to the sensational creation of a new taxon Lophotrochozoa

(HALANYCH et al. 1995; COHEN et al. 1998). However, I cannot understand this measure with regard to the following aspect. If the comparable sequences of ribosomal DNA should have originated once as apomorphous agreements – which would have to be justified as proof for the phylogenetic relationship – then the agreements in the genetic information would have to be reflected in some way in the appearance, in the morphology. However, in the discussed example of the "Lophotrochozoa", this is not the case. Agreements in the molecular patterns of genetic information remain arbitrary for phylogenetic research as long as they cannot be justified as synapomorphies between certain organisms – and then in unquestioned harmony with corresponding feature evaluations from other levels of organization.

Life Cycle

As a consequence of its widespread occurrence among the Radialia, it is first of all not problematic to postulate a biphasic life cycle with a pelagic larva and a benthic adult for the ground pattern of the Radialia – as an autapomorphy in comparison with the sister group (p. 66).

Secondly, it is no problem to specify the pelagic larva for the ground pattern of the Deuterostomia. This is with acceptable certainty the planktotrophic dipleurula with a uniform band of cilia around the mouth (p. 94).

The three entities of the "Tentaculata" must be discussed separately, but here also we will arrive at clear statements that will be justified later. The planktotrophic cyphonautes larva can be set in the ground pattern of the Bryozoa. With equal certainty, the planktotrophic actinotrocha larva belongs in the ground pattern of the Phoronida. In the Brachiopoda the primary larval form is a lecithotrophic swimming larva (LÜTER 1997); however, I do not consider the phenomenon of larval lecithotrophy to be primitive, but rather an autapomorphy that arose in the stem lineage of the Brachiopoda (p. 91).

Let us now return to the ground pattern of the Radialia. As a result of the mentioned distribution, we may hypothesize a planktotrophic larva. Are there additional, convincing arguments to postulate certain elements of the cyphonautes larva, the actinotrocha or dipleurula for the larva of the stem species of the Radialia? I cannot find many, but can refer to two phenomena. (1) If the protonephridia of the actinotrocha originate from the ground pattern of the Bilateria (p. 68), then they must have been a characteristic of the Radialia larva. (2) In the cyphonautes larva the ribs of the filter apparatus bear three rows of cilia (p. 78). This is exactly the ciliation of one half of the tentacle of the adult in the ground pattern of the Radialia and could have been taken over by the larva from the adult in the evolution of the Radialia. It must be said, however, that this process could also have occurred first in the stem lineage of the Bryozoa.

The preceding justification of the monophyly of an entity Radialia of "Tentaculata" + Deuterostomia is opposed by molecular phylogenetic analyses with increasingly aggressive claims to hold the truth. "Recognition that brachiopods and phoronids are close genealogical allies of protostome phyla such as molluscs and annelids, but are much more distantly related to deuterostome phyla such as echinoderms and chordates, implies either (or both) that the morphology and ontogeny of blastopore, mesoderm and coelom formation have been widely misreported or misinterpreted, or that these characters have been subject to extensive homoplasy. This inference, if true, undermines virtually all morphology-based reconstructions of phylogeny during the past century or more" (COHEN 2000, p. 225)[4].

If true – the time does not seem ripe for an unemotional recognition of all characteristics from the molecule to the whole organism and their consequent inclusion in the methodology of phylogenetic systematics.

Bryozoa

The Bryozoa (Ectoprocta) are sessile, colonial organisms. We begin with the distribution of their adelphotaxa, the Phylactolaemata and Gymnolaemata, in fresh and sea water as well as assigned alternatives in the tentacle apparatus – a condition that we know from experience will encounter difficulties.

The limnetic Phylactolaemata with ca. 50 species have a horseshoe-shaped lophophore and a flap – the epistome – over the opening of the mouth (Fig. 24). The marine Gymnolaemata with several thousand species, in contrast, all have a ring-shaped, closed tentacle crown and a naked throat (Fig. 25).

What is the **primary state of the tentacle apparatus**? The agreements between the lophophore of the Phylactolaemata as well as the corresponding filter apparatus of the Phoronida and Brachiopoda must, in the most economical explanation, be considered as homologies (p. 68); I cannot see any conflicts in this interpretation.

The ostensible difficulties now arise. How can a tiny number of limnetic Bryozoa for which fresh water is most certainly a secondary habitat exhibit the primitive state of the tentacle apparatus? The answer is: the stem species of the Bryozoa consisted of marine individuals with a horseshoe-shaped lophophore and an epistome. The stem lineages of the Phylactolaemata and the Gymnolaemata separated in the sea. Stem lineage representatives of the Phylactolaemata with the primitive lophophore migrated into fresh water and did not have the possibility for larger evolutionary diversi-

[4] cf. COHEN et al. (1996, 1997, 1998).

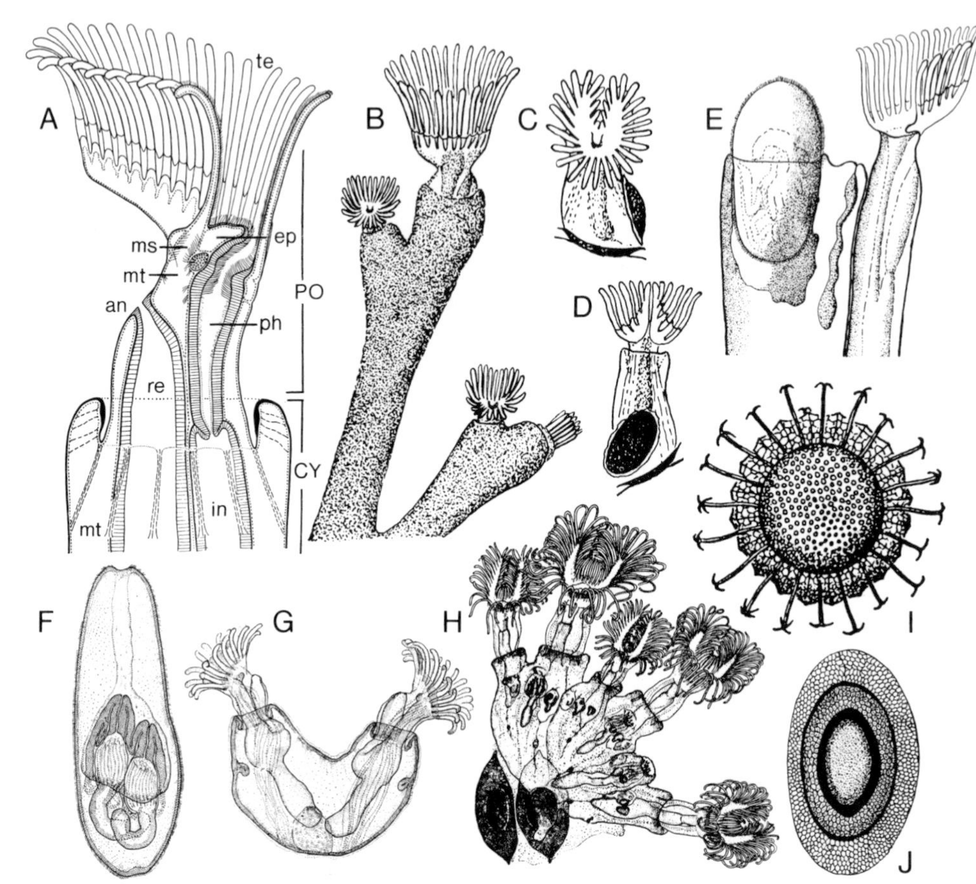

Fig. 24. Phylactolaemata (Bryozoa). **A** *Plumatella fungosa.* Sagittal section with division of the zooid into polypid (*PO*) and cystid (*CY*). **B** *Fredericella sultana.* Colony with polypid buds. **C** and **D** *Fredericella sultana.* First zooid from statoblast with cleft between the arms of the lophophore. **E–G** *Plumatella fungosa.* Sexual reproduction with brood care. **E** Emergence of a young colony from the embryo sac. **F** Free swimming ciliated colony founder with two invaginated polypids. **G** Attachment with extension of the polypids. **H–J** Asexual reproduction via statoblasts. **H** *Lophophus crystallinus.* Young colony from a statoblast, the opened shells of which are visible on the ground. **I** *Cristatella mucedo.* Statoblast with outer circle of hooks. **J** *Plumatella casmiana.* Statoblast with chambered, air-filled annulus acting as a float. *an* Anus; *ep* epistome; *in* intestine; *ms* mesocoel; *mt* metacoel; *ph* pharynx; *re* rectum; *te* tentacle. (**A, F–H** Brien 1960; **B–E** Marcus 1926; **I** Ryland 1970; **J** Rao 1973)

fication in this biotope-poor milieu. In the stem lineage of the Gymnolaemata, the transformation of the lophophore into a closed tentacle crown occurred in the sea – and here the widely varying habitats provided the impetus for intensive speciation. Only single members of the Gymnolaemata with the apomorphous tentacle apparatus immigrated later into fresh water – for example, the now widely distributed species *Paludicella articulata*. JEBRAM (1973) formulated similar considerations.

Furthermore, there is an impressive example for the direction of evolution from a lophophore to a tentacle crown within the Phylactolaemata. At a first glance, the filtration apparatus of the fresh water species *Fredericella sultana* appears to be closed to a crown. In fact, it still consists of the two arms of a lophophore which are now only separated by a slit; this is particularly evident in juvenile individuals (Fig. 24C, D).

One may accuse me of drawing too broad comparisons - a glance at the basal split of the Arthropoda seems to me to be useful at this point for an understanding of evolutionary routes. Representatives of the stem lineage of the Onychophora crawled on to land with a series of primitive characteristics and have retained some of them in the terrestrial milieu - for example, the subepidermal body wall musculature or segmental nephridia spread over the entire trunk. The Euarthropoda as the sister group remained at first in the sea and developed in the original marine milieu prominent apomorphies such as the plate skeleton with abandonment of the subepidermal muscular tube, the jointed extremities or the head with several segments and the compound eyes (Vol. II, p. 79).

Returning to the Bryozoa, sessility and formation of colonies led in the sea to a unique apomorphous body construction with division of the individual into **cystid** and **polypid**. The firmly attached cystid forms a protective shell and is at the same time responsible for budding and growth. The mobile polypid carries the tentacle apparatus as feeding organ. It can be extended and withdrawn into the cystid.

The cystid wall is primarily a purely organic body cover of protein and chitin. In the Phylactolaemata its consistence ranges from solid sclerotization (*Plumatella*) to semifluid gel (*Lophopus*). A solidification of the cystid wall by calcareous impregnation occurred among the Gymnolaemata - independently in the Cyclostomata and in the Cheilostomata.

No apomorphous characteristics can be derived from the structure of the **coelom**. A mesocoel as a ring-shaped lophophore coelom at the base of the lophophore with supply of the tentacles as well as a spacious metacoel in the cystid belong in the ground pattern of the Bryozoa. Mesocoel and metacoel are only incompletely separated. There is a cavity lined with epithelium and lacking pores to the outside in the epistome of the Phylactolaemata; in addition, the epistome coelom opens widely into the mesocoel. As a result of these features, the earlier postulated homology with the protocoel of the Deuterostomia was discarded (NIELSEN 1995; MUKAI et al. 1997).

However, the **funiculus** as a derivative of the coelom is a characteristic autapomorphy of the Bryozoa. In the primitive form it is a separate, individual organ in every zooid of the colony (Phylactolaemata, Cyclostomata). The tubular funiculus passes through the coelom from the stomach to the ventral body wall; peritoneal cells and muscles enclose a central lumen.

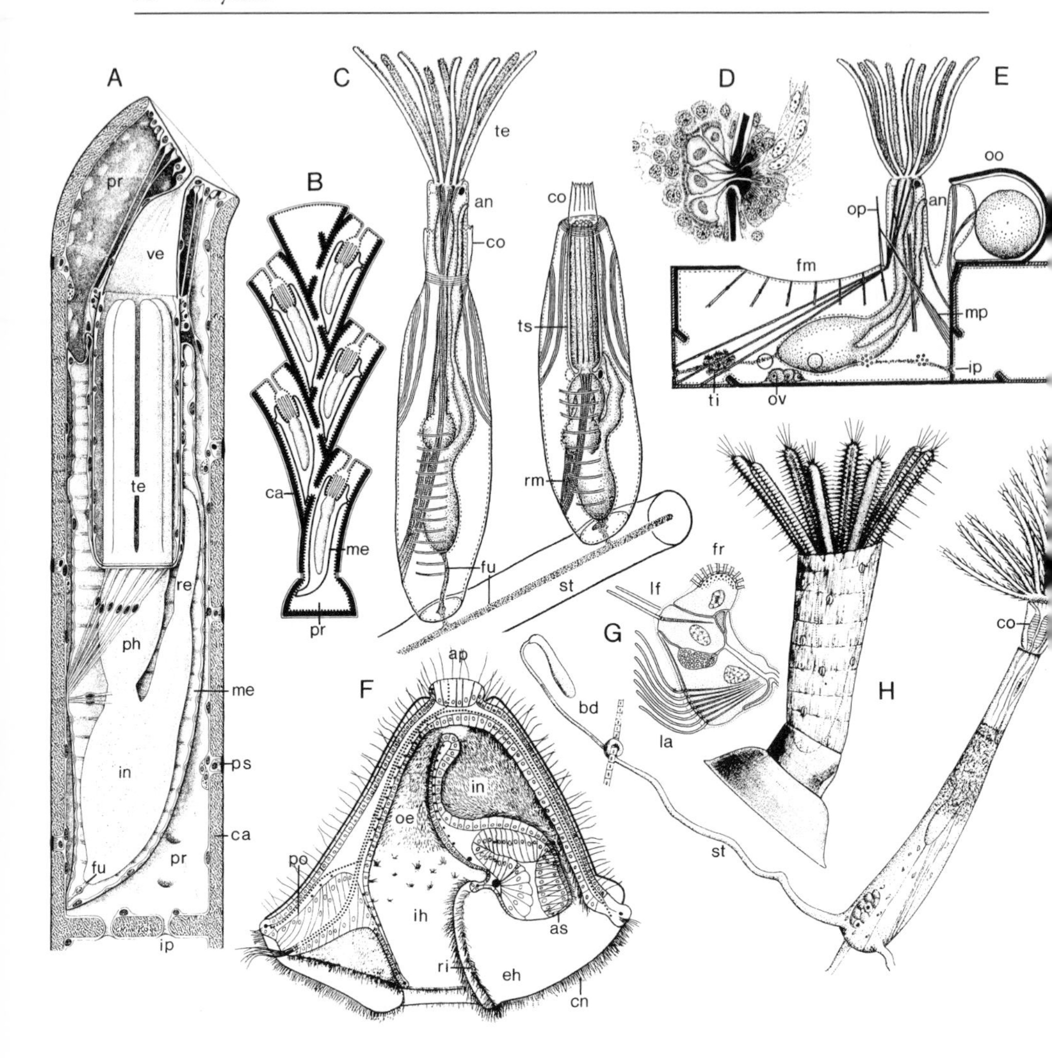

In contrast, the funiculus in the Eurystomata (Ctenostomata + Cheilostomata) has developed into a central vessel for transporting fluids through the colony (CARLE & RUPPERT 1983). This is an apomorphous state. The funiculus links zooids of the colony through pores in the cystid walls. Complexes of cells that probably control the flow of material within the colony occur at these pores; they are designated as rosettes, septula or communication organs.

Irrespective of its various forms, the funiculus is the site at which the testes of all Bryozoa are formed, while the ovaries develop on the body wall.

The hypothesis of a homology between the funiculus and the blood vessel system of the Phoronida and Brachiopoda is controversial (CARLE &

◀ **Fig. 25.** Gymnolaemata (Bryozoa). **A** *Crisia eburnea* (Cyclostomata). Zooid with retracted polypid. Lateral view. **B** Colony of *Crisia eburnea* (Cyclostomata). Demonstration of calcified wall under the periostracum, the primary body cavity and the membranous sac (secondary body cavity). Ectoderm as *dotted line* under the cystid wall. **C** *Bowerbankia* (Ctenostomata). Zooid with extended and zooid with partially retracted tentacle apparatus. The zooids originate from a stolon by which they are linked through strands of the funiculus. **D** *Bowerbankia imbricata* (Ctenostomata). Rosette of cells and tissue of the funiculus at an interzooidal pore. These pores are openings in the cystid wall of neighboring zooids. **E** *Membranipora* (Cheilostomata). Box-shaped zooid with open operculum and expanded tentacle apparatus. The zooid has formed an ooecium to receive eggs (brood care). **F** *Electra pilosa* (Cheilostomata). Schematic side view of cyphonautes larva. **G** *Membranipora*. Cyphonautes larva. Cross section through a rib of the food filter with three rows of cilia. **H** *Crisia eburnea* (Cyclostomata). Ancestrula (colony founder) with one tentacle apparatus; 7 days after attachment of the lecithotrophic larva. **I** *Monobryozoon limicola* (Ctenostomata). Secondary, solitary member of the Bryozoa. *an* Anus; *ap* apical organ; *as* adhesive sac; *bd* bud; *ca* calcified cystid wall; *cn* corona; *co* collar; *eh* outflow chamber; *fm* frontal membrane; *fr* frontal cilia; *fu* funiculus; *ih* inflow chamber; *in* intestine; *ip* interzooidal pore; *la* lateral cilia; *lf* laterofrontal cilia; *me* membranous sac; *mo* mouth opening; *mp* constrictor of the operculum; *oe* oesophagus; *oo* ooecium; *op* operculum; *ov* ovary; *ph* pharynx; *po* pyriform (*pear-shaped*) organ; *pr* primary body cavity; *ps* pore; *re* rectum; *ri* ciliated rib; *rm* retractor muscle; *st* stolon; *te* tentacle apparatus; *ti* testis; *ts* tentacle sheath; *ve* vestibulum. (**A, B** Nielsen & Pedersen 1979; **C, E** Ryland 1970; **D** Bobin 1979; **F** Nielsen 1995; **G** Nielsen 1971; **H** Nielsen 1970; **I** Franzén 1960)

RUPPERT 1983; NIELSEN 1995). According to a general methodological point of view, however, I must point out that the funiculus itself can be interpreted as an autapomorphy of the Bryozoa and that the mentioned change may be used as justification of a monophylum Eurystomata without having to make a decision on the hypothesis of its origin.

A biphasic life cycle with a planktotrophic larva is taken over as a plesiomorphy from the ground pattern of the Radialia and does not offer very much as such. In the stem lineage of the Bryozoa, however, the evolution of a plankton larva in the state of the **cyphonautes** (Fig. 25 F) occurred; in other words, it may be interpreted as an autapomorphy of the Bryozoa. In fact, the well-known triangular cyphonautes larvae are only found in a few members of the Ctenostomata and Cheilostomata (NIELSEN 1971, 1995, 1998a; STRICKER et al. 1988). Even so, we must still place the cyphonautes in the ground pattern of the Bryozoa simply because it is the sole planktotrophic larva in the entity. Many forms of lecithotrophy have developed within the Ctenostomata and the Cheilostomata, but are not relevant in the present context. Two circumstances, however, are important for the justification of subtaxa of the Bryozoa. In the Cyclostomata, ontogenesis always proceeds through lecithotrophic larvae without a gut; they are incubated in gonozooids (modified zooids). In the limnetic Phylactolaemata there are neither planktotrophic nor lecithotrophic larvae; the larval phase of the primary life cycle is completely reduced. Small, ciliated spheres emerge from the embryo sacs of adults

(Fig. 24E–G). These are not larvae, but rather young animals with an invaginated polyp or also even small colonies with the anlage of two to three zooids.

Autapomorphies (Fig. 21 → 2)

- Sessile colony.
  The stem species of the Bryozoa was a firmly attached, colonial organism. The colony arises by budding after attachment of the plankton larva and its metamorphosis to the colony founder (ancestrula). The zooids of the colony are each able to feed independently, but remain in bodily contact throughout their lives.
- Division of the body into cystid and polypid.
  The individual zooid is divided into the firmly attached cystid as envelope for the mobile polypid with tentacle apparatus.
- Multiciliated cells on the tentacles.
  In the ground pattern of the Radialia the ciliated tentacle cells each carry one cilium (p. 69). In the Bryozoa the frontal and lateral ciliary bands of the tentacles are formed from multiciliated cells (Fig. 22B, C).
- Pharynx with triradiate lumen.
  Pharynx wall of myoepithelial cells. Contractions of the radially directed, striated myofilaments extend the pharynx lumen.
- Funiculus.
  Tube-like cord of peritoneal and muscle cells that extends from intestine to the ventral side of the cystid. In the ground pattern an individual organ in each zooid.
- Lack of nephridial excretory organs.
  Protonephridia are realized in the actinotrocha larva of the Phoronida, composite metanephridia in the adult Phoronida and Brachiopoda. We must hypothesize both systems for the ground pattern of the Radialia (p. 68) and accordingly interpret the absence in the Bryozoa as a secondary state.
- Cyphonautes larva (Fig. 25F).
  Laterally compressed, triangular body in bivalve shell. Apical organ linked by a nerve cord to a pear-shaped tactile organ for attachment. This pyriform organ is the anterior part of the corona that surrounds the larva on the underside as a ciliary band. An extensive vestibulum is separated into inflow and outflow chambers by two ribs with three rows of cilia; however, the cilia arranged in rows may originate from the ground pattern of the Radialia (p. 72). Stiff laterofrontal cilia probably serve as a mechanical filter (STRATHMANN & McEDWARD 1986; NIELSEN 1995). Attachment sac on the roof of the outflow chamber.

Systematization

Due to the lack of an understanding of the postulates of phylogenetic systematics, the traditional division of the Bryozoa into three classes of equal rank is followed even in modern handbooks (BOARDMAN et al. 1983; WOOLLACOTT et al. 1997) and textbooks. The classification then has the following appearance:

Classis Phylactolaemata
Classis Cyclostomata (Stenolaemata)
Classis Gymnolaemata
  Ordo Ctenostomata
  Ordo Cheilostomata

Kinship relationships between the three "classes" cannot be seen from this division. This is, however, possible when we consequently ask about the highest-ranking sister groups of the Bryozoa at the first level of subordination. Then, namely, we can take up the already mentioned, very old division into the limnetic Phylactolaemata and the marine Gymnolaemata (ALLMAN 1856; see HYMAN 1959) and justify these two entities as equal-ranking monophyla. For the Gymnolaemata, we can then provide good arguments of an adelphotaxa relationship between the Cyclostomata (Stenolaemata) and an entity already named by MARCUS (1938) as Eurystomata in the union of Ctenostomata and Cheilostomata. With reservations as to the difficulties in the relationship between the latter two taxa – which will be described later – we reach the following phylogenetic systematization.

Bryozoa
  Phylactolaemata
  Gymnolaemata
    Cyclostomata
    Eurystomata
      Ctenostomata
      Cheilostomata

We will present justifications and open problems group for group in the following survey.

Adelphotaxa Phylactolaemata – Gymnolaemata

Phylactolaemata (Fig. 24)

Plumatella and *Fredericella* with tube-shaped, branched growth. *Lophopus, Pectinatella* and *Cristatella* as compact, massive colonies.

Plesiomorphies. Horseshoe-shaped lophophore with two arms – epistome lies as a flap over the mouth opening. The individual zooids of the colony each possess a separate, cord-like funiculus.

Autapomorphies. Fresh water inhabitants – complete reduction of the primary plankton larva – development of young colonies in the embryo sacs of adult organisms – asexual formation of statoblasts on the funiculus to survive periods of unfavorable weather.

The ostensible crown-shaped tentacle apparatus of *Fredericella* is a lophophore with two arms (p. 75) – modified within the Phylactolaemata.

The mentioned apomorphies in the development and the existence of statoblasts provide support for the monophyly of the Phylactolaemata and against the hypothesis of a twofold invasion of fresh water (BARTOLOMAEUS & GROBE 2000).

Gymnolaemata (Fig. 25)

Plesiomorphies. Marine organisms – biphasic life cycle with planktotrophic cyphonautes larva. This larva was taken over from the ground pattern of the Bryozoa into the stem lineage of the Gymnolaemata – continuing separate funiculus cord.

Autapomorphies. Ring-shaped, closed crown of tentacles around the opening of the mouth – loss of epistome above the mouth.

Adelphotaxa Cyclostomata – Eurystomata

Cyclostomata (Stenolaemata; Fig. 25 A, B)

Crisia, Tubulipora, Hornera

Plesiomorphies. Separate funiculus cord per zooid as in the Phylactolaemata continued from the ground pattern of the Bryozoa (and beyond the ground pattern of the Gymnolaemata) – the individual zooids are linked by open pores (apomorphous alternative in the Eurystomata) – circular opening of the cylindrical cystid without closing apparatus.

Autapomorphies. Unique construction of the coelom. Metacoel as a membranous sac that is clearly lifted from the cystid wall (NIELSEN & PEDERSEN 1979). Circular muscles on the outside wall of the sac; their contraction effects the extension of the tentacle crown. The cavity between the membranous sac and the body wall is interpreted as a primary body cavity – lack of frontal cilia on the tentacles (p. 70) – calcification of the body wall – gonozooids. Development of the eggs to larvae in breeding compartments as modified individuals (zooids) of the colony – lecithotrophic larvae develop under polyembryony in the zooids.

Eurystomata (Fig. 25C–H)

Plesiomorphies. Primary uncalcified cystid wall – planktotrophic cyphonautes larva taken over from the ground pattern of the Bryozoa and Gymnolaemata – no gonozooids.

Autapomorphies. No musculature on the body wall. Parietal muscles that pass through the cystid control the extension of the cystid – cellular rosettes occupy the pores in the cystid wall between zooids – branched funiculus vessels without longitudinal muscles; they make contact with the rosettes and effect transfer of fluids through the colony via the pores.

The **Eurystomata** have **closing apparatus** at the opening of the cystid. In comparison to the Cyclostomata, this is at first certainly an apomorphous state. However, the closing systems of the Ctenostomata and the Cheilostomata are completely different. In the Ctenostomata, the apparatus is a purely cuticular collar (Fig. 25C), in the Cheilostomata a mobile cover (operculum; Fig. 25E). There are principally two possibilities to interpret this situation.

1. They are alternative apomorphies that evolved independently from a condition without closure in the stem lineages of the Ctenostomata and Cheilostomata. This is difficult to validate.
2. One of the two systems is primitive and belongs in the ground pattern of the Eurystomata.

Findings from the Japanese species *Labiostomella gisleni* have been put forward in support of the second possibility (SILÉN 1942, 1944). *Labiostomella* has a lip under the laterally displaced opening of the cystid; it is firmly bound with the cystid wall. Furthermore, *Labiostomella* has a structure designated as a collar – but in contrast to the Ctenostomata it is thick, not folded and composed of two cell layers (Fig. 26D,E). If a homology should exist between these two cases, then the complex lip/operculum should form the apomorphous homology and the compared collar struc-

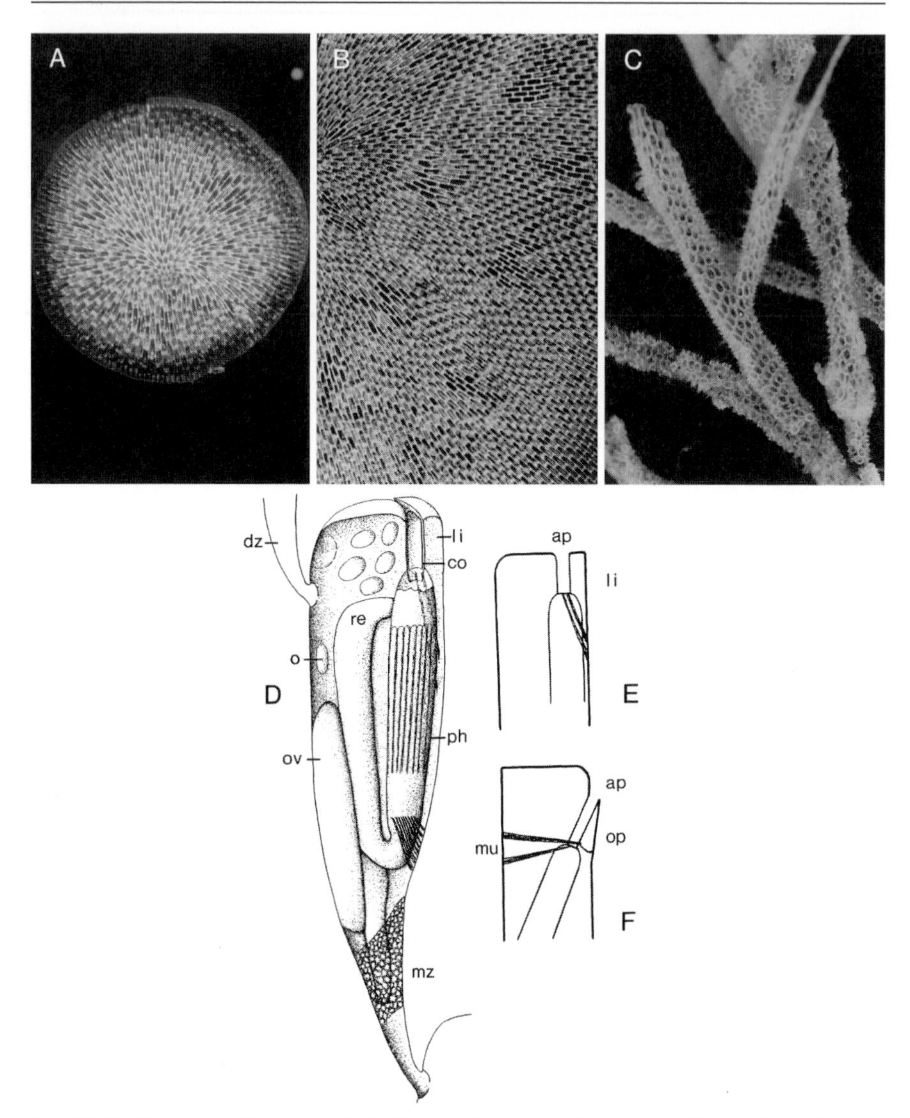

Fig. 26. Cheilostomata (Bryozoa). **A, B** *Membranipora membranacea*. **A** Young, flat colony on the brown alga *Macrocystis*. The colony grows in the periphery by sprouting new zooids. **B** Section of the colony with complete array of joined, box-like cystids. **C** *Electra pilosa*. Colony with oval cystids. As a result of coverage of filamentous algae, tube-like closed colonies form. **D, E** *Labiostomella gisleni*. **D** Side view of zooid. Distal part cut in the sagittal plane. **E** Longitudinal section through the opening of a cystid with firmly fixed lip under the laterally displaced aperture. **F** General scheme of the Cheilostomata for comparison. Displacement of the aperture to the ventral side and mobile operculum. *ap* Aperture; *co* collar; *dz* daughter zooid; *li* lip; *mu* muscle; *mz* mother zooid; *o* egg; *op* operculum; *ov* ovary; *ph* pharynx; *re* rectum. (**A–C** Originals; **A, B** Puget Sound, Washington, USA; **C** Helgoland, North Sea; **D–F** Silén 1944)

tures a plesiomorphous agreement. This would have two consequences: (1) *Labiostomella gisleni* belongs to the Cheilostomata; (2) the collar is a ground pattern feature of the Eurystomata and, as such, is not suitable for the justification of a monophylum Ctenostomata. The appreciable differences in both cases led SILÉN to make highly cautious formulations about the homology question which were simply ignored in later summaries.

A convincing clarification of this situation down to the ultrastructural level is needed. I can only, with reservation, set the collar and operculum as respective autapomorphies in the justification of the taxa Ctenostomata and Cheilostomata. The collar may be the primitive state of the closure apparatus of the Eurystomata, in which case the Ctenostomata would form a paraphylum.

Ctenostomata – Cheilostomata

Ctenostomata

Alcyonidium, Flustrella, Paludicella, Bowerbankia, Monobryozoon as a taxon with secondary, solitary species in soft sediments (Fig. 25 I).

Plesiomorphy. Cystid wall uncalcified.

Autapomorphy. (?) Collar (without *Labiostomella*) as a thin cuticular membrane that protrudes from the diaphragm. When the polypid is extended, the collar surrounds the lower part of the tentacle sheath. With retracted polypid the collar folds itself to a cone that closes the vestibulum. Finally, the diaphragm is closed by sphincter muscles at the base of the collar (Fig. 25 C).

Cheilostomata

Membranipora, Electra, Flustra, Bugula

Autapomorphies. At least the vertical zooid walls are calcified – operculum. Opening of the cystid closable with a mobile cover (possibly only a firmly attached lower lip as in *Labiostomella gisleni* belongs in the ground pattern of the Cheilostomata).

Fig. 27. Phoronida. **A–E** *Phoronis muelleri.* Organization and development. **A** Actinotrocha larva. Ventral view. Crown-like arrangement of the larval tentacles. Ventral pore of the metasome diverticulum. Next to it, the openings of the protonephridia. **B** Metamorphosis. Expulsion of the metasome diverticulum leads to a U-shaped gut. Degradation of larval tentacles and a part of the episphere. **C** Young organism after metamorphosis. Short adult tentacle. **D** Adult. "Anterior end". Epistome without cavity over the opening of the mouth. Mesocoel canals and blood vessels extend into the tentacles. Double layer of myoepithelial cells separates the mesocoel from the extensive metacoel. **E** Actinotrocha. Sagittal section. Cavity of the episphere with gel-like ECM. Extensive blastocoel in the region of the larval tentacles. Mesocoel anlage basal on growing adult tentacles. Metasome diverticulum not drawn. *EP* Episphere; *TE* tentacle region; *HY* hyposphere. **F** *Phoronis psammophila.* ♀ Top view of the horseshoe-shaped lophophore. With location of anus and pores of the metanephridia. **G** *Phoronis australis.* Lophophore in the form of a double spiral. *an* Anus; *at* adult tentacle; *bl* blastocoel; *bv* blood vessel; *es* epistome of the adult without coelom; *in* intestine; *lo* lophophore organ; *lt* larval tentacle; *md* metasome diverticulum; *mo* mouth; *ms* mesocoel; *mt* metacoel; *pm* pore of the metasome diverticulum; *po* pore of the metanephridium; *pp* pore of the protonephridium. (**A–E** Bartolomaeus 1993, 2001; **F, G** Emig 1977)

Phoronida

Small entity with a dozen marine species in the traditional division into two supraspecific taxa – *Phoronis* and *Phoronopsis.* However, *Phoronis* is apparently not a monophyletic entity (EMIG 1974, 1985).

The Phoronida are inhabitants of self-made tubes that are mainly deposited in soft sediments (sand, mud) of the sea bottom. The tubes of the solitary organisms stand vertically in the substrate; they are secreted by glands of the ectoderm (EMIG 1982). The anterior end with the tentacle apparatus is extended from the tube to filter food. The functional posterior part serves as an inflatable end bulb for anchoring in the tube.

In spite of the small number of species, the lophophore apparatus exhibits an extraordinary diversity in its shape. The spectrum ranges from an oval form in *Phoronis ovalis* (? secondary simplified), through the common horseshoe (*Phoronis muelleri, P. hippocrepia, P. psammophila*) and spirals in *Phoronis australis* (Fig. 27 G) to a helical roll in the taxon *Phoronopsis* (Fig. 28 A).

A planktotrophic larva with a pair of protonephridia can be postulated without conflict for the ground pattern of the Phoronida. The protonephridia go back to the ground pattern of the Radialia and further to the ground pattern of the Bilateria. On the other hand, it cannot be determined to what extent features that are only known from Phoronida larva can be taken over as plesiomorphies. We must identify the following elements of the actinotrocha as well as the mechanism of their metamorphosis into the adult as apomorphies of the Phoronida (Fig. 27).

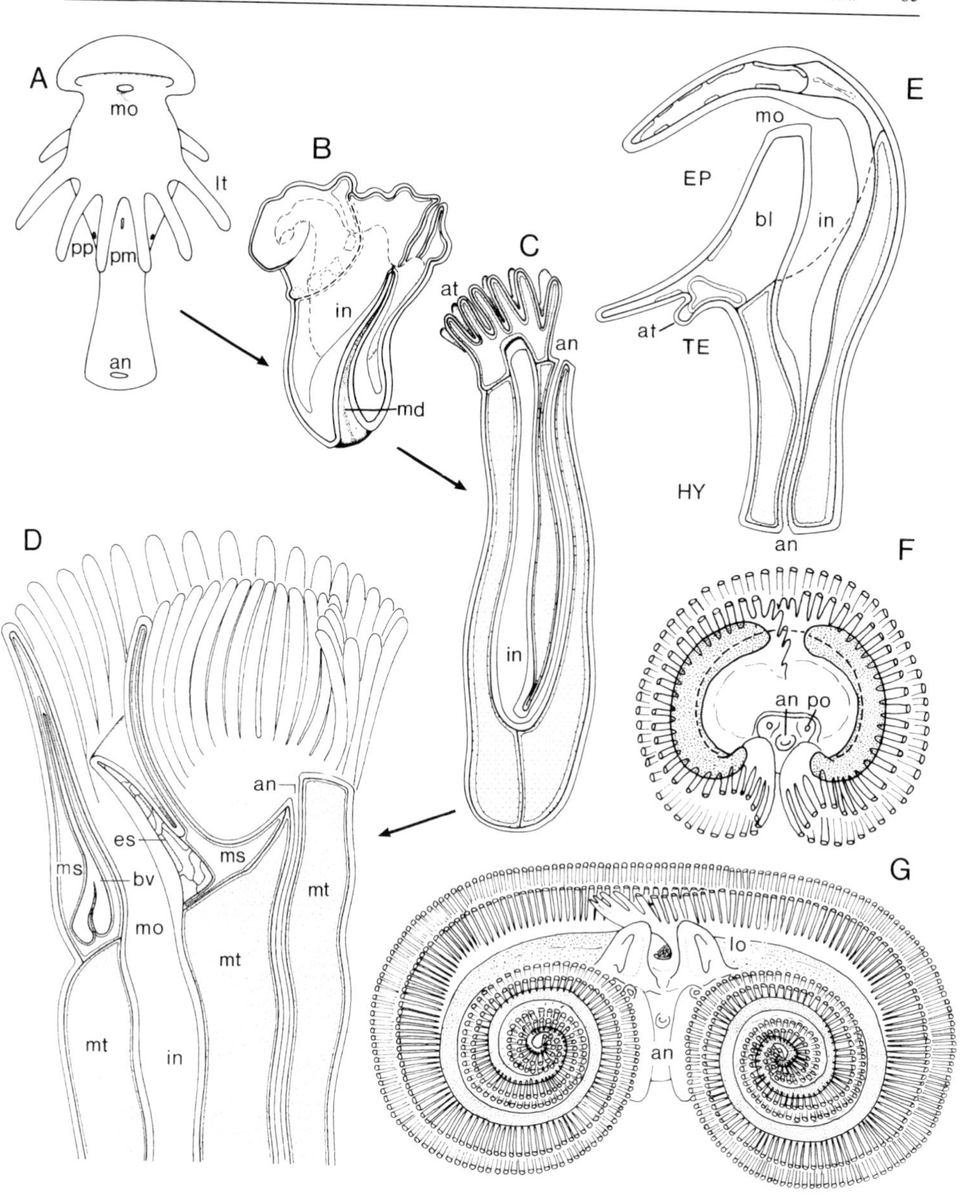
A
mo
lt
pp
pm
an
B
in
md
C
at
an
in
D
es
ms
bv
mo
ms
an
mt
mt
mt
in
E
mo
EP
bl
in
at
TE
HY
an
F
an po
G
lo
an

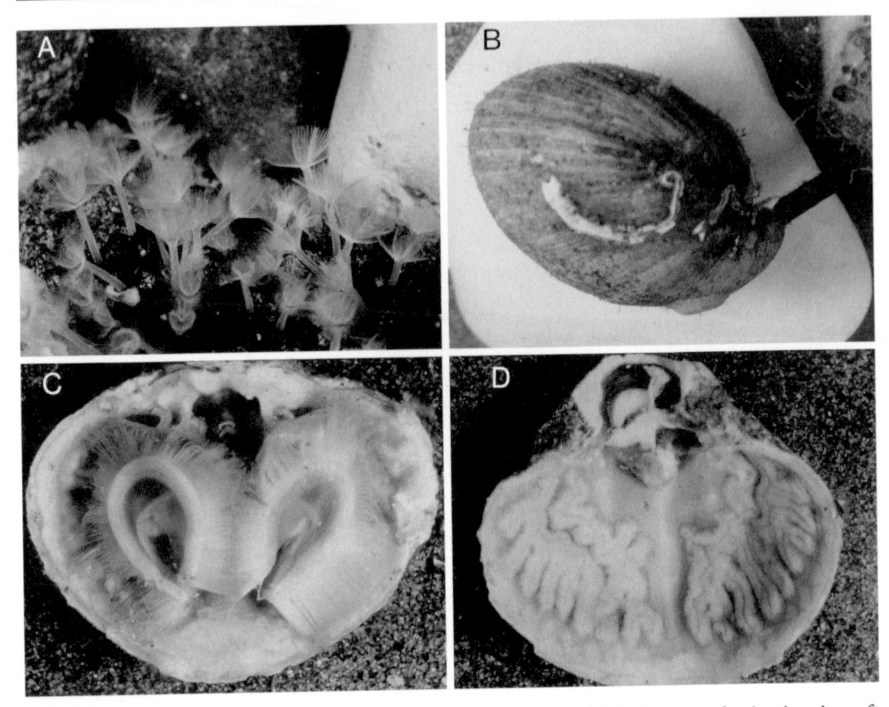

Fig. 28. Phoronida and Brachiopoda. **A** *Phoronopsis* (Phoronida). Dense colonization in soft sediment. The helically rolled tentacle crown protrudes out of the dwelling tube. **B** Testicardines (Brachiopoda) in morphological orientation. Side view (anterior end left). Flat dorsal shell and bulbous ventral shell. The foot with which the animal adheres to hard substrates emerges upwards from the posterior end of the ventral shell. **C, D** *Terebratalia transversa* (Brachiopoda). Shells separated, views of the inside. **C** Dorsal shell. An arm skeleton supports the lophophore with tentacles. **D** Ventral shell. Gonads grow out under the shell over metacoel canals into the mantle. (Puget Sound, Washington, USA. Originals)

Autapomorphies (Fig. 21 → 3)

- Features of actinotrocha larva.
  Umbrella-like episphere (preoral tube).
  Radial arrangement of larval tentacles.
  Metasome diverticulum (or pouch) as an invagination of the ventral side of the larva in the region of the mesosome-metasome boundary.
- Metamorphosis.
  Everting the metasome diverticulum. Drawing the larval alimentary canal into the pouch and further development to the U-shaped intestine of the adult. The end of the pouch transforms to the functional posterior end of the body. Degradation of the episphere and the larval tentacles. The tentacles of the adult develop.

Brachiopoda

The Brachiopoda are solitary organisms of the sea bottom – vagile tube inhabitants in soft sediments or sessile surface colonists on hard substrates.

The small number of ca. 330 recent species contrasts with the knowledge of over 12,000 fossil species from the Paleozoic and Mesozoic eras. This is due to the petrified shells, the most conspicuous feature of the Brachiopoda. These consist of a dorsal and a ventral shell valve and constitute a fundamental difference to the lateral shell valves of the Bivalvia.

The following discussion concentrates on the features that are relevant for the elaboration of apomorphies in the ground pattern of the Brachiopoda and the evaluation of kinship relationships within the entity. We are concerned with four high-ranking, presumably monophyletic taxa (p. 92) and consider their bonds to the substrate (Fig. 29).

Lingulida (*Lingula, Glottidia*). Tube inhabitants in soft sediments. With long stalk.
Discinida (*Discinisca*). On hard ground. Very short stalk that emerges from a slit in the ventral valve.
Craniida (*Crania*). On hard ground. No stalk. Fastened with the ventral valve.
Testicardines (*Lacazella, Tegulorhynchia, Terebratulina, Argyrotheca, Terebratella*). On hard ground. Anchoring with stalk; outlet through a hole in the ventral valve.

We will start with the **shell**. Even though shell-carrying organisms are lacking for an outgroup comparison in the neighborhood of the Brachiopoda, we may identify a shell without a hinge as a common plesiomorphy of the Lingulida, Discinida and Craniida. In the ground pattern, the shell valves are held together only by musculature. The evolution of the hingeless shell from a state with articulation through teeth and sockets is hard to imagine. Rather, a shell with hinge can be interpreted without conflict as an evolutionary novelty of the Testicardines. Another, interesting alternative arises from the chemical composition of the shell. In the Lingulida and Discinida, the shell is composed of calcium phosphate (75–94%) and an organic portion of chitin and protein, in the Craniida and Testicardines of calcium carbonate (88–99%) and protein, but no chitin (HOLMER et al. 1995). Again, we have no possibility for a direct outgroup comparison. However, we may consider that in any case calcium carbonate is the major hard substance of animal skeletons. This may be an indication that the calcium phosphate-chitin shells of the Lingulida and Discinida are to be interpreted as an apomorphy.

The shells lie on the **mantle** of the soft body from which they are secreted. The dorsal and ventral lobes of the mantle cover a filter cavity that

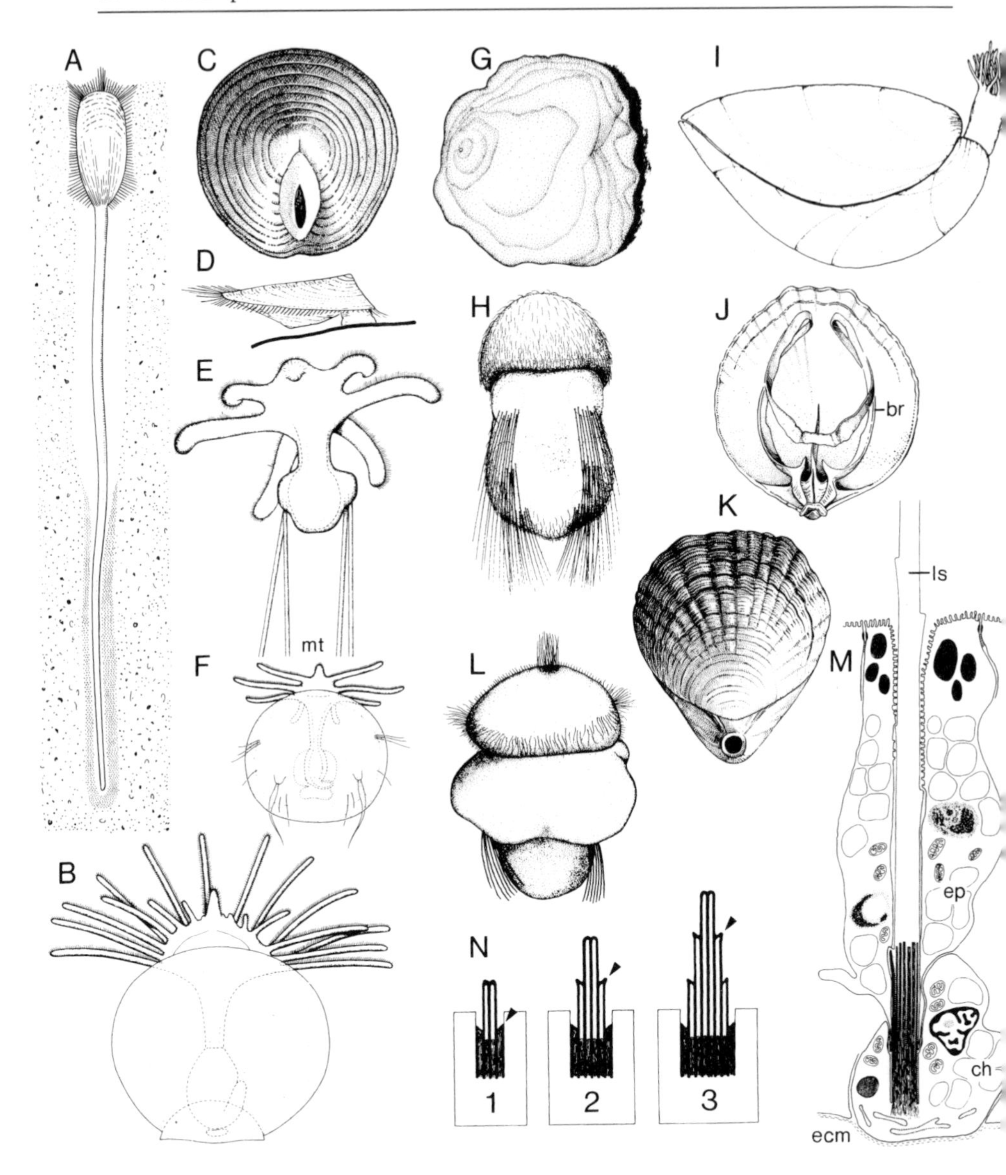
A
B
C
D
E
F
mt
G
H
I
J
br
K
L
M
ls
ep
ch
ecm
N
1
2
3

◀ **Fig. 29.** Brachiopoda. **A, B** Lingulida. **A** *Lingula* in the tube. Setae of the mantle edge protrude out of the shell. Long stalk deep in the sediment. **B** *Lingula*. Pelagic, planktotrophic juvenile stage with paired adult tentacles and unpaired median tentacle. **C–F** Discinida. **C** *Discinisca lamellosa*. Ventral shell with cutout for the stalk to pass through. **D** *Discinisca lamellosa*. Side view. Anchored with short stalk to hard substrate. **E** *Discinisca*. Lecithotrophic, pelagic larva with larval setae. **F** *Discinisca*. Planktotrophic juvenile stage with adult tentacles and unpaired median tentacle. **G, H** Craniida. **G** *Crania anomala*. View of the dorsal shell. Ventral shell fused with hard substrate. **H** *Crania anomala*. Lecithotrophic larva from the pelagic zone with bundles of larval setae. **I–M** Testicardines. **I** *Terebratulina retusa*. Side view. Stalk protrudes from a hole in the ventral shell. **J** *Magellania flavescens*. Dorsal shell with arm skeleton (brachidium). Inside view. **K** *Magellania flavescens*. Inside view of ventral shell. **L** *Notosaria nigricans*. Lecithotrophic larva. Three-divided stage with apical ciliary circlet and long larval setae. **M** *Notosaria nigricans*. Larval seta with chaetoblast that is followed outwards by an indented epidermis cell. The epidermis cell has no contact with the seta. **N** *Lingula anatina*. Adult seta. Diagram of the formation of a horsetail structure. The chaetoblast forms new microvilli with projecting tips at distances around the central seta canals. The secreted material forms circles about the seta. The *arrowheads* point to the upper circle at times 1, 2 and 3. *br* Brachidium; *ch* chaetoblast; *ecm* extracellular matrix; *ep* indented epidermis cell; *ls* larval seta; *mt* median tentacle (**A** Kozloff 1990; **B, E, F, H** Nielsen 1991; **C, J, K** Beauchamp 1960; **D** Hyman 1959; **I** Lüter 1995; **L–N** Lüter 1998 a)

is divided into an inflow chamber and an outflow chamber by the lophophore apparatus. The evolution of mantle and shell are inseparably linked. Accordingly, the mantle cannot be considered as a separate autapomorphy of the Brachiopoda.

The **stalk** or **pedicle** is a second conspicuous feature in the construction of the Brachiopoda. Strangely, it is described even in the recent literature as an analogous organ with convergent evolution in the stem lineages of the hinge-less Ecardines and the hinge-carrying Testicardines (WILLIAMS et al. 1997). There are certainly serious differences. The stalk of the Lingulida is only joined to the ventral valve; for movement in soft sediments and for fixing in the burrow it has an extensive coelom as hydrostatic skeleton with antagonistic musculature in the periphery. The stalk of the Testicardines is fused with the mantle all round; the central connective tissue and the thick cuticle form basic elements of the anchoring organ that, after adhesion, do not allow any further change of location.

However, even when I take the differences into account, I cannot imagine the evolution of a sedentary stem species with two shell valves without a stalk. In addition, of course, we can refrain from the idea of a stalk-less stem species of the Brachiopoda fastened by a shell as realized by the Craniida and Thecidacea within the Testicardines.

If we postulate homology, then the situation in the Lingulida with emergence of the stalk between two similarly shaped shells must be interpreted as plesiomorphy and set in the ground pattern of the Brachiopoda. A consequence is the hypothesis of a primary lifestyle in soft sediment.

There is a convincing model for the evolutionary route from an inflatable, mobile stalk to a rigid anchoring organ. In the hinge-less Discinida as adelphotaxon of the Lingulida (p. 93), a short adhesive organ with basal structural agreement to the stalk of the Lingulida has evolved (WILLIAMS et al. 1997). The opposite route is just as improbable as the idea of the abandonment of the shell hinge in favor of a hinge-less shell (see above).

The **coelom** with a division into mesocoel and metacoel is taken over from the ground pattern of the Radialia. The complete separation of the compartments in the Craniida is possibly the plesiomorphous state (p. 93).

The lack of **protonephridia** must be considered as secondary in comparison with the Phoronida and thus be interpreted as an autapomorphy of the Brachiopoda.

Metanephridia of ectoderm and mesoderm together with the two-part coelom belong in the ground pattern of the Radialia (p. 66). Thus, they represent a plesiomorphy for the Brachiopoda.

In the Brachiopoda, epidermal **setae** exist in two forms. The different states correlate with the two stages of the biphasic life cycle (LÜTER 1998 a, 2000 a); they are both interpreted as autapomorphies of the Brachiopoda (Fig. 29 M, N).

Larval setae belong to the lecithotrophic swimming larvae; they are arranged in bundles. The larval seta is formed in a chaetoblast that is situated inwards of a single invaginated epidermis cell. There are no follicle cells. The entire material of the seta is produced by the chaetoblast.

Adult setae are formed at the edge of the mantle after metamorphosis. The chaetoblast of the setae lies at the base of a deeply invaginated setal follicle. The follicle cells participate in the formation of the seta by secreting an enamel layer. Adult setae are absent in the Craniida. The larval and adult setae of the Brachiopoda are characterized by regular rings of spine-like projections that are created by peripheral microvilli of the chaetoblasts.

With mention of the setae we come to the **life cycle** of the Brachiopoda. A biphasic cycle with a lecithotrophic larva in the pelagic zone (Fig. 29 L) and a planktotrophic adult in the benthic zone can be hypothesized for the ground pattern (LÜTER 1997, 1998 a).

This biphasic life cycle is realized in the Craniida and the Testicardines. The swimming larva with larval setae do not take up any food; for metamorphosis they anchor themselves with the posterior end of the body.

Significant evolutionary changes have occurred in the Discinida and the Lingulida. In *Discinisca* (Fig. 29 E, F) the lecithotrophic swimming larva is followed by an – also pelagic – juvenile stage with few tentacles, a shell and already with the planktotrophic feeding of the sessile adult (CHANG 1977). In the stem lineage of the Lingulida, the route leads further to complete suppression of the lecithotrophic larva. In *Lingula* a pelagic, planktotrophic young animal with several tentacles hatches directly from the egg (Fig. 29 B). The juvenile planktotrophic stages of the Discinida and Lingulida possess an unpaired median tentacle that is absent in the adult.

We must pursue our argumentation about the postulated lecithotrophic swimming larva in the ground pattern of the Brachiopoda a step further. The lecithotrophic larva of the Brachiopoda cannot be followed back to the ground pattern of the Radialia because a planktotrophic swimming larva can be justified for the Radialia (p. 72). In other words, the change from a planktotrophic larva to a lecithotrophic larva must have occurred in the stem lineage of the Brachiopoda. The larval lecithotrophy thus turns out to be an autapomorphy in the ground pattern of the Brachiopoda – irrespective of the evolution of a new planktotrophic juvenile stage in the stem lineage of the Lingulata (see above).

Autapomorphies (Fig. 21 → 4)

- Shell with dorsal and ventral valves as product of corresponding mantle folds. The following subfeatures (significant changes within the Brachiopoda in parentheses) belong in the ground pattern:
  Uniform, flat shell valves (more strongly developed ventral shell with overhang at the posterior end).
  No hinge (with terminal joint). Emergence of a stalk between the shell valves (evolution of outlet openings in the ventral valve).
  Chemical composition from calcium carbonate and protein (calcium phosphate + chitin + protein).
- Stalk (pedicle).
  Inflatable organ with central coelom and peripheral musculature for burrowing and anchoring in soft sediment.
- Lack of protonephridia.
- Setae.
  Lecithotrophic swimming larva with bundles of larval setae. Formed solely by the chaetoblast.
  Adult setae in rows at the edges of the mantle. Composite formation from material of the chaetoblast and wall cells of the setal follicle.
  Larval and adult setae with rings of spine-like projections.
  For the often discussed agreement with the capillary bristles of the Annelida, the hypothesis of a convergent evolution is demanded a posteriori (Vol. II, p. 44; LÜTER & BARTOLOMAEUS 1997).
- Lecithotrophic swimming larva.
  Ciliated larva that do not take up food in the pelagic zone. Component of a biphasic life cycle. Evolution from a planktotrophic larva in the stem lineage of the Brachiopoda.

Systematization

In the framework of a textbook, we will follow the problem of kinship relationships among the four above-mentioned taxa Lingulida, Discinidia, Craniida and Testicardines. A prerequisite for this is their justification as monophyla that will be established in the beginning on the basis of selected autapomorphies.

Lingulida: Lack of a lecithotrophic swimming larva.
Discinida: Slot in the ventral shell valve for the emergence of the stalk (a hole in the ventral shell of the Testicardines evolved independently).
Craniida: Adhesion of the ventral valve through a thickened epithelial region with microvilli. Lack of a stalk.
Testicardines: Hinge at the posterior end of the shell; with teeth in the ventral valve and sockets in the dorsal valve (hinge teeth-dental sockets mechanism). Hole or foramen for emergence of the stalk in the ventral, overhanging shell. Intestine without anus. Rostrally directed reflection of the mantle valves during or after metamorphosis.

Even in the recent literature, there are three competing hypotheses about the relationships of the three hinge-less entities among themselves and with the Testicardines (e.g., NIELSEN 1991; POPOV et al. 1993; CARLSON 1995; HOLMER et al. 1995; WILLIAMS et al. 1996; COHEN & GAWTHORP 1997; LÜTER 1998a).

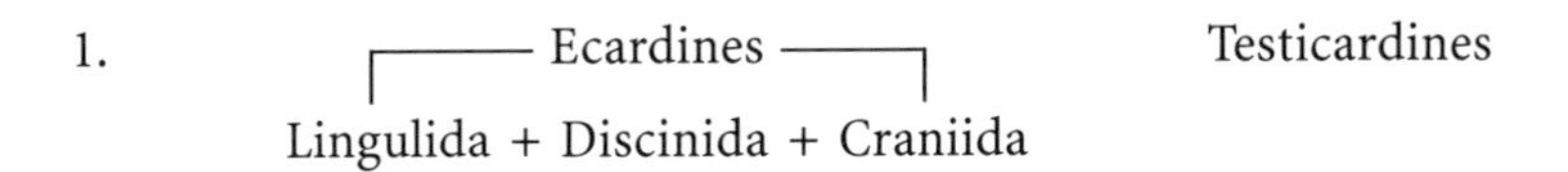

The traditional division of the Brachiopoda into Ecardines and Testicardines remains an option because it possibly reflects an adelphotaxa relationship. Of course, the alternative "absence or presence of a hinge" is not relevant for this; the former state represents a plesiomorphy. Nevertheless, there is remarkable agreement in the development of the lophophore in the Lingulida, Discinida and Craniida. In the early growth of the apparatus, an unpaired, temporary median tentacle is formed (WILLIAMS et al. 1997); this could be an autapomorphy of the Ecardines (WILLIAMS et al. 1996). Findings on the genome are consistent with the grouping of the three "inarticulated lineages" in a single taxon (COHEN et al. 1996, 1997).

2. Lingulata — Lingulida + Discinida; Calciata — Craniida + Testicardines

In the second hypothesis, the Lingulata and the Calciata form the highest-ranking sister groups of the Brachiopoda (HOLMER et al. 1995).

The combination of the Lingulida and Discinida in the monophylum Lingulata is not controversial. The common calcium phosphate + chitin shell represents a convincing autapomorphy. The planktotrophic juvenile stage with tentacles must be mentioned as a further autapomorphy.

In the possible sister group Calciata, the existence of a calcium carbonate shell forms the plesiomorphy. A common apomorphy of the Craniida and the Testicardines, on the other hand, may be the displacement of the gonads from a central position in the metacoel to peripheral canals under the shell – developed once in the stem lineage of an entity Calciata.

3. Craniida ┌── N. N. ──┐
Lingulida + Discinida + Testicardines

On the basis of two possibly primitive features, the Craniida were finally interpreted as sister group of all other Brachiopoda (LÜTER 1998a). They alone lack setae in the adult. Furthermore, the Craniida are the only Brachiopoda with complete separation of the coelom into mesocoel and metacoel. This corresponds to the state in the Phoronida and can be taken as a plesiomorphy by outgroup comparison. The existence of adult setae and a secondary, uniform coelom are, in other words, possibly autapomorphies of an unnamed entity consisting of the Lingulida + Discinida + Testicardines.

Deuterostomia

The name-giving new formation of the mouth in the ontogenesis is certainly expected at the beginning of the unique features of the Deuterostomia. Much more convincing as an autapomorphy, however, is the axial complex in the protosome, the evolution of which as a new excretory organ in the stem lineage of the Deuterostomia can be postulated.

Autapomorphies (Fig. 33 → 1)

- Blastopore → anus.
  The anus in the adult arises from the blastopore at the posterior end of the embryo. The mouth is a new formation ventral in the anterior body.
- Lack of proto- or metanephridia.
- Axial complex in protosome.
  Components: (1) Heart. Section of the blood vessel system without endothelium in the protocoel. (2) Pericardium. A contractile vesicle that covers or encloses the heart. (3) Glomerulus. Highly branched section of

the vessel system emerging from the heart. Covered with podocytes from the epithelium of the protocoel (NIELSEN 1995).
Function: Primary urine is transported from the blood vessel system through ultrafiltration by podocytes into the protocoel.

- Trimery.
  Division of the body into three parts: protosome, mesosome and metasome with an unpaired protocoel as well as paired states of meso- and metacoel (in the Echinodermata the three sections of the coelom were previously known as axocoel, hydrocoel and somatocoel).
- Dipleurula.
  Planktotrophic larva. Ventral opening of the mouth is surrounded by a closed ciliary band for locomotion and feeding.

With the presented autapomorphies we have limited the extent of the monophylum Deuterostomia to the three traditional "stems" Echinodermata, Hemichordata (Pterobranchia + Enteropneusta) and Chordata. The Chaetognatha are excluded; they are treated as a taxon of the Bilateria with unclarified relationships at the end of this volume. The already mentioned Phoronida and Brachiopoda are, in contrast to NIELSEN (1995), not included; the name Neorenalia proposed by this author is thus a synonym for Deuterostomia.

Three "stems" in the sense of descent communities with equal rank are not possible in the phylogenetic system of organisms. Besides, only two of the three usual stems of the Deuterostomia can be justified as monophyla. The "Hemichordata" form a paraphylum and are accordingly disbanded. Within the "Hemichordata" the "Pterobranchia" are then considered as a paraphyletic species group.

Larva in Life Cycle

Does a planktotrophic or a lecithotrophic larva belong in the ground pattern of the Deuterostomia? Let us review a few facts.

Among the Echinodermata, planktotrophic larvae exist in the four large taxa of the Eleutherozoa – the bipinnaria in the Asteroida, the auricularia in the Holothuroida as well as pluteus larvae in the Ophiuroida and Echinoida. Regardless of group-specific differences, larval development begins with a "dipleurula" stage (Fig. 30). This is an early larva in which the ventral opening of the mouth is surrounded by a single, closed ciliary band serving for locomotion and feeding. In contrast, among the Crinoida – the adelphotaxon of the Eleutherozoa – only lecithotrophic doliolaria larvae are known; here four or five separate ciliary rings encircle the barrel-shaped body at right angles to the longitudinal axis.

In the paraphyletic "Pterobranchia" (p. 136) *Rhabdopleura* and *Cephalodiscus* each have simple, lecithotrophic larvae without ciliary bands or

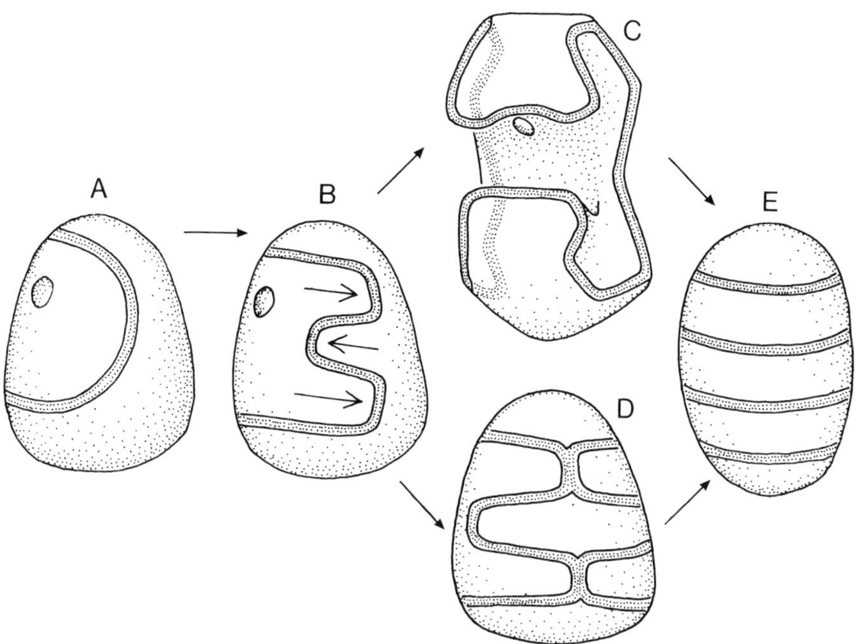

Fig. 30. Echinodermata. General sketch of the formation of complex larval ciliary bands from a simple pattern. **A** Dipleurula-stage with a simple circumoral band in the early development (e.g., Echinoida). **B** Tortuous ciliary band that is created by local proliferation or new arrangement of cells (*arrows*). **C** Further differentiation to the patterns of the auricularia (Holothuroida). **D** Initial pattern in the larva of *Florometra serratissima* (Crinoida), dorsal view. Fusion of the loops at two positions on the back. **E** Transformation into doliolaria-stage with separate ciliary rings. (Lacalli & West 1988)

rings. Rather, the larvae are ciliated on all sides; they swim for only a short time and do not take up any food.

The widely distributed larva of the Enteropneusta is the planktotrophic tornaria (Figs. 31 C, 53 E–H). Similar to the dipleurula of the Echinodermata, a single ciliary band arises around the mouth in the early development; the pre- and postoral regions of the band grow later to comparably complicated ciliary loops. The tornaria develop the telotroch as a unique element; this is a separate ciliary ring around the posterior end of the larva with locomotory function. However, in the Enteropneusta there are also lecithotrophic larvae with a uniform ciliation; they also have a telotroch (Fig. 53 C).

We start the interpretation in the Echinodermata with the question of whether the dipleurula of the Eleutherozoa or the doliolaria of the Crinoida represents the original, plesiomorphous larval form of the entity. The answer is obtained by a more detailed comparison of larval development in the Holothuroida and Crinoida. In the ground pattern of the Holothuroida, the closed ciliary band of the auricularia (Fig. 30 C) breaks into sec-

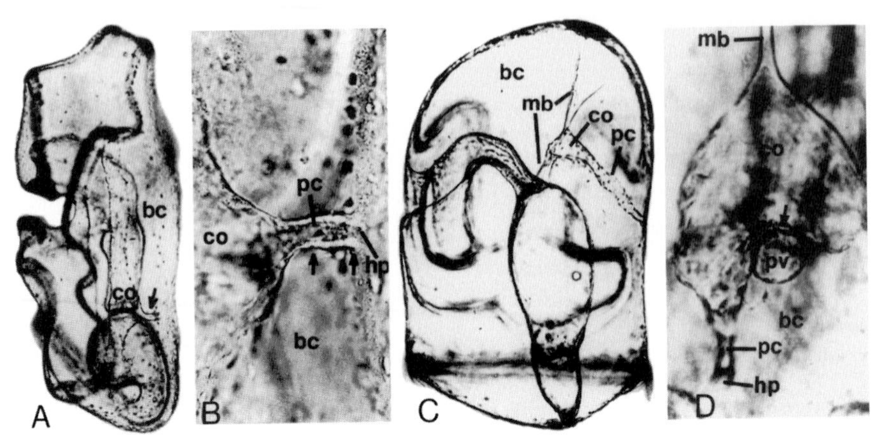

Fig. 31. Nephridial organ (pore canal – hydropore complex) of larvae of Echinodermata and Enteropneusta. **A, B** Bipinnaria of *Asterias forbesi*. **A** Left lateral view of an early larva with left coelom, blastocoel and pore canal (*arrow*). **B** Highly magnified pore canal and dorsal hydropore. The *arrows* delineate the ectodermal part of the pore canal. **C, D** Tornaria of *Schizocardium brasiliense* (Enteropneusta). **C** Left lateral view of early larva with anterior coelom on muscle bands, blastocoel and pore canal. **D** Dorsal view of the complete nephridial organ. A pulsating bladder is joined laterally with the coelom wall; it encloses a small cavity which passes over to the blastocoel. In the Echinodermata there is a corresponding contractile bladder. *bc* Blastocoel; *co* coelom; *hp* hydropore; *mb* muscle band; *pc* pore canal; *pv* pulsatile vesicle. (Ruppert & Balser 1986)

tions which then rearrange to several separate, ciliary rings that run transversely; the auricularia is transformed into a doliolaria stage during the development.

Correspondingly, the doliolaria pattern of the Crinoida does not develop directly, but proceeds through an initial, temporary pattern in the form of a single, tortuous ciliary band (LACALLI & WEST 1986; LACALLI 1988). In *Florometra*, the loops extend upwards, meet at two positions on the dorsal side of the larva and thus exhibit the separate stripes of doliolaria (Fig. 30 D).

The evaluation seems unambiguous. Lecithotrophic doliolaria stages are apomorphous, repeated convergently evolved larvae. A planktotrophic dipleurula belongs in the ground pattern of the Echinodermata (JÄGERSTEN 1972; STRATHMANN 1988 a; NIELSEN 1998 a).

Now we must discuss the derivation of tornaria in the Enteropneusta. The agreements between dipleurula and tornaria with the uniform pre- and postoral running ciliary band are so large that, 150 years ago, JOHANNES MÜLLER interpreted the first tornaria as a larva of the Echinodermata. The circle of arguments was recently closed with the identification of an identical larval excretory organ in the bipinnaria of the Asteroida and the tornaria (RUPPERT & BALSER 1986). The nephridial complex has a pulsatile vesicle. The ultrafiltration of coelom fluid proceeds via podocytes and the filtrate is transported by a ciliated pore canal to the dorsal hydropore (Fig. 31). The hypothesis of a homology of dipleurula and tornaria as planktotrophic larvae cannot be better justified. The consequences of this hypothesis are presented in the following considerations (Fig. 32).

1. If a benthic, lecithotrophic larva should represent the original developmental mode of the Enteropneusta, then the planktotrophic tornaria must have arisen from it and from this, in turn, the dipleurula under reduction of the telotroch. Such ideas contradict functional considerations (STRATHMANN 1988b). In the bipinnaria and auricularia, only one ciliary band is available for swimming and feeding; a functional compromise between the two tasks must exist. In the tornaria, the telotroch constitutes a separate locomotory band that releases the former band from the requirements of movement. In other words, an evolutionary interpretation is only reasonable when it postulates the transformation of the dipleurula with one ciliary band to a tornaria with an additional telotroch. This planktotrophic larva belongs in the ground pattern of the Enteropneusta. Secondary lecithotrophic larvae have taken the telotroch over from the tornaria.
2. The hypothesis of a homology between dipleurula and tornaria, however, does not necessarily follow the hypothesis of an adelphotaxa relation between the Echinodermata and Enteropneusta. Rather, the branchial region of the trunk with U-shaped gill slits represents a complex, apomorphous agreement between the Enteropneusta and the Chordata on the basis of which these two entities can be combined to the monophylum Cyrtotreta (Fig. 33).

As will be validated later, there are furthermore no arguments for a sister group relationship of the Echinodermata with the Cyrtotreta (Enteropneusta + Chordata), with *Rhabdopleura* or with the Cephalodiscida. There is a possibility that the Echinodermata stand as adelphotaxon to all other Deuterostomia – and for this case, we have the answer to the question posed above. A planktotrophic larva is to be set in the ground pattern of the Deuterostomia.

This conclusion demands that the lecithotrophic, circumferentially ciliated larvae of *Rhabdopleura* and *Cephalodiscus* must be of a secondary nature. There are models for this. In species of the taxa *Astropecten* and *Ctenopleura* (Asteroida), transformed bipinnaria exist as barrel-shaped larvae with complete ciliation, but without a trace of ciliary bands (KOMATSU et al. 1988). On the other hand, I can find no evidence for an evolution from lecithotrophic to planktotrophic larvae in the Deuterostomia.

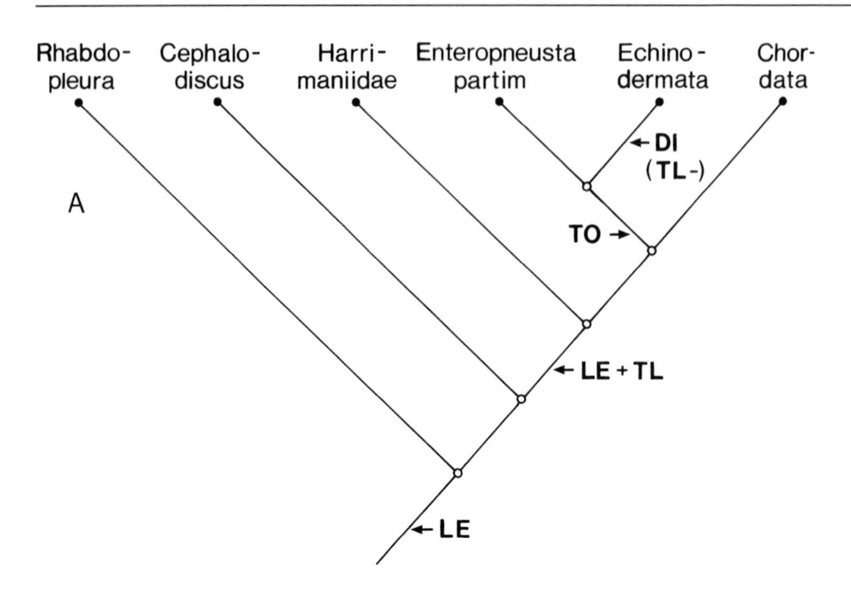
Rhabdo-
pleura
Cephalo-
discus
Harri-
maniidae
Enteropneusta
partim
Echino-
dermata
Chor-
data
A
DI
(TL-)
TO
LE + TL
LE

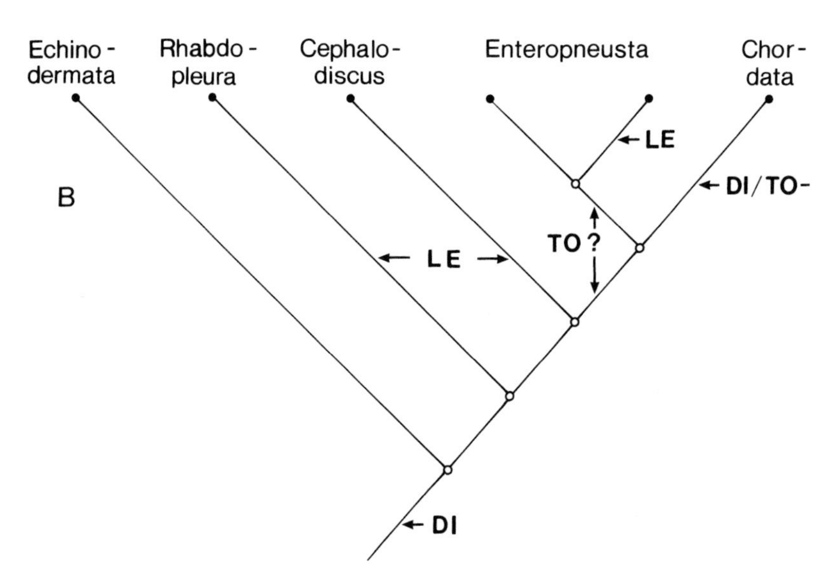
Echino-
dermata
Rhabdo-
pleura
Cephalo-
discus
Enteropneusta
Chor-
data
B
LE
DI/TO-
TO ?
LE
DI

◀ **Fig. 32.** Deuterostomia. The two extreme possibilities for the evolution of a planktotrophic larva with circumoral ciliary band under the prerequisite of a homology between dipleurula (without telotroch) and tornaria (with telotroch). **A** The stem species of the Deuterostomia possessed a lecithotrophic larva. A telotroch developed in the lecithotrophic larva of an entity from the Harrimaniidae (Enteropneusta) + the remaining Enteropneusta + Echinodermata + Chordata. The planktotrophic tornaria evolved in the stem lineage of the Enteropneusta partim + Echinodermata and finally a dipleurula under reduction of the telotroch in the stem lineage of the Echinodermata. **B** The dipleurula belongs in the ground pattern of the Deuterostomia and was taken over from there into the stem lineage of the Echinodermata. The tornaria with telotroch evolved from the dipleurula in the stem lineage of the Enteropneusta + Chordata or first in the stem lineage of the Enteropneusta (? In diagram). Lecithotrophic larvae arose secondarily in the stem lineage of *Cephalodiscus*, *Rhabdopleura* and some Enteropneusta (Harrimaniidae). The planktotrophic (dipleurula or tornaria) larva was reduced or transformed in the stem lineage of the Chordata. (cf. Salvini-Plawen 1998, Figs. 1a and 2). *Dl* Dipleurula; *LE* lecithotrophic larva; *TL* telotroch; *TO* tornaria

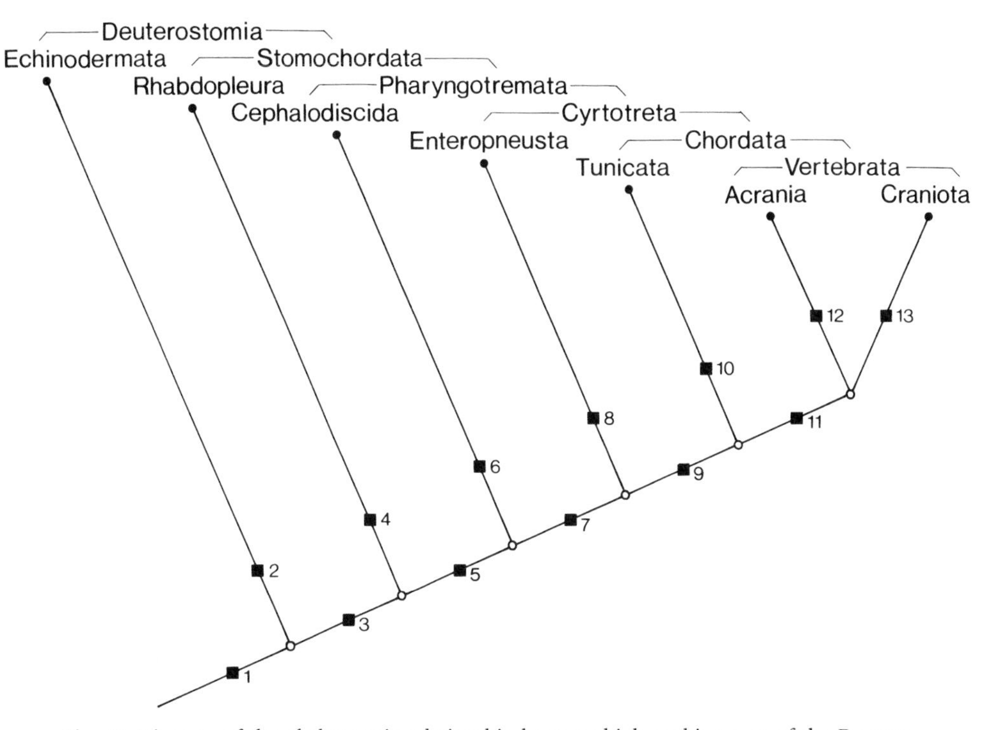

Fig. 33. Diagram of the phylogenetic relationship between high-ranking taxa of the Deuterostomia

Systematization

Deuterostomia
- **Echinodermata**
- **Stomochordata**
  - **Rhabdopleura**
  - **Pharyngotremata**
    - **Cephalodiscida**
    - **Cyrtotreta**
      - **Enteropneusta**
      - **Chordata**
        - **Tunicata**
        - **Vertebrata**
          - **Acrania**
          - **Craniota**

In the diagram of relationships (Fig. 33), the autapomorphies with which the presented systematization will be validated from entity to entity are numbered. At this point, we must address the controversial problem of the position of the Echinodermata in the phylogenetic system of the Deuterostomia. There are two "soft" arguments to place the Echinodermata as the sister group alongside all other taxa unified under the name Stomochordata.

1. There is no specific agreement by which an adelphotaxa relationship of the Echinodermata with a specific subtaxon of the remaining Deuterostomia could be identified.
2. The stomochord as extension of the intestine into the protosoma is absent in the Echinodermata with pentameric symmetry. Is this possibly a plesiomorphous condition?

Let us ask ontogenesis. In the development of the Enteropneusta, the stomochord occurs during metamorphosis of the planktonic tornaria into the trimeric benthic organism; a stomochord has also been observed in late larval stages (VAN DER HORST 1939).

If the absence of a stomochord in the Echinodermata would be of secondary nature, then a rudimentary anlage in the comparable stage of dipleurula would be conceivable. However, this is not the case, which speaks for a primary lack.

However, this argument is open to attack. In the trimeric Deuterostomia, the stomochord "supports" the axial complex as excretory organ and thus could belong together with the latter in the ground pattern of the Deuterostomia. This possible function was dropped on transformation to radial symmetry with the result of a complete reduction of the stomochord in the adults and larvae.

Echinodermata – Stomochordata

Echinodermata

The central autapomorphy of the Echinodermata derives from the pentameric radial symmetry. Its secondary nature has been unequivocally documented on the basis of the bilaterally symmetrical larvae. Five radii form the dominant elements of the body construction. Its expression in the habitus is reflected in a corresponding internal organization of the water-vascular system.

When we concentrate on the outstanding features we must include the calcareous skeleton with stereome structure and the extremely changeable connective tissue.

Autapomorphies (Fig. 33 → 2; 36 → 1)

- Pentamery.
  Five radii (ambulacra) and five interradii (interambulacra) are arranged about a central main axis which is directed perpendicular towards the mouth opening. The oral side about the mouth is primarily oriented towards free water; the aboral side (apical side or apical surface) is opposite.
- Water-vascular system (ambulacral system).
  The ground elements of the ambulacral system are derivatives of the protocoel and the left mesocoel. The protocoel opens to the outside via a hydropore; it is linked on the other side by the stone canal to the mesocoel. The mesocoel forms a ring canal around the foregut and five radial canals with ambulacral tentacles in the radii. The tentacles serve primarily for microphagic nutrition.
- Stereom.
  Calcite skeleton of mesodermal origin with intracellular genesis. Individual stereoblasts (skeleton-forming cells) fuse to syncytia in the vacuoles of which singular calcareous sclerites are formed. The sclerites then combine to a three-dimensional network. Connective tissue migrates into the pore spaces of the stereom.
- Mutable collagenous tissue.
  "Echinoderms possess the most unusual collagenous structure yet discovered in the Animal Kingdom" (WILKIE & EMSON 1988) with extremely variable expansivity of the tissue. Short times down to fractions of a second are sufficient to change the strain state from solid rigidity to extensive relaxation. On the basis of nervous control, this ability is important for the rapid change from an energetically favorable rigid state of entire body parts to their use in locomotion.

Fossils from the Cambrian period allow us to make statements on the time course in the evolution of central features of the Echinodermata (PAUL & SMITH 1984; SMITH 1988 a, 1990; Fig. 34).

A calcareous skeleton with stereom pattern was already present in the † Carpoida († *Cothurnocystis*). This represents the first autapomorphy of the Echinodermata detected in the fossil remains.

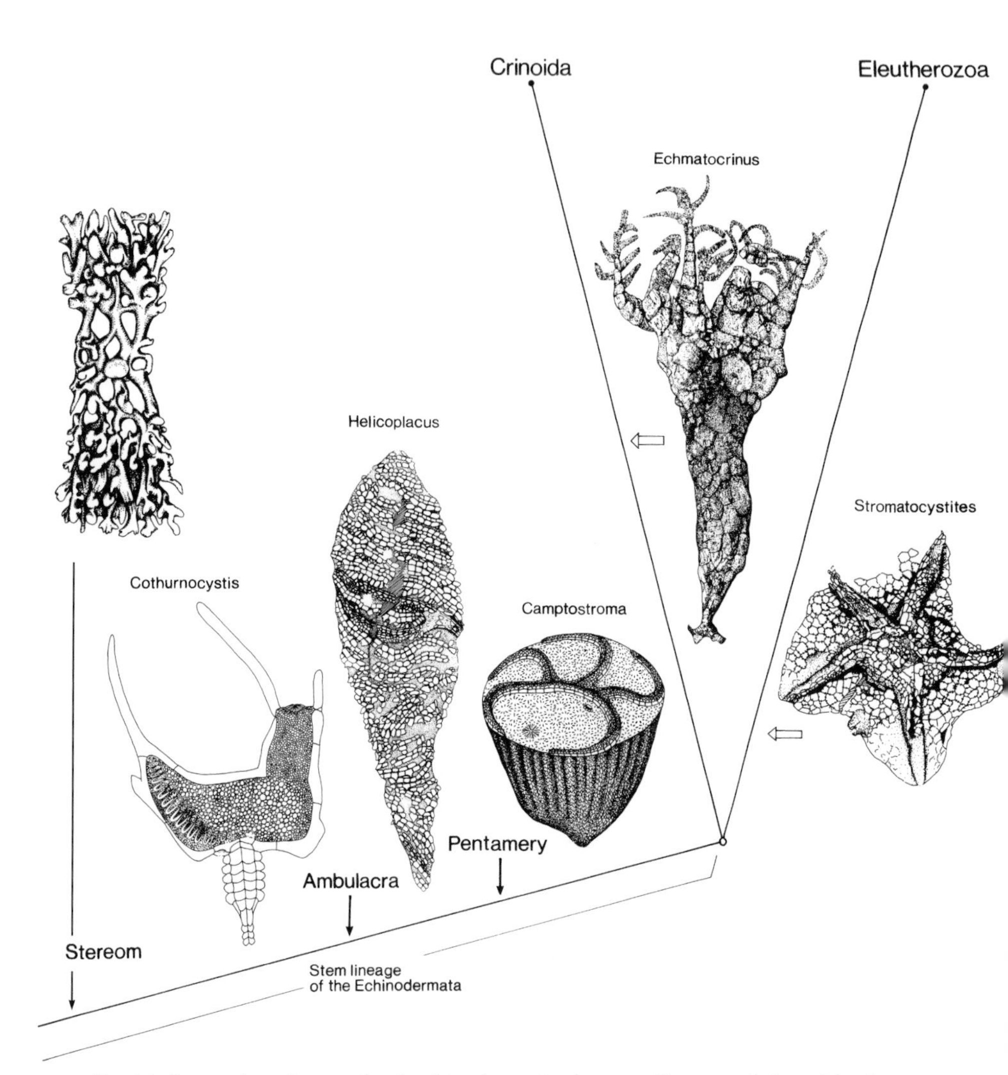

Fig. 34. Integration of some fossil echinoderms in the stem lineages of the Echinodermata, the Crinoida and the Eleutherozoa. Time course in the evolution of the stereome, the ambulacra and the pentameric symmetry in the stem lineage of the Echinodermata. Explanations in text (outline according to Paul & Smith 1984; Smith 1988). An element from the calciferous ring of *Rhabdomolgus ruber* (Holothuroida) is chosen as an example of the stereome structure. (Menker 1970)

The water-vascular system apparently evolved before the pentaradial symmetry. The Helicoplacoida exhibit a triradial symmetry in the arrangement of ambulacra. They were interpreted as sessile suspension feeders with radial canals of a water-vascular system under the ambulacral feeding grooves.

In † *Camptostroma* the pentamery of the Echinodermata was then realized. *Camptostroma* appears as the youngest representative of the stem lineage of the Echinodermata before the division into the adelphotaxa Crinoida and Eleutherozoa. On the oral side of the spherical body, there are five ambulacra that are separated by broad, interambulacral fields. The mouth is central in the oral side, a distinct periproct zone lies in the periphery. *Camptostroma*, like † *Helicoplacus*, was estimated to be a sedentary suspension feeder, with the aboral surface probably embedded in soft sediment.

In the broad spectrum of divergent opinions on the asymmetrical † Carpoida, it even remains controversial what is the front and what is the back. On the basis of the stereom structure, PAUL & SMITH (1984) interpreted the Carpoida as early stem lineage representatives of the Echinodermata. In contrast, JEFFERIES (1986) did not consider the stereom skeleton to be an autapomorphy of the Echinodermata; in his opinion it was already evolved in the stem species of an entity consisting of Echinodermata+ Chordata. In the Calcichordata hypothesis (JEFFERIES), different Carpoida belong to different taxa of the Chordata. The attempt to order specific Carpoida in the stem lineages of the Acrania, the Tunicata and the Vertebrata implies a multiple, independent reduction of the stereom skeleton in these stem lineages. This inference is improbable. Anyway, the controversies (PETERSON 1994, 1995; NIELSEN 1995; GEE 1996; JEFFERIES 1997) are not relevant for our systematization because the kinship relationships among the recent Deuterostomia are independent of the position of the fossil Carpoida.

Systematization

The five common "classes" of the Echinodermata – Crinoida, Asteroida, Ophiuroida, Echinoida and Holothuroida – can each be considered unequivocally as a monophylum. This will be justified in more detail below.

For reasons of optimal presentation, I will start with a judgement on the phylogenetic relationship between the five entities. Some features must be emphasized, the evolutive evaluation of which can only be made later in another context.

We will start with the two sister group relations that are accepted today with extensive agreements.

Crinoida – Eleutherozoa

The Crinoida are primarily sessile organisms with the mouth and ambulacral grooves oriented to free water. On the other hand, the Eleutherozoa in a union of all other echinoderms are vagile organisms with the mouth directed towards the substrate. The most economical hypothesis of a single change to a freely moving species in the stem lineage of an entity Eleutherozoa is supported by further autapomorphies (p. 112).

Echinoida – Holothuroida

The large extension of the oral side with the ambulacra under reduction of the apical side to a tiny perianal field is considered as the major synapomorphy. Echinoida and Holothuroida are combined in the monophylum Echinozoa.

The two hypotheses form the cornerstones of the modern phylogenetic systematization of the Echinodermata (SMITH 1984b; LITTLEWOOD et al. 1997, 1998a)[5]. Consequently, there can only be three competing hypotheses on the relationships between the Asteroida, Ophiuroida and Echinozoa (Fig. 35).

1. Asteroida – Cryptosyringida (Ophiuroida + Echinozoa)

The major apomorphous agreement between the Ophiuroida and Echinozoa results from the evolution of epineural canals inside the body in which the radial nerves of the ectoneural nerve system are embedded (SMITH 1984b). In contrast, in the open ambulacral grooves of the Asteroida, the nerves run in the epidermis.

2. Ophiuroida – N.N. (Asteroida + Echinozoa)

A single argument from the construction of the water-vascular system is available for discussion. The lack of internal ampullae in the Crinoida and Ophiuroida can only be considered as a plesiomorphy, their existence along the radial canals in the Asteroida, Echinoida and Holothuroida as the apomorphous alternative. However, the agreement is not available as an autapomorphy of an entity encompassing these organisms. Internal bladders are absent in all Asteroida of the Ordovician period. They could have first occurred in the stem lineage of the Asteroida and must a posteriori be understood as the result of a convergent evolution in comparison with the ampullae of the Echinozoa.

[5] The assumption of an adelphotaxa relationship between the Holothuroida and all other Echinodermata (SMILEY 1988) is not accepted.

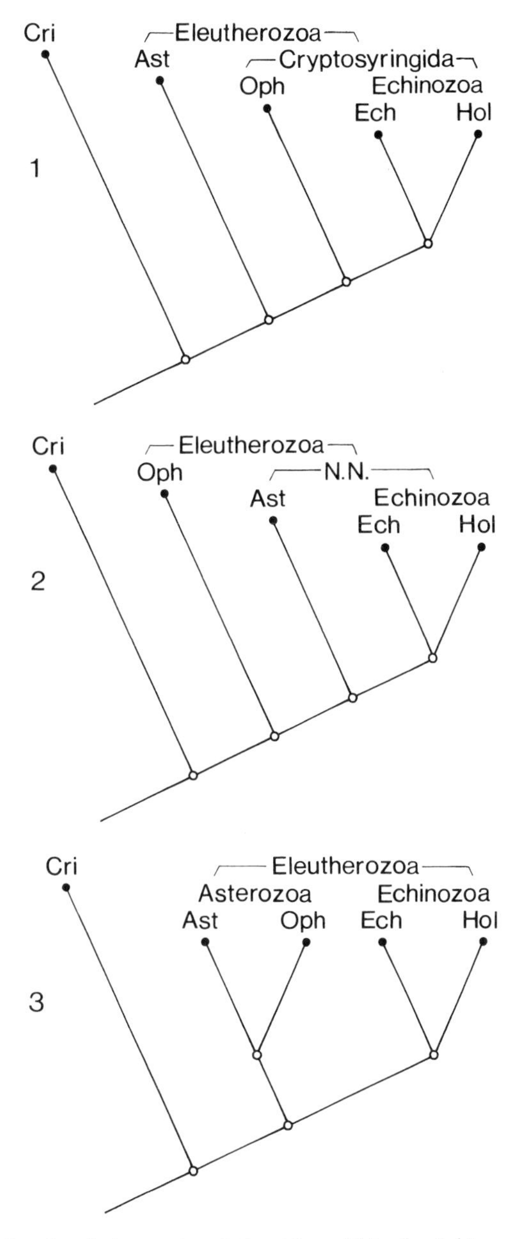

Fig. 35. Three competing hypotheses for the phylogenetic relationships within the Echinodermata. The well-established sister group relationships Crinoida – Eleutherozoa and Echinoida – Holothuroida are shown in identical form in all three diagrams. **1** The Asteroida form the sister group of all other Eleutherozoa. **2** The Ophiuroida form the sister group of all other Eleutherozoa. **3** Asterozoa and Echinozoa represent equal-ranking sister groups of the Eleutherozoa. *Ast* Asteroida; *Cri* Crinoida; *Ech* Echinoida; *Hol* Holothuroida; *Oph* Ophiuroida

3. Asteroida – Ophiuroida

A union of the two taxa to a monophylum Asterozoa has long been under discussion (FELL 1963, 1982) – recently also on the basis of mitochondrial gene arrangements (M.J. SMITH et al. 1993).

The common occurrence of five arms is probably a plesiomorphy taken over from the ground pattern of the Eleutherozoa.

The tube-like intestine with loops can be interpreted as a plesiomorphy of the Crinoida and Echinozoa, the sac-like intestine of the Asteroida and Ophiuroida as apomorphy. However, the presence of an anus in the ground pattern of the Asteroida (p. 116) compared to its absence in the Ophiuroida is a major difference.

We turn to the **ontogenesis**. No fruitful arguments for an analysis of the relationships within the Echinodermata arise from the divergent development of planktotrophic larvae in the discussed taxa (STRATHMANN 1988 a). The auricularia of the Holothuroida is taken as plesiomorphous larva in the pattern of features of the stem species of recent Echinodermata (p. 135). The agreements with the bipinnaria of the Asteroida are of plesiomorphous nature and thus of little relevance for the current discussion.

For the Crinoida we urgently need concepts about the ground pattern larva from analyses of sessile species. As yet, only the derived larval form of doliolaria from secondarily vagile feather stars is known.

How do phylogenetic systematics judge the conspicuous agreements between the pluteus larvae of the Ophiuroida and the Echinoida? Homology as larvae with long projections supported by calcareous spicules would be conceivable under the following circumstances: (1) the Ophiuroida and Echinoida form adelphotaxa. The three presented relationship hypotheses do not favor this assumption. (2) The pluteus larva evolved in the stem lineage of the Cryptosyringida and was later reduced in the Holothuroida. (3) The pluteus arose in the stem lineage of the Eleutherozoa and was later independently abandoned twice in the Asteroida and Holothuroida. There is also no evidence in favor of these two notions. Accordingly, the hypothesis of an independent occurrence of pluteus larvae must be favored – the evolution of the ophiopluteus with four projections in the stem lineage of the Ophiuroida as well as the evolution of the echinopluteus with eight arms in the stem lineage of the Echinoida.

Result

The second hypothesis of an adelphotaxa relation between Asteroida and Echinozoa is improbable and will not be followed further.

The first hypothesis with a taxon Cryptosyringida as well as the third hypothesis with a monophylum Asterozoa remain competitors.

After evaluation of all available morphological and molecular data, the highest probability is assigned to the first hypothesis with the Asteroida and Cryptosyringida as sister groups (LITTLEWOOD et al. 1997, 1998a). The resultant phylogenetic system is shown as a relationship diagram (Fig. 36) and in the following hierarchical tabulation.

Echinodermata
- **Crinoida**
- **Eleutherozoa**
  - **Asteroida**
  - **Cryptosyringida**
    - **Ophiuroida**
    - **Echinozoa**
      - **Echinoida**
      - **Holothuroida**

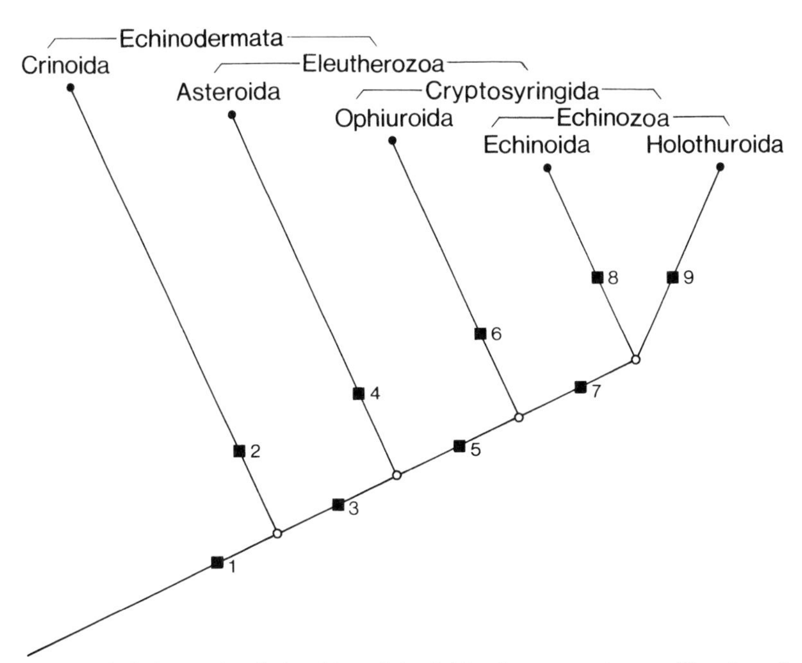

Fig. 36. Diagram of phylogenetic relationships of the Echinodermata under consideration of the favored hypothesis with the Asteroida and Cryptosyringida as highest-ranking adelphotaxa of the Eleutherozoa

Crinoida – Eleutherozoa

Conventional classifications operate with a confrontation of the Pelmatozoa (Crinozoa) and the Eleutherozoa as two equal-ranking subphyla of the Echinodermata. In the Pelmatozoa are placed a series of high-ranking fossil taxa and the "class" Crinoida as sole entity with recent members. Even SMITH (1984b) still considered the fossil † Cystoidea as the sister group of the Crinoida.

The consequent phylogenetic system, however, does not recognize sister group relationships between an entity with recent species and a purely fossil entity. Fossils must always be placed in the stem lineages of entities with recent members.

Therefore, in the stem lineage concept (Vol. I, p. 43), the fossil Cystoidea are merely stem lineage members of the monophylum Crinoida. This statement holds equally for all other fossil echinoderms for which at least one autapomorphy of recent Crinoida can be demonstrated.

From this point of view, a taxon Pelmatozoa or Crinozoa (FELL 1982) superordinated to the Crinoida must be abandoned; the names are superfluous. The highest-ranking adelphotaxa of the Echinodermata are the Crinoida and the Eleutherozoa.

Crinoida

Sessility as well as a division of the body into crown and stalk are outstanding features of the ground pattern of the Crinoida – realized in the sea lilies (*Hyocrinus, Bathycrinus, Endoxocrinus*). Vagile feather stars evolved within the Crinoida under abandonment of the stalk (*Antedon, Florometra, Heterometra*); with the reputed pentacrinus stage they recapitulate a short sessile phase in the life cycle (Fig. 37D, E).

Today, sessility and division of the body are correlated features, however, in the phylogenesis of the Echinodermata, they evolved one after the other. Sessility arose in the stem lineage of the Echinodermata, was present in the last common stem species of all recent Echinodermata (p. 101), and is therefore included in the ground pattern of the Crinoida as a plesiomorphy. The division of the body into a crown and stalk, on the other hand, occurred first in the stem lineage of the Crinoida. The first tendencies to a division into crown and stalk can be recognized in the Cambrian taxa † *Lepidocystis* and † *Echmatocrinus* (Fig. 34). The stalk section is covered by plates; stalk joints were not yet present.

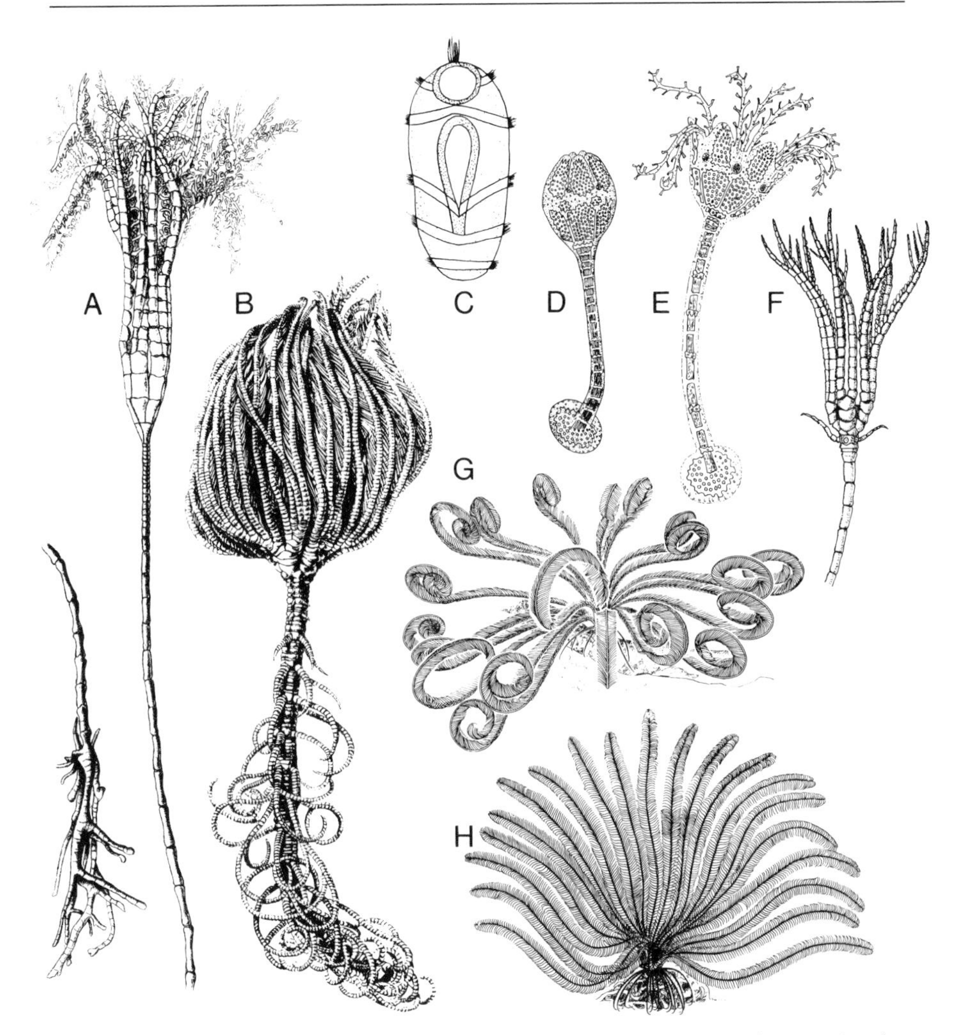

Fig. 37. Crinoida. **A, B** Stalked sea lilies. **A** *Bathycrinus carpenteri.* Stalk without cirri. *Below left*: end of stalk with root. **B** *Endoxocrinus parrae.* Stalk with belts of cirri. **C–H** Comatulida. Secondary, freely moving feather stars. A remnant of the stalk carrying a circle of cirri on the calyx. **C–F** Developmental stages of *Antedon.* **C** Doliolaria larva. **D, E** Young pentacrinus stages with calyx, stalk and adhesive disk. **F** Later pentacrinus stage before detachment of the calyx from the stalk. Cirri have developed at the uppermost section of the stalk. **G, H** *Heterometra savignyi.* **G** Spreading out of the arms in still water. **H** Arrangement of the arms to a filtration fan in tidal flow. The aboral arm side is directed against the flow direction; food particles trapped by the tentacles are passed on in flow shadow to the ciliated feeding grooves of the pinnules. (**A** Ludwig & Hamann 1907; **B** Rasmussen 1978; **C–F** Breimer 1978; **G** Magnus 1965; **H** Magnus 1964)

Fig. 38. Crinoida. **A–C** *Florometra* with ten arms. Secondary vagile feather star. Adult after discarding the stalk. **A** View of the oral side. **B** Side view (main axis is vertical). Conserved uppermost section of the stalk with circle of cirri for anchoring to hard substrates. **C** Oral side highly magnified. Five ambulacral grooves radiate from the central mouth; they demonstrate the primary pentameric symmetry. With branching the ciliary grooves then run out on the ten arms, flanked on the lateral sides by ambulacral tentacles. The pinnules insert alternately in the arms. Anus next to the mouth in interradius (*left* in picture). **D** *Ptilocrinus pinnatus.* Species with five unbranched arms and a stalk without cirri. **E** Position of the gonads in the pinnules (majority of the freely moving Comatulida). The genital cord of the coelom stretches to the tips of the arms. However, gonads only occur in a few proximal pinnules. **F** Scheme of the organization, shown in vertical sections: *left* through a radius with start of the arm, *right* through an interradius. Oral side up. Coelom *lightly dotted*, skeletal elements *thickly dotted*, nerve system *black*. *am* Aboral canal of the metacoel; *an* aboral nerve mass; *as* axial sinus; *ba* brachialia; *bn* brachial nerve; *ca* calyx; *ce* centrodorsale; *ci* cirrus; *co* chambered organ of the right metacoel; *gc* genital coelom; *gp* pinnules with gonads; *hp* hydropore; *hyr* hyponeural ring; *i* section through intestine; *m* metacoel (as in the *right* of the picture degraded into numerous clefts; not shown); *nr* nerve ring; *oe* oesophagus; *om* oral canal of the metacoel; *pb* primibrachiale; *p+m* fusion product of protocoel and left metacoel; *ra* radial canal; *rd* radiale; *ri* ring canal. (**A–C** Originals; Puget Sound, Washington, USA; **D** Rasmussen 1978; **E, F** Heinzeller & Welsch, in R. W. Harrison, F. S. Chia (eds.) (1994) Microscopic Anatomy of Invertebrates, Vol. 14: Echinodermata, pp. 18, 22. Reprinted by permission of John Wiley & Sons Inc.)

Autapomorphies (Fig. 36 → 2)

- Division into crown and stalk.

  Crown

  1. Calyx with oral tegmen and theca (outside wall).
  2. Primarily five arms (brachia) with arm sections (brachialia) on the edge of the calyx (Fig. 38D); secondary increase within the Crinoida.
  3. Pinnules as finger-like appendices with alternating insertion on the sides of the arms.

  Stalk (columna)

  Primarily constructed from uniform sections with cylindrical vertebrae (columnalia). Terminal differentiation to an anchor (Fig. 37 A).

  Cirri do not belong in the ground pattern. They first evolved as appendices on the stalk within the Crinoida.
- Coelom (Fig. 38 F).

  (HEINZELLER & WELSCH 1994)

  1. Protocoel and left mesocoel fused in the calyx to a labyrinth of communicating clefts. The axial sinus around the axial organ is the only larger cavity.
  2. Unpaired pore of the protocoel of the larva has multiplied in the adult to numerous, separate hydropores. Openings on the oral side.
  3. Numerous stone canals arise from the ring vessels of the mesocoel and open in the left metacoel.
  4. Chambered organ. Swellings of tubes of the right metacoel in the cavity of the sclerite centrodorsale. Compartmentalization through folds of the coelothelium.

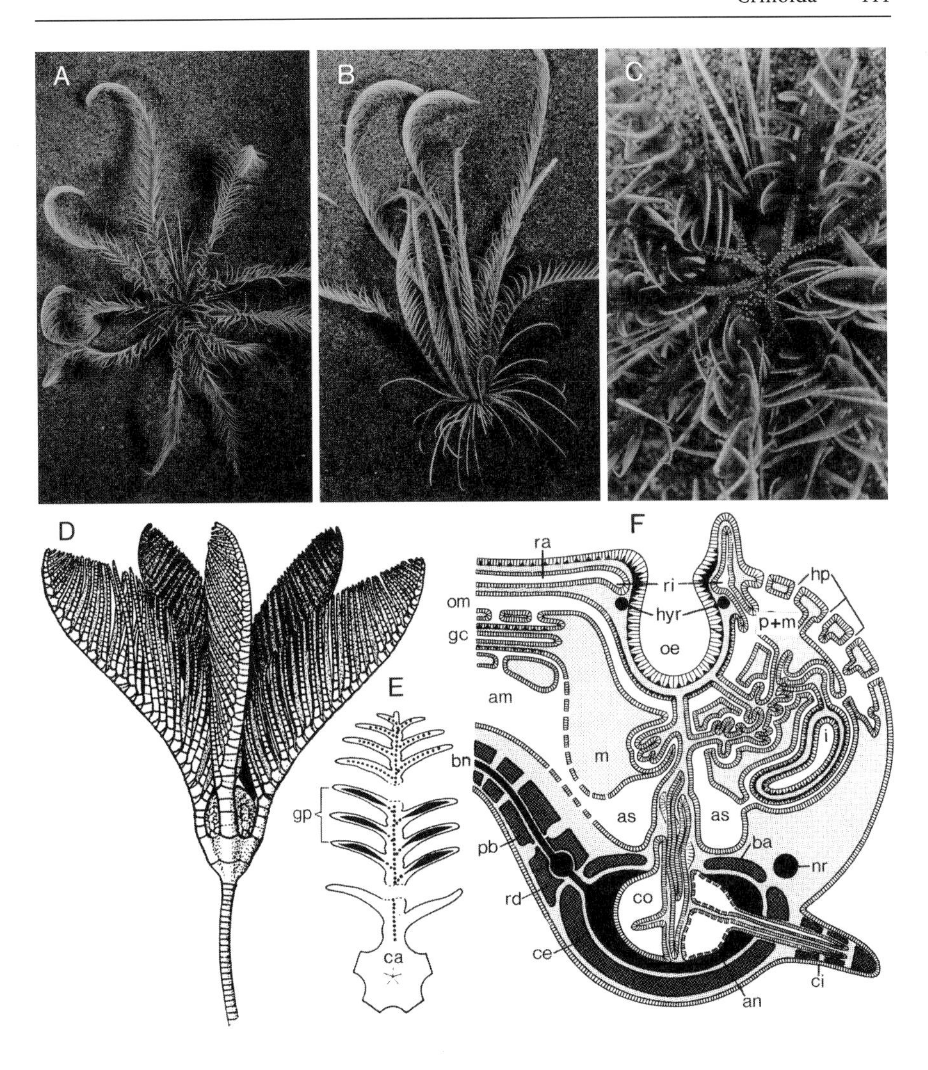

5. Canal-like genital coelom in arms and pinnulae.

- Gonads.

  Displacement over the genital canals in the arms (stalked sea lilies) and further in pinnulae near the calyx (vagile feather stars).

- ? Doliolaria.

  The doliolaria of vagile feather stars with the evolution of separated ciliary strips from a primarily closed ciliary band (Fig. 30) is certainly an apomorphous larva. Until we know more about the development of stalked sea lilies, however, we cannot judge whether a doliolaria belongs in the ground pattern of all Crinoida.

Further plesiomorphous features are correlated with the original sessility. Mouth and anus lie next to each other on the tegmen. Ciliated ambulacral grooves run on the surface of the body from the arms over the tegmen and further to the central mouth opening. They effect the transport of particulate food that is trapped on the arms and pinnules by the ambulacral tentacles. According to this task in the framework of a microphage feeding, the ambulacral tentacles have neither bladders inside nor peripheral adhesive disks.

A phylogenetic system of the recent Crinoida does not yet exist. Sessile sea lilies with primarily five arms and a stalk without cirri (*Hyocrinus, Ptilocrinus*) and the secondarily vagile feather stars with cirri as devices for periodic anchoring to the ground represent opposite extremes of life forms.

Eleutherozoa

The central process in the phylogenesis of the Eleutherozoa was the change from a sessile to a freely moving organism. Associated with this is the new orientation of the oral side to the substrate. From the ambulacral tentacles serving primarily for microphage feeding, ambulacral tube feet with mainly locomotory function have evolved.

Autapomorphies (Fig. 36 → 3)

- Vagile stem species.
- Oral side oriented to ground – apical surface in free water.
- Conically running ambulacral tube foot.

  Locomotory tube feet without adhesive disks or suckers have been hypothesized for the ground pattern of the Asteroida (p. 116) and are realized in all Ophiuroida. As a consequence of this distribution, they are to be included in the ground pattern of the Eleutherozoa.

  Ambulacral tube feet with adhesive disks must have occurred independently twice in the Eleutherozoa; once within the Asteroida and then again in the stem lineage of the Echinozoa (Echinoida + Holothuroida).
- Supply of the ambulacral tube feet.

  The radial canals of the water-vascular system have lateral branches that supply the tube feet (Fig. 41). In the Crinoida, they originate directly from the radial canals.
- Polian vesicles.

  Bladder-like appendices on ring canal of the water-vascular system. They are absent in the Crinoida (plesiomorphy) and, strangely, they are absent in the Echinoida. If we wanted to consider the latter as primitive too, the consequence would by a multiple, independent occurrence of polian vesicles within the Eleutherozoa. More economical is the hypothesis of a single evolution in the stem lineage of the Eleutheroza and a reduction in the Echinoida.

- Madreporic plate.
  The hydropore of the protocoel exits through a calcified body, the madreporic plate (madreporite; Figs. 39 D, 40 A). Its absence in the Crinoida may be interpreted as a plesiomorphy.
- Motile spines.
  Reduced together with the peripheral calcareous skeleton in the Holothuroida.

With † *Stromatocystites* from the lower Cambrian (Fig. 34) we refer to an early stem lineage member of the Eleutherozoa (PAUL & SMITH 1984; SMITH 1988 a) in which the reorientation in the biotope was not yet realized. *Stromatocystites* had a pentagonal outline, a slightly curved oral side and a flat apical side; as in *Camptostroma* (stem lineage of the Echinodermata), the anus occurs in an interradius of the oral side. *Stromatocystites* probably laid freely on the substrate with the oral side up and was a suspension feeder.

Moreover, from fossils like *Stromatocystites*, we can develop notions about the habitus of the last common stem species of the recent Eleutherozoa. I consider a pentaradiate body with mildly differentiated arms and flowing transition to a central section to be probable. The Asteroida may come close to the ground pattern with the majority of its species. The separation of five thin arms against a round, central disk is a derived state of the Ophiuroida. With expansion of the oral side under curving of the body in the stem lineage of the Echinozoa, the arms were fused in the new, round to sac-like body.

Asteroida – Cryptosyringida

Asteroida

The interpretation of three character alternatives is of major importance for the evaluation of the ground pattern of the Asteroida.

1. Larval development solely with bipinnaria – development through bipinnaria and brachiolaria.
2. Conically running ambulacral tube feet without an adhesive disk – ambulacral tube foot ends in a sucker.
3. Intestine with apical anus – intestine blindly closed.

The planktotrophic bipinnaria (Fig. 40 C) of seastars can be understood as a continuation of a dipleurula/auricularia, as is retained in the Holo-

Fig. 39. Asteroida. **A** *Astropecten irregularis*. Burrowing starfish in the act of burying itself in sandy soil. **B** *Astropecten irregularis*. Oral side. Arm with ambulacral tube feet without suction disk. **C** *Crossaster papposus*. Sun star with secondary multiplication of arms. **D** *Asterias rubens*. Common starfish. Apical side with a light madreporic plate. **E** *Asterias rubens*. Oral side with central opening of the mouth. Ambulacral feet with suction disks in double rows on both sides of the ambulacral groove. **F** *Pygnopodia helianthoides*. Large increase in the number of arms. (**A–C** Heligoland, North Sea; **D, E** Sylt, North Sea; **F** Puget Sound, Washington, USA. Originals)

thuroida as plesiomorphy. Evolutionary changes involve the anterior end. Fusion of the rostral tips of the ciliary cord leads to separation of a band that runs closed around a preoral field of the ventral side. A bipinnaria is realized at the start of metamorphosis to pentameric adult in free water in species of the taxa *Astropecten* and *Luidia*.

In general, however, brachiolaria larvae arise from bipinnaria. At the anterior end an adhesion apparatus consisting of three brachiolar arms and an attachment organ develops – for anchoring to the ground during metamorphosis (Fig. 40 D).

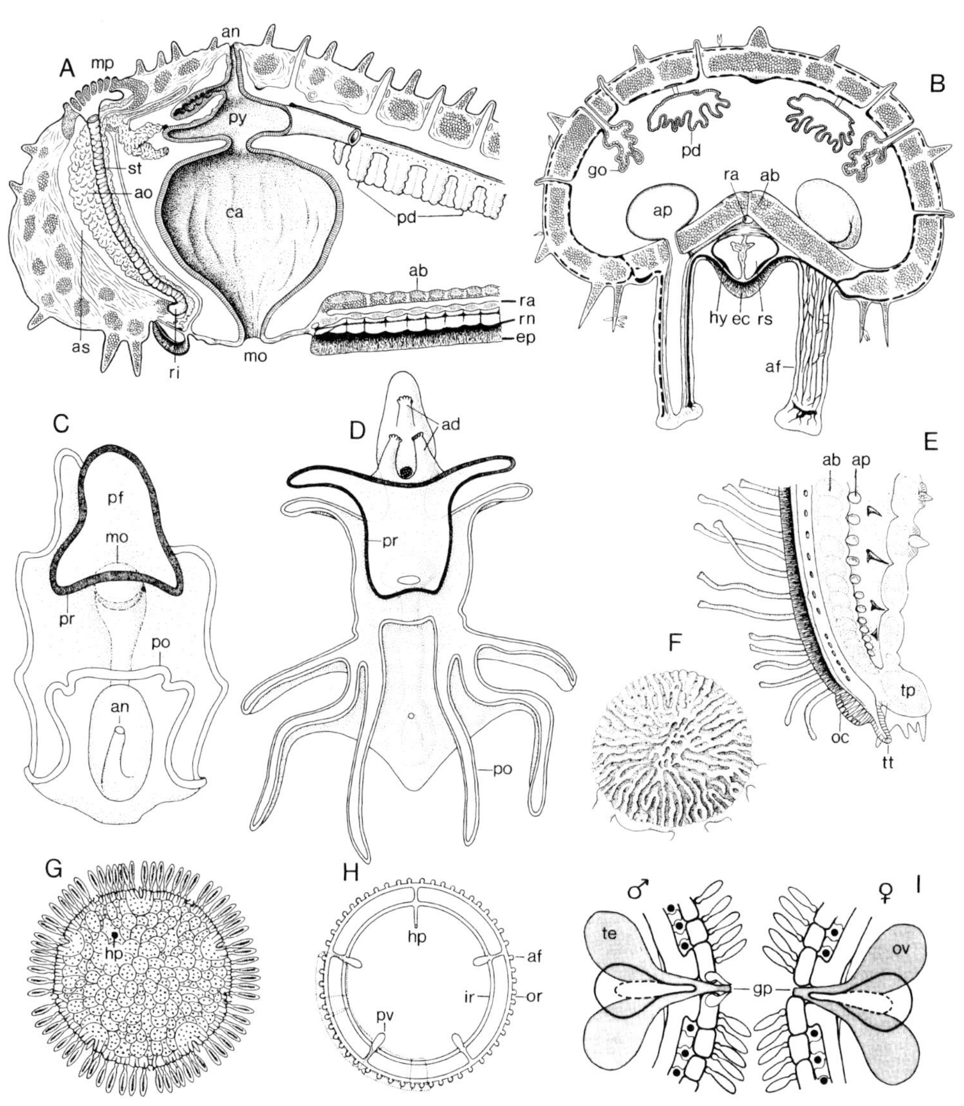

Fig. 40. Asteroida. **A** Organization. *Left* Section through the central part; *right* through a radius with arm base. **B** *Asterias.* Cross section of an arm. **C, D** Larvae. **C** Bipinnaria of *Astropecten polyacanthus.* Ventral view. **D** Brachiolaria of *Asterias* with three adhesive arms at the anterior end. **E** Longitudinal section through the tip of an arm of *Asterias glacialis.* **F** Madreporic plate of *Asterias.* **G, H** *Xyloplax medusiformis.* **G** Apical view of the round body (diameter: ca. 6 mm) with a hydropore. **H** Water-vascular system. With an internal and an external circular vessel. **I** *Xyloplax.* Genital organs with sexual dimorphism. Male with penile papilla that is supported by spines. *ab* Ambulacral plate; *ad* arms with adhesive papilla; *af* ambulacral tube foot; *an* anus; *ao* axial organ; *ap* ampulla; *as* axial sinus; *ca* cardia; *ec* ectodermal nerve system; *ep* epithelium; *go* gonad; *gp* genital pore; *hp* hydropore; *hy* hyponeural nerve system; *ir* inner ring canal; *mo* mouth; *mp* madreporic plate; *oc* ocellus; *or* outer ring canal; *ov* ovary; *pd* pylorus diverticulum; *pf* preoral field; *po* postoral ciliary band; *pr* preoral ciliary band; *pv* polian vesicle; *py* pylorus; *ra* radial vessel; *ri* ring canal; *rn* radial nerve; *rs* radial sinus; *st* stone canal; *te* testes; *tp* terminal plate; *tt* terminal tentacle. (**A, B, E, F** Marinelli 1960; **C** Cuénot 1948; **D** Ruppert & Barnes 1991; **G, H** Baker et al. 1986; **I** Rowe et al. 1988)

The following conclusion must be drawn from the existence of species that possess solely the bipinnaria. Only bipinnaria larvae belong in the ground pattern of the Asteroida (WADA et al. 1996). Larval adhesive systems are an evolutionary new acquisition within the entity. Widespread speculations in which the adhesion of a brachiolaria should be the "recapitulation" of the former sessility of the Echinodermata become invalid with this interpretation.

Also with the second feature, *Astropecten* and a few further taxa stand apart from the bulk of the seastars with suckers on the ambulacral tube feet. *Astropecten* species have pointed feet with which they can rapidly bury themselves in soft ground (Fig. 39 B).

Crinoida and Ophiuroida do not possess adhesive disks. The outgroup comparison demands a clear answer. The lack of suckers was taken over as a plesiomorphy from the ground pattern of the Eleutherozoa into the ground pattern of the Asteroida. Adhesive disks on the ambulacral tube feet evolved first within the monophylum. Furthermore, it is reasonable to assume a connection between the evolution of larval adhesive systems in the brachiolaria and the invasion of hard ground as living space. Moreover, the hypothesis of a convergent evolution of adhesive disks in the stem lineage of the Echinozoa (Echinoida + Holothuroida) is a posteriori mandatory; however, with a calcified rosette they are also constructed differently in the latter (p. 124).

We now come to the last alternative. Various seastars (e.g., *Porcellanaster, Eremicastor, Ctenodiscus, Luidia*) have a blindly closed intestine. In contrast, there are apparently species in the taxon *Astropecten* which, like the majority of the Asteroida, possess an anus (LAFAY et al. 1995). Since the Crinoida and the Echinozoa possess an intestine with anus, one may set this condition in the character pattern of the stem species of the Eleutherozoa. Thus, the anus is probably a plesiomorphy in the ground pattern of the Asteroida, its lack within the entity is a secondary state.

A habitus with five arms and broad transition into the central part is probably taken over from the ground pattern of the Eleutherozoa. A surprising palette of body forms has evolved within the Asteroida, from deep sea inhabitants with extremely thin, long arms (*Brisinga, Freyella*) to five-cornered structures in the form of flat plates (*Anseropoda*) or blown-up cushions (*Culcita*).

Exceptions to the pentamery are documented by the sun star *Crossaster papposus* with 8–15 arms or the sunflower starfish *Pygnopodia helianthus* that can carry up to 24 arms (Fig. 39).

Autapomorphies (Fig. 36 → 4)

- Light-sensory organ at the tips of the arms (Fig. 40 E).
  Collection of pigment-containing optic cushions in the epidermis. A calcareous plate and the terminal ambulacral tube foot (end tentacle) cover the eyespot.
- Paired intestinal diverticula in the arms (Fig. 40 A, B).
  Five diverticula arise from the apical region of the intestine. They run in the arms and split there into two branches.
- Gonads in the arms (Fig. 40 B).
  The anlagen of the five gonads split in the central disk into two branches. One branch each is inserted into the adjacent arms and lies lateral to the intestinal diverticulum there.
- Bipinnaria larva.
  As mentioned above, only the bipinnaria with metamorphosis in free water belong in the ground pattern of the Asteroida. In comparison with the dipleurula/auricularia, the separated preoral ciliary band on the ventral side is an autapomorphous subfeature.

Various drafts for a phylogenetic system of the Asteroida are the subject of controversial discussion (GALE 1987; BLAKE 1987; LAFAY et al. 1995; WADA et al. 1996). Here, I will only point out that LAFAY et al. (1995) hypothesized the species-rich taxon *Astropecten* as the sister group of all other Asteroida. However, a strict evaluation of the above-mentioned character alternatives as well as the justification of *Astropecten* as a monophyletic species group are necessary.

An informative example for the fundamental difference between a traditional, essentialistic classification and the phylogenetic systematization is the diverging judgement of the recently discovered deep sea species *Xyloplax medusiformis* and *Xyloplax turnerae* (Fig. 40 G–I).

Under the subjective measures of traditional classifications ***Xyloplax*** is interpreted as "one of the most significant echinoderm discoveries this century..." (ROWE et al. 1994, p. 149). A series of characters was set as being essential for the separation from the Asteroida and used to justify the creation of an independent echinoderm "class" Concentricycloida (BAKER et al. 1986; ROWE 1988; ROWE et al. 1988; HEALY et al. 1988; ROWE et al. 1994). These characters include the tiny, medusoid body; a uniserial ring of ambulacral tube feet on an outer ring canal; an additional inner ring canal of the ambulacral system; sexual dimorphism with five interradial papilla in the male and the suspicion of copulation; thread-like sperms with inserting cilia far to the front in connection with internal fertilization. This is all very interesting. However, the differences to other "classes" of Echinodermata are irrelevant when we leave the procrustean bed with orders, classes, and phyla. Now we are no longer concerned with the creation of categories on the basis of subjectively chosen differences, but rather must undertake an objective search for apomorphous agreements to find the sister group.

SMITH (1988b) made a corresponding judgement of *Xyloplax*. *Xyloplax* shares a series of agreements that can be evaluated as apomorphies with the also disk-shaped, deep sea seastar *Caymanostella*:

Circle of peripheral inframarginalia with club-like spines; large peristomial field with a ring of trapeziform mouth plates; madreporite as solitary pore opposite a plesiomorphous sieve plate; ambulacral tube foot without adhesive disk. They justify an adelphotaxa relation between *Xyloplax* and *Caymanostella*. Since its creation (BELYAEV 1974) through to the most recent review with the description of a "genus" *Belyaevostella* (ROWE 1989), the "family" Caymanostellidae has been and is a taxon of the Asteroida. Logically, the same must hold for *Xyloplax* in the interpretation as sister group of the Caymanostellidae (*Caymanostella* + *Belyaevostella*).

Cryptosyringida

Autapomorphies (Fig. 36 → 5)

- Epineural canals in the arms.
  Folds of the oral epidermis with radial nerves of the ectodermal nerve system close to form tubes or canals that are named epineural canals (ectoneural canals). The inwardly displaced radial nerves now lie in the aboral wall of the canals. The open ambulacral grooves of the Asteroida with a peripheral, unprotected position of the radial nerves in the skin form the plesiomorphous state (Fig. 41).

Ophiuroida – Echinozoa

Ophiuroida

At first glance, serpent stars are impressive due to the thin, completely mobile arms and their sharp contrast to the small, central disk.

Their unique locomotion is based on a series of vertebra that are linked in the arms by joints and flexor muscles. The nature of the joints determines the mode of locomotion. In the zygospondylic articulation of the Ophiurina, median projections of one ossicle mesh in a joint groove on the adjacent vertebra (Fig. 43H). In principle, this allows only a backward and forward motion of the arms in the horizontal plane (examples: *Ophiura*, *Ophiocoma*; Figs. 42A, 43E). The streptospondylic articulation of the Euryalina operates

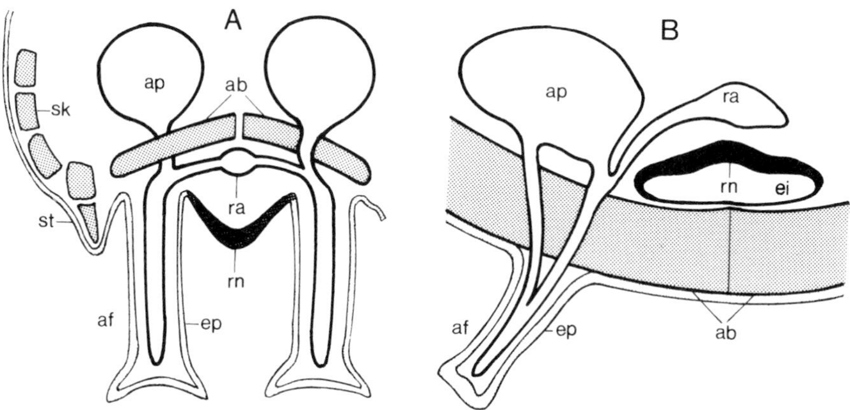

Fig. 41. Demonstration of the displacement of the ectoneural nerve system in the Cryptosyringida to within the body comparing seastars with sea urchins. The presentation extends only over the radial nerves and the water-vascular system. **A** Cross section of the arm of *Asterias* shows the "open" ambulacral grooves of the Asteroida. The radial nerves lie unprotected in the epidermis. The radial canals of the water-vascular system are also outside the ambulacral plates; the connections to the internal ampulla run between the parts of the skeleton. **B** Section through two ambulacral plates of *Paracentrotus* demonstrates the inwardly displaced radial nerves of the Echinoida. The nerve is enclosed in the aboral wall of the epineural canal. Together with the nerve canal, the radial canal of the water-vascular system now also runs internally from the ambulacral plates. Two canals run through the ambulacral plate (autapomorphy of the Echinoida) to join the foot with the ampulla. *ab* Ambulacral plate; *af* ambulacral tube foot; *ap* ampulla; *ei* epineural canal; *ep* epidermis; *ra* radial vessel; *rn* radial nerve; *sk* skeletal plate; *st* spine. (Kaestner 1963, modified)

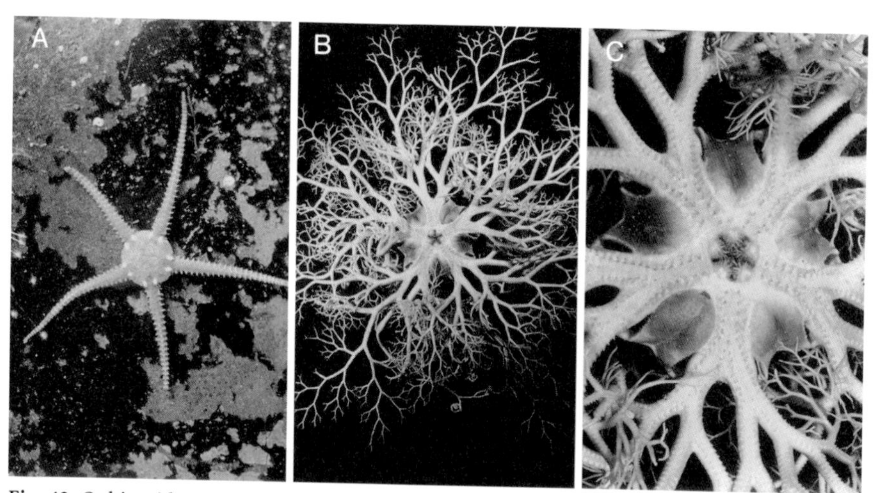

Fig. 42. Ophiuroida. **A** *Ophiura albida* (Ophiurina) from the apical side. The cylindrical arms, sharply separated against the central disk can only move horizontally. **B, C** *Gorgonocephalus eucnemis* (Euryalina). Medusa's head, from the oral side. **B** Complete view with highly dichotomous branching of the five arms that are rotatable in all directions. **C** Central disk with mouth opening and origin of the arms. The first branching already begins in the periphery of the disk. To the side of the arms are clefts leading to the ectodermal bursae (respiratory organs). (**A** Heligoland, North Sea; **B, C** Puget Sound, Washington, USA. Originals)

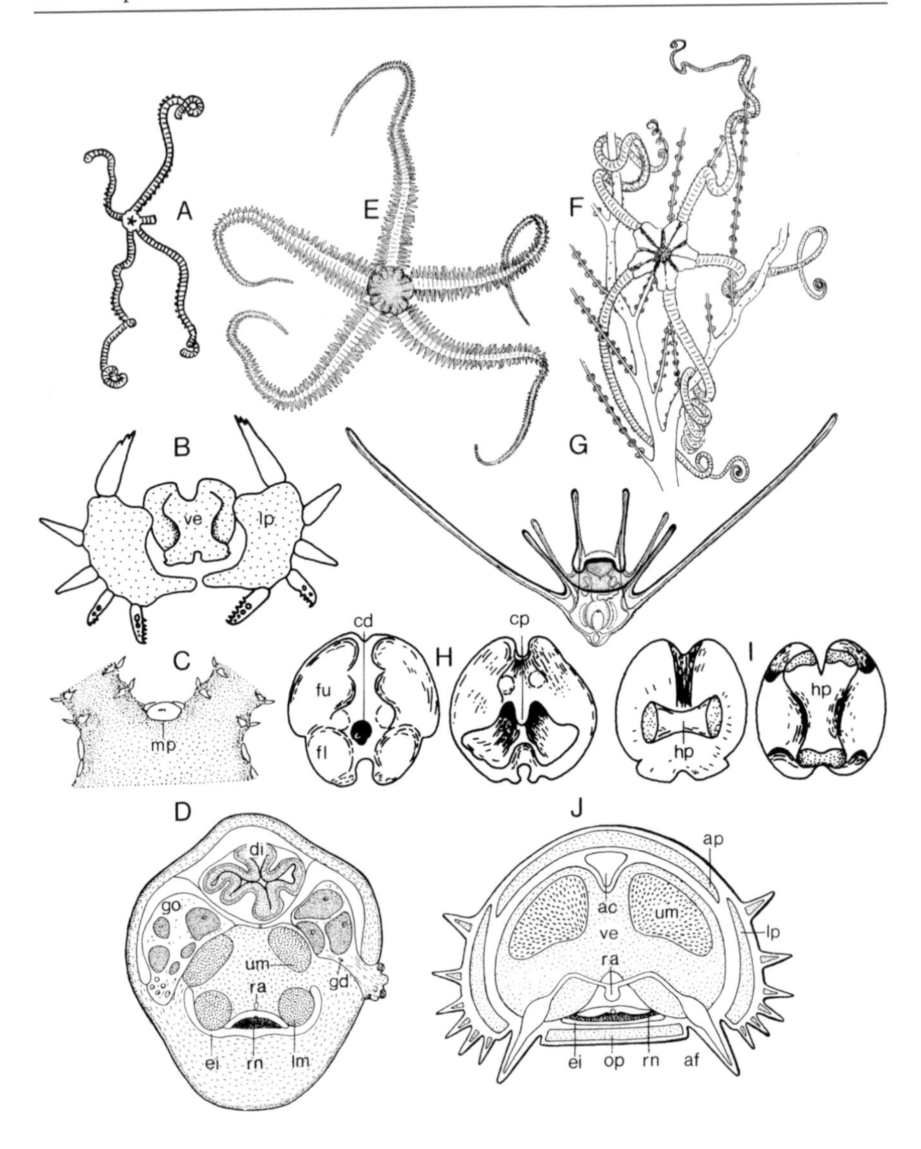

with hour-glass joints. Constrictions in the joint collars of consecutive vertebrae are at right angles to each other (Fig. 43I). In this way, the arms can be turned in various directions; an anchoring in the biotope by twining around objects is possible (examples: *Asteronyx* with five simple arms, *Gorgonocephalus* with strong branching of the arms – Figs. 42B, C, 43F).

◀ **Fig. 43.** Ophiuroida. **A–D** *Ophiocanops fugiens.* **A** Habitus. **B** Skeleton of the arms. Cross section. In the *center* the vertebra, to the *sides* the two lateral plates with spines; aboral and oral plates are absent. **C** Aboral view of the disk with interradial position of the madreporic plate at the edge. **D** Cross section of an arm. Soft body. Unpaired intestinal diverticulum and two rows of gonads extend along the aboral side of the arm. The individual gonads have separate gonoducts. With two pairs of intervertebral muscles. **E–J** Neophiuroida. **E** Ophiocoma (Ophiurina). All Ophiurina have simple, unbranched arms. **F** *Asteronyx excavata* (Euryalina) with primitive, simple arms that are wound around a thorny coral (Antipatharia). **G** Ophiopluteus larva of *Ophiomaza cacaotica.* **H** Zygospondylic articulation of the vertebra in Ophiurina in the example of *Ophiolepis.* Proximal vertebral surface (*left*) with central pit, distal surface (*right*) with corresponding joint tubercule. **I** Streptospondylic articulation of vertebra in Euryalina in the example of *Astrophyton.* Distal vertebral surface (*left*) with transverse tapering of the joint collar. Proximal surface (*right*) with vertical hour-glass constriction. **J** Arm cross section with four peripheral plates; of the intervertebral muscles, the oral pair is omitted to expose the canals to the ambulacral tube feet. *ac* Arm coelom; *af* ambulacral tube foot; *ap* aboral plate; *cd* central depression (joint pit); *cp* central projection (joint tubercule); *di* intestinal diverticulum; *ei* epineural canal; *fl* pit for lower intervertebral muscle; *fu* pit for upper muscle; *gd* gonoduct; *go* gonad; *hp* joint collar with central constriction (hour-glass principle); *lm* lower intervertebral muscle; *lp* lateral plate; *mp* madreporic plate; *op* oral plate; *ra* radial vessel of the ambulacral system; *rn* radial nerve; *um* upper intervertebral muscle; *ve* vertebra. (**A** Cuénot 1948; **B, C** Fell 1963; **D, J** Marinelli 1960; **H, I** Meglitsch & Schram 1991; **G** Dawydoff 1948 a)

Autapomorphies (Fig. 36 → 6)

- Habitus.
  Five thin arms, strictly separated from a round, central disk.
- Vertebrae or vertebral ossicles in the arms (Fig. 43 B).
  The single vertebra develop from two adjacent calcareous bodies. In the ground pattern of the serpent stars, these fuse completely only in the proximal parts. In the distal sections of the arms they remain primarily separated (plesiomorphy in *Ophiocanops fugiens*, p. 123), The vertebrae are linked by two pairs of strong muscles.
- Calcareous plates in the periphery of the arms.
  Only two lateral plates per vertebral section belong in the ground pattern of the Ophiuroida. This plesiomorphous state is realized in *Ophiocanops fugiens* (Fig. 43 B). Unpaired aboral and oral plates additionally occurred first in the Neophiuroida (Fig. 43 J).
- Aboral-oral transfer of the madreporic plate.
  An apomorphy in comparison with the aboral position of the madreporite in Asteroida. The position at the edge of the disk in *Ophiocanops fugiens* represents the first stage of the process (Fig. 43 C).
- Absence of an anus.
  An intestine with anus is found in all Crinoida, Echinoida and Holothuroida; furthermore, an anal opening can be postulated for the ground pattern of the Asteroida (p. 116). This circumstance supports a secondary lack in the serpent stars; we can hypothesize the loss of the anus in the stem lineage of the Ophiuroida.

- Ophiopluteus.
  A planktotrophic larva with four pairs of long projections supported by calcareous spicules (Fig. 43G) belongs in the ground pattern of the Ophiuroida. There are extensive agreements in the existence of larval arms and their equipment with skeletal rods with the echinopluteus of the Echinoida. The previously mentioned kinship hypotheses (p. 106), however, demand the assumption of a convergent evolution of pluteus larvae in the stem lineages of the Ophiuroida and Echinoida.

Ophiocanops fugiens – Neophiuroida[6]

"Our analysis of morphological data clearly shows that previous classifications have been based on grades, not clades[7], and the higher taxonomy of ophiuroids is in need of complete revision" (SMITH & PATERSON 1995, p. 227).

Even so, the highest-ranking adelphotaxa relationship seems to be clarified. *Ophiocanops fugiens* from Indonesia with a small disk and very long arms (MORTENSEN 1933) has been discussed as the adelphotaxon of all other Ophiuroida (FELL 1962, 1963; SMITH & PATERSON 1995). In other words, a single species stands opposed to the entire Neophiuroida[8] – just as the crustacean *Cyclestheria hislopi* stands opposed to the entire Cladocera (water fleas), the salmon leech *Acanthobdella peledina* to all other Hirudinea or *Mastotermes darwiniensis* to the remaining Isoptera (termites) (Vol. II, pp. 69, 154, 292). In each case, one recent species is the sister species of a species-rich, supraspecific taxon with recent members.

These examples are mentioned in order to validate the following interpretation. In a consequent phylogenetic system *Ophiocanops fugiens* cannot be hidden in an order or subclass of otherwise fossil Oegophiurida "as the only extant member" (FELL 1962, 1963; SMITH & PATERSON 1995). Such a measure once again illustrates the pitfalls on dealing with categories. If we eliminate them, the shoe is then on the other foot. The fossil relatives are to be ordered in a stem lineage at the end of which the recent species *Ophiocanops fugiens* stands as the adelphotaxon of the Neophiuroida.

---

[6] tax. nov.

[7] Grades = paraphyla; clades = monophyla.

[8] In order to avoid the very similar, but confusing naming of equal-ranking subtaxa of the Ophiuroida, I have chosen the new name Neophiuroida.

We present some feature alternatives between *Ophiocanops fugiens* and the Neophiuroida with evaluations.

Ophiocanops fugiens	**Neophiuroida**
Marginal position of the madreporic plate at the edge of the disk (Fig. 43 C). Step on the way of an aboral to oral displacement	Madreporic plate shifted to the oral side of the disk
Plesiomorphy	Autapomorphy
Paired anlagen of the vertebral ossicles remain separated in the distal part of the arm	Fusion of the anlagen to solid calcareous bodies along the length of the arms
Plesiomorphy	Autapomorphy
In the periphery of the arms only two lateral plates on each side of the vertebra. No aboral or oral plates	Two lateral as well as one aboral and one oral plate per "vertebral section"
Plesiomorphy	Apomorphy, but possibly evolved first within the Neophiuroida
Lack of bursa and bursa clefts	Bursa as sac-like invaginations of the epidermis. Clefts on the oral side of the disk to the side of the base of the arm
Plesiomorphy	Apomorphy (Fig. 42 C). Possibly evolved first within the Neophiuroida
Numerous gonads in paired, serial arrangement in the arms; with separate outlets (Fig. 43 D)	Gonads mainly or completely limited to the disk; outlets through bursa clefts
Long, unpaired appendices of the intestine in the arms above the vertebral bodies (Fig. 43 D)	Intestine mainly or only developed in the disk

The position of the gonads and intestinal diverticula in the arms of *Ophiocanops fugiens* was evaluated as being original by FELL (1963); however, it could also be the result of a secondary extension due to a lack of space in the tiny disk (MORTENSEN 1933).

There is one last word in this context. *Ophiocanops fugiens* has hourglass joints, leading back to the above-described joints of the arm vertebrae. The Neophiuroida are divided into two distinct groups (SMITH & PATERSON 1995) (? adelphotaxa) – the streptospondylic Euryalina (*Asteronyx, Asteroschema, Euryale, Gorgonocephalus*) and the zygospondylic Ophiurina (majority of the ca. 2000 serpent star species). I do not know of an evolutionary evaluation of the alternative joint systems. If hour-glass joints should constitute the apomorphy and, in addition, have arisen only once, then *Ophiocanops fugiens* would have to be ordered in the Euryalina; the features presented here as plesiomorphies would then have to be discussed anew (? reductions).

Echinozoa

Autapomorphies (Fig. 36 → 7)

- Habitus.
  Strong extension of the oral side with the ambulacra under aborally directed bulging of the body. The apical or aboral surface is reduced by this process to a tiny field about the anus.
- Internal skeletal system about the oesophagus.
  A homology between the chewing apparatus of the Echinoida (Aristotle's lantern) and the calcareous ring of the Holothuroida seems reasonable (LITTLEWOOD et al. 1997), but is also disputed (HAUDE 1994; SMILEY 1988, 1994).
  Even so, there is a Devonian member of the stem lineage of the Holothuroida (SMITH 1984b), *Rotasaccus dentifer* (HAUDE & LANGENSTRASSEN 1976), which possesses a chewing apparatus with five jaws and a nonarmoured body wall with wheel-like sclerites (Fig. 44A,B).
  Lantern and calcareous ring are closely associated with the circumoral water vessel and nerve, and they are both encased in the peripharyngeal coelom.
- Ambulacral tube foot with adhesive disk.
  Synapomorphy of the Echinoida+Holothuroida; convergence of evolution of tube foot with sucker within the Asteroida (p. 116).
- Sclerites on the ambulacral tube foot.
  Wall of the foot with calcareous spicules; adhesive disk with end plate in the form of a rosette (Fig. 44C,D).
- Perianal coelom.
- Well developed haemal system with rete mirabile.

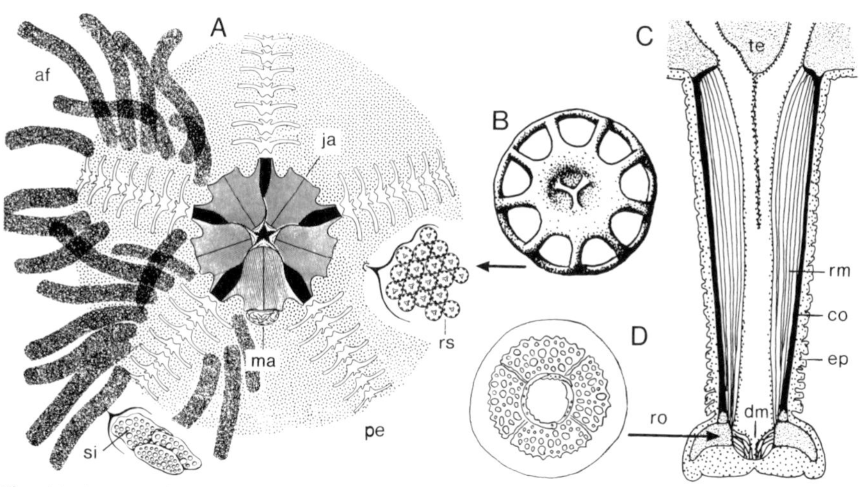

Fig. 44. Apomorphous features from the ground patterns of the Echinozoa, Echinoida and Holothuroida. **A–B** † *Rotasaccus dentifer.* Member of the stem lineage of the Holothuroida from the Devonian period. **A** Diagram of the oral side. Five jaws in the center (ground pattern of the Echinozoa). Sclerites in the wall of the ambulacral tube foot (ground pattern of the Echinozoa). Wheel-like sclerites in the unarmoured body wall (ground pattern of the Holothuroida). **B** Reconstruction of a sclerite. **C** Schematic longitudinal section through an ambulacral tube foot of regular Echinoida. Sclerite plate in the form of a rosette in the terminal adhesive disk (ground pattern of the Echinozoa). Connection of the ambulacral tube foot to the inner ampulla through two canals in the shell (ground pattern of the Echinoida). **D** Adhesive disk with rosette of *Echinoneus.* View from the inside. *af* Ambulacral tube foot; *co* collagen; *dm* adhesive disk muscle; *ep* epidermis; *ja* jaw; *ma* madreporic plate; *pe* perradial elements; *rm* retractor muscle; *ro* rosette; *rs* wheel-like sclerite; *si* sieve platelet; *te* test. (**A, B** Haude & Langenstrassen 1976; **C** Smith 1984; **D** Hyman 1955)

Echinoida – Holothuroida

Echinoida

Autapomorphies (Fig. 36 → 8)

- Corona.
  Rigid globular armour made up from five double rows of ambulacral plates and five double rows of interambulacral plates in a regular sequence (Fig. 46 D, F).
- Two pores per ambulacral tube foot in the ambulacral plates.
  Each ambulacral tube foot is connected to the internal ampulla through two small canals running through the armour (Fig. 41 B). This is unique among the Eleutherozoa.
- Peristome.
  A flexible, connective tissue on the oral side forms a broad peristomial membrane (Fig. 47 B).

- Apical system with periproct (Fig. 46B).
  The aboral or apical system consists of a ring of ten plates. One plate each is at the end of a double row of the corona. Through the five terminal ambulacral plates (ocular plates) five tentacles emerge as the ends of the radial canals. There are five gonopores in the five interambulacral plates. One of these genital plates transforms to the madreporic plate. The apical ring of plates is followed by the periproct; this is a flexible membrane around the anus.
- Aristotle's lantern (Fig. 45).
  Cone-shaped, pentameric jaw apparatus in the main axis of the corona. The orally directed tip is formed from the chewing sections of the teeth. Forty calcareous elements – five rigid teeth and 35 stereom skeletons – are arranged in five identical units. The single unit consists of (1) a two-halved pyramid (jaw), (2) one tooth, (3) a pair of epiphyses, (4) a rotula and (5) a two-part compass.
  The pyramids hold and guide the teeth. The epiphyses are apically joined rigidly to the pyramids, but are folded over joints with the rotulae (p. 129). The compass parts lie on the rotulae; they have nothing to do with the movements of the jaw for feeding.
  The lantern is moved by 60 muscles. Strong interpyramidal muscles close the lantern in the periphery. Protractors and retractors pull to the edge of the peristome.
- Perignathic girdle.
  Internal calcareous element as muscle insertions on the oral edge of the corona. In the ground pattern the girdle probably consists only of interambulacral apophyses (p. 129).
- Lack of polian vesicles.
  The existence of polian vesicles on the ring canal of the ambulacral system is justified for the ground pattern of the Eleutherozoa (p. 112). According to this hypothesis, their absence in the Echinoida must be a secondary state.
- Echinopluteus.
  Planktotrophic larva. In the ground pattern with eight arms that are supported by calcareous spicules. Evolution in the stem lineage of the Echinoida in convergence to the pluteus larva of the Ophiuroida (pp. 106, 122).

A roundish shell of 20 rows of plates in regular sequence consisting of two rows each of ambulacra and interambulacra has been postulated for the last common stem species of the recent Echinoida (LITTLEWOOD & SMITH 1995). This hypothesis, however, in no way justifies the system entity "Regularia" of traditional classifications. The described, regular structure of the shell represents a plesiomorphous state in the character pattern of the last common stem species of the recent sea urchins or – in other

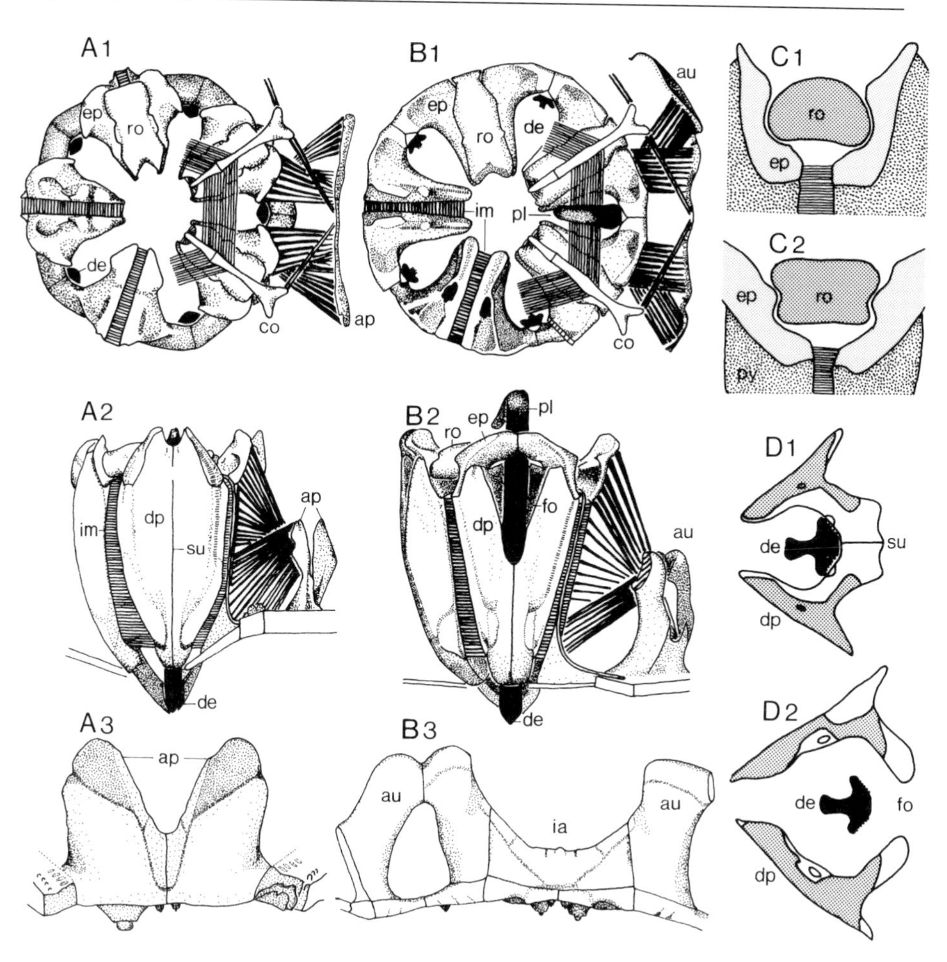

Fig. 45. Echinoida. Aristotle's lantern. **A** Socket-joint lantern of the Cidaroida, *Eucidaris tribuloides* as example. **B** Hinge-joint lantern of the Euechinoida, *Echinus acutus* as example. **A1+B1** Apical views. Lantern musculature is only shown on one pyramid. Removed in upper radius the compass, in the next the rotula and in the following radii the epiphyses (in **B1** one tooth with plumule). **A2+B2** Side views (interradius). Musculature with insertion in the perignathic belt only shown for the pyramid on the *right*. **A3+B3** Perignathic belt. **A3** Interamubulacral apophysis with two tips. **B3** Interambulacral crest and ambulacral auricle. **C** Joint system between epiphysis and rotula. Transverse section. **C1** Socket joint of the Cidaroida (apomorphy). **C2** Hinge joint of the Euechinoida (plesiomorphy). **D** Cross-cut through pyramid and tooth of *Paracentrotus lividus* (Euechinoida). Tooth surrounded by the halves of the pyramid. Hemipyramids in **D1** linked by suture, in **D2** widely separated by a foramen. *ap* Apophysis; *au* auricle; *co* compass; *de* tooth; *dp* hemipyramid; *ep* epiphysis; *fo* foramen; *ia* interambulacral crest; *im* interpyramidal muscle; *pl* plumule; *py* pyramid; *ro* rotula; *su* suture. (**A–C** Märkel 1979; **D** Märkel 1976)

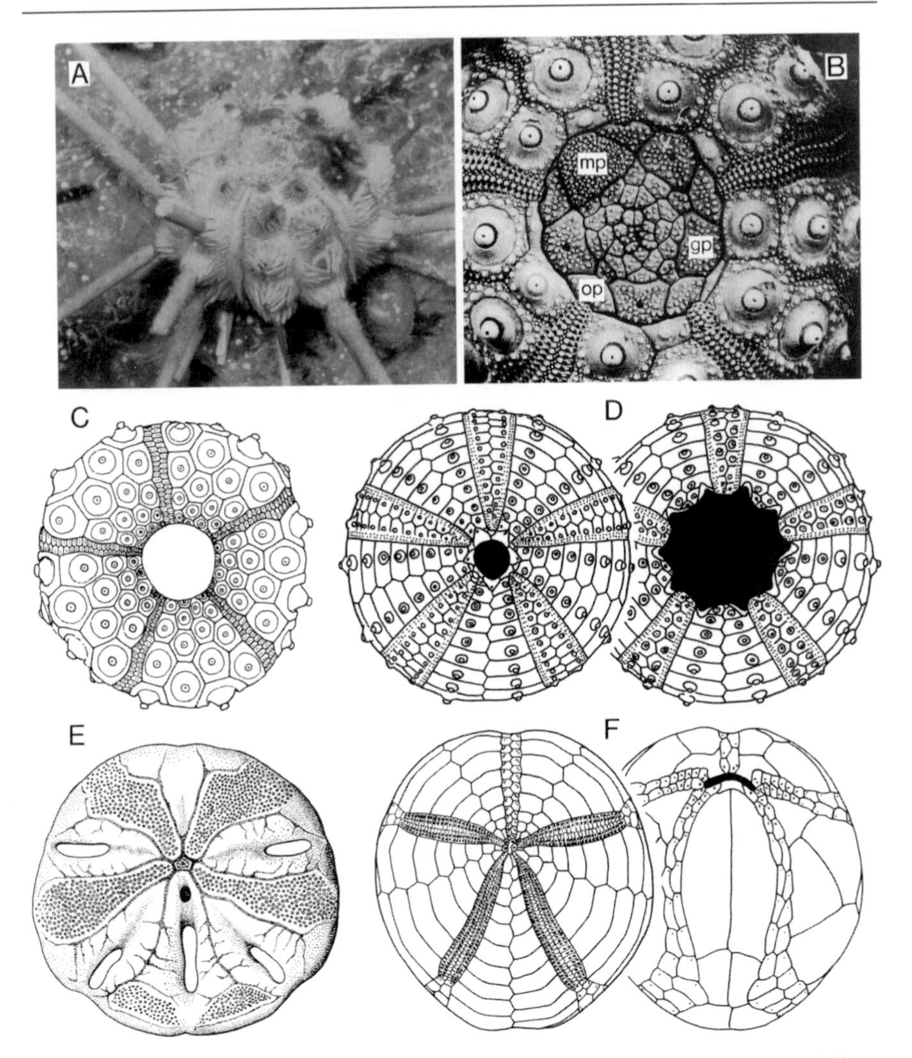

Fig. 46. Echinoida. **A** *Cidaris cidaris* (Cidaroida). The long primary spines are partially or completely broken off. **B** *Eucidaris metularia* (Cidaroida). View of the apical surface with periproct. A ring of ten plates in the periphery: five ocular plates at the end of the double rows of ambulacralia; five genital plates in continuation of the double row of interambulacralia. **C** † *Archaeocidaris* (member of the stem lineage of the Echinoida). With four rows of interambulacral plates between the double rows of ambulacral plates. **D** *Echinothiara* (Euechinoida). Shell of a regular sea urchin, apical view (*left*) with periproct and oral view (*right*) with peristomal field (*black*). **E, F** Irregular sea urchins. **E** *Mellita testudinata* (Irregularia, Clypeastroida). Oral side. The anus is close behind the mouth opening. **F** † *Linthia* (Irregularia, Spatangoida). Anus displaced from the apical field (*left*) to the posterior end of the bilaterally symmetrical body. Peristome with mouth opposed shifted forwards. *gp* Genital plate; *mp* madreporic plate; *op* ocular plate. (**A** Banyuls-sur-Mer, Mediterranean Sea. Original; **B–D, F** Smith 1984 a; **E** Cuénot 1948)

words – in the ground pattern of the entity Echinoida; it evolved in the stem lineage of the Echinoida in the Paleozoic era. Paleozoic members of the stem lineage of the Echinoida with considerable variations in the number of ambulacral and (or) interambulacral plate rows are known.

On the basis of fossil findings the development of the regular shell plates in the ground pattern of the recent Echinoida can be followed more precisely.

The **Cidaroida** and the **Euechinoida** (union of all other sea urchins) are today interpreted as the highest-ranking sister groups of the Echinoida, as will be justified below. Fossils such as † *Miocidaris* (lower Carboniferous, Permian-Liassic) with a regular shell of ten double rows of calcareous plates can be readily ordered in the Cidaroida. † *Archaeocidaris* (lower Carboniferous, Permian) with several rows of interambulacral plates (Fig. 46C), on the other hand, stands apart from the character pattern of the stem species of the recent Echinoida; the taxon is accordingly placed as an "advanced stem group member" (LITTLEWOOD & SMITH 1995, p. 215) in the stem lineage of the Echinoida before its splitting into the adelphotaxa Cidaroida and Euechinoida. This interpretation has far-reaching consequences. The stem species of the recent sea urchins must have been an organism with the essential features of the recent Cidaroida. Features discussed purely and simply as excellent diagnostic organizational traits of the Cidaroida such as the extremely thin, tortuous rows of ambulacra or the existence of a single main spine per interambulacral plate are then mandatory items of the ground pattern of the Echinoida and therefore remain as plesiomorphies of the Cidaroida (Fig. 46A, B).

As for the corona, the state of the chewing apparatus in the ground pattern can be defined in more detail by an analysis of † *Archaeocidaris* (LEWIS & EMSON 1982). Here, the nature of the joint between the epiphysis and the rotula plays a decisive role. In the socket-joint lantern of the Cidaroida, the main joint consists of a hemispherical condyle of the rotula and a corresponding cavity in the epiphysis; the latter surrounds the condyle on all sides (Fig. 45, C1). By contrast, in the hinge-joint lantern of the Euechinoida a crista of the epiphysis is encompassed by two protrusions on the rotula (Fig. 45, C2). *Archaeocidaris* possesses a hinge joint – just like the living Euechinoida. In other words, this state belongs in the ground pattern of the Echinoida as a plesiomorphy; the socket joint of the Cidaroida is the apomorphy of the alternative.

Further feature alternatives correlate with the different joint systems. The interpretation for the foramen magnum of the pyramids seems to be unambiguous. In the Euechinoida, the semipyramids of the jaw deviate upwards and wide apart to form the foramen here (Fig. 45, B2); in the Cidaroida the joining suture is closed up to the top (Fig. 45, A2). Since the jaws of *Archaeocidaris* exhibit a deep foramen, this state is to be set in the ground pattern of the Echinoida.

We now come to the perignathic girdle. It is absent in the fossil † *Archaeocidaris.* The lantern muscles probably insert directly on the interambulacral plates around the peristome. The Cidaroida have only interambulacral apophyses with two tips (Fig. 45, A3). In the Euechiniscoida additional auricles with an opening lie over the ambulacra (Fig. 45, B3). Even when *Archaeocidaris* is not available for comparison, the direction of evolution seems to be well-founded. The Cidaroida represent the primitive state. Through growth of the laterally extending tips of neighboring apophyses over the ambulacralia the auricles of the Euechinoida developed around a central hole.

Thus, the ground has been prepared for the validation of the Cidaroida and Euechinoida as the highest-ranking sister groups of the Echinoida on the basis of alternative apomorphies (JENSEN 1981; SMITH 1981, 1984a; LEWIS & EMSON 1982; LITTLEWOOD & SMITH 1995).

For the **Cidaroida** (*Cidaris, Stylocidaris*), we have named two features from the construction of the chewing apparatus that can be evaluated as **autapomorphies.** These are the socket joint between the epiphysis and the rotula (socket-joint lantern) and the firm connection of the pyramid halves over practically their entire length.

For the **Euechinoida,** the evolution of the interambulacral auricle in the peripheral girdle may be considered as an **autapomorphy** from the field of the chewing apparatus. Furthermore, gills (protrusions on the edge of the peristomial membrane) as well as sphaeridia as static sensory organs (tiny swollen spines on adoral ambulacral plates) are only known among the Euechinoida and can be readily assigned as **autapomorphies** of the entity.

The Euechinoida encompass a whole series of high ranking monophyla of regular sea urchins whose mutual kinship relations have not been sufficiently clarified for presentation in a textbook. However, the Euechinoida also include the well-known irregular sea urchins with secondary bilateral symmetry. I will only make the following statement. It is completely wrong to discard the taxon Irregularia together with the necessary dissolution of the paraphyletic “Regularia”, as has been propagated in some modern textbooks.

Various special features can be interpreted without conflict as **autapomorphies** of a monophylum **Irregularia** (Figs. 46E, F, 47C):

1. Evolution of a new, bilaterally symmetrical body. The anus migrates from the circle of the apical genital plates in an interradius on the edge of the shell and further to the oral side of the shell.
2. With the displacement of the anus the genital plate of the corresponding interradius is reduced; the Irregularia have only four terminal interambulacral plates with gonopores in their ground pattern.
3. Aristotle’s lantern in the ground pattern of the Echinoida is a gripping lantern – realized in all regular sea urchins (MÄRKEL 1979); the teeth

Fig. 47. Echinoida and Holothuroida (Echinodermata). **A** *Psammechinus miliaris* (Echinoida). Purple-tipped sea urchin. View of the oral side. **B** *Echinus esculentus* (Echinoida). Edible sea urchin. Oral side. The tips of the five teeth extend into the oral cavity. Ten dark mouth tentacles intersperse the soft peristomal membrane around the mouth. The membrane is densely occupied by tridentate pedicellaria. **C** *Echinocardium cordatum* (Echinoida). Heart urchin. Sand inhabitant with bristle-like spines as digging tools. **D** The sea cucumber *Cucumaria planci* (Holothuroida). Tree-like branched tentacles around the mouth. Five radii each with two rows of ambulacral tube feet. **E** *Stichopus regalis*. Royal sea cucumber (Holothuroida). Life form differentiation at the top and underside. Mouth with shield-shaped tentacles subterminal on the underside. **F** *Leptosynapta* (Holothuroida). Worm-shaped, naked sand inhabitant. Radial canals and ambulacral tube feet reduced. (**A** Sylt, North Sea; **B, C** Heligoland, North Sea; **D, E** Banyuls-sur-Mer, Mediterranean Sea; **F** Puget Sound, Washington, USA. Originals)

grasp from outside to within and create pentaradiate bite wounds. Against this, the compact grinding lantern belongs in the ground pattern of the Irregularia (MÄRKEL 1978); it probably originated in connection with the change to microphage feeding in the stem lineage of the Irregularia. Within the Irregularia the grinding lantern is present throughout the life cycle in members of the subtaxon Clypeastroida (*Clypeaster, Dendraster, Melitta*), while in the Cassiduloida (*Cassidulus, Echinolampas*) it only develops in the juvenile phase. The Spatangoida (*Spatangus, Echinocardium*) – standing as adelphotaxon of the Clypeastroida + Cassiduloida – are more strongly derived in two aspects. The opening of the mouth of the heart urchins is shifted forwards in the interradius of bilateral symmetry and the jaw apparatus has completely disappeared.

Holothuroida

Sea cucumbers stand out because of their soft, unprotected bodies. A peripheral calcareous skeleton only exists in the form of microscopic sclerites in the connective tissue beneath the epidermis. In comparison to the extensive calcareous skeleton of the other Eleutherozoa, this must be interpreted as an outstanding apomorphous state.

As for the internal organization, the unpaired gonad with one interradial genital pore must be emphasized, whereby the pentameric symmetry is overlaid by a novel bilateral symmetry. The interpretation of this phenomenon is a subject of controversy. If the state of a single gonad is judged as a plesiomorphy, this would have the following consequence. Either the Asteroida + Ophiuroida + Echinoida form a monophylum with five gonads as a ground pattern apomorphy or gonads with a pentameric arrangement have occurred repeatedly and convergently within the Eleutherozoa. Neither assumption has a convincing validation. I tend to an evaluation of the single gonad as an apomorphy of the Holothuroida – namely under the following aspects. On locomotion in the ground pattern of the sea cucumbers, the main axis of the pentameric symmetry is oriented parallel to the ground, with the opening of the mouth at the anterior end and the anus at the posterior end, i.e., not like the other Eleutherozoa where it is directed perpendicularly through the mouth to the substrate. Perhaps three rows of locomotory ambulacral tube feet (trivium) also preferentially belong in the primitive movements. In any case, the limitation to one gonad and outlet on the interradius directed towards free water seems to be understandable with the new orientation of the main axis.

Autapomorphies (Fig. 36 → 9)

- Peripheral microsclerites (ossicles).
  Restriction of the calcareous skeleton to microscopically small calcareous bodies of manifold structures under the epidermis (Figs. 48 B, 49 A).
- Peripheral musculature of circular and longitudinal muscles.
  Evolution of a peripheral musculature in connection with the reduction of the calcareous skeleton. A uniform, outer circular muscle layer is followed inwards by five isolated, longitudinal muscle bands in the radii of the body.
- Calcareous ring around the pharynx.
  Composed of ten calcareous ossicles – five radial and five interradial sclerites (Figs. 48 A, 49 F). They form insertion sites for the mentioned longitudinal muscles by which the mouth field with tentacles can be retracted.
  The possible derivation from an apparatus with five jaws was discussed above (p. 124).
- Tentacles.
  Differentiation of ambulacral feet around the mouth to serve for microphage feeding (Fig. 49 D–F).
- Respiratory trees.
  Two invaginations from the hindgut with strong branching inside the body. Ventilation by pumping movements of the hindgut.
  Evolution of the respiratory tree can be explained in connection with the occurrence of the leather-like, impermeable connective tissue under the skin. Respiratory trees are reduced in "thin-skinned" sea cucumbers such as the Apoda (p. 135).

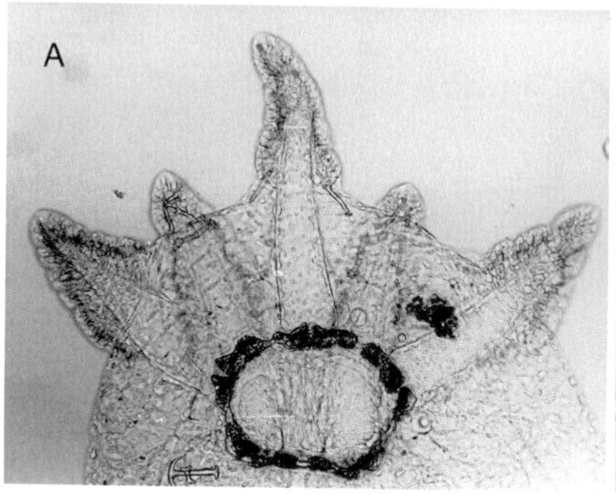

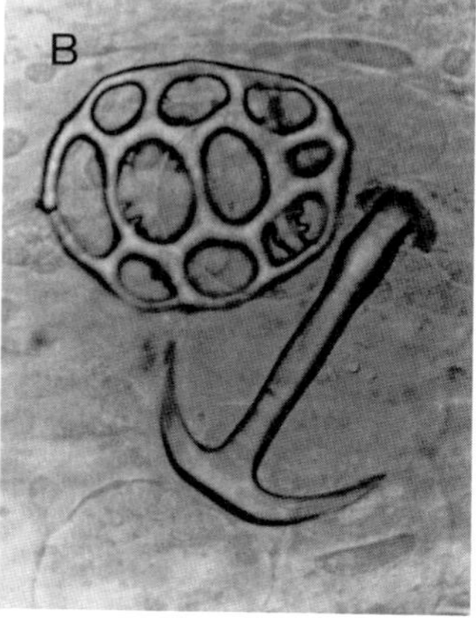

Fig. 48. Holothuroida. Synaptidae. **A** *Leptosynapta*. Anterior end with tentacle and calcareous ring. **B** *Leptosynapta minuta*. Characteristic skin sclerites of the apodal Synaptidae with anchor and associated anchor plate. (Originals: **A** Argeles, French Mediterranean coast; **B** Heligoland, North Sea)

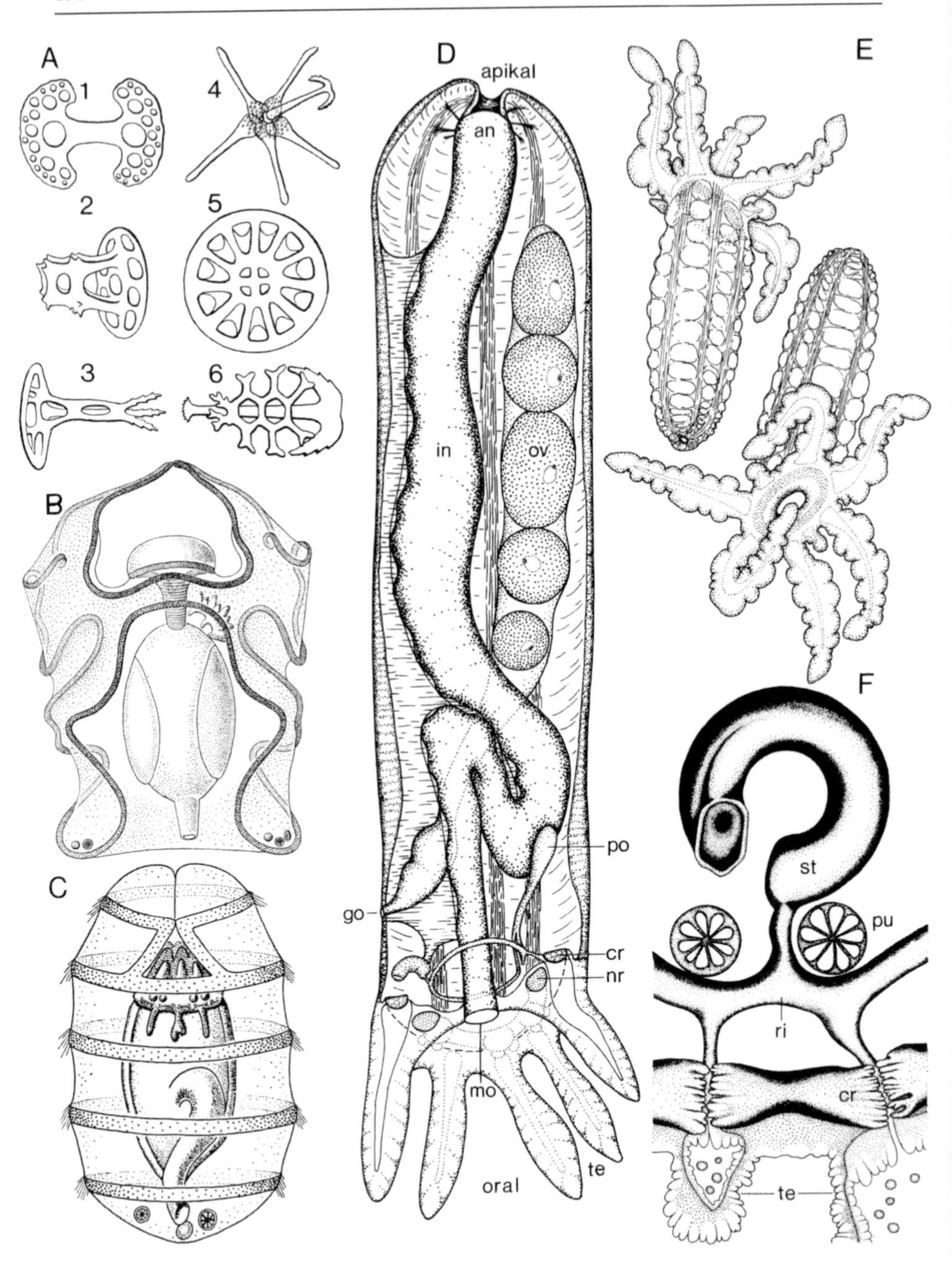
A
1
2
3
4
5
6
B
C
D
apikal
an
in
ov
po
go
cr
nr
mo
te
oral
E
F
st
pu
ri
cr
te

◀ **Fig. 49.** Holothuroida. **A** Selection of subepidermal sclerites (ossicles). *1 Oneirophanta affinis.* Two-part grid plate. *2 Holothuria impatiens.* Stool. *3 Holothuria multiceps.* Stool with rods. *4 Ankyroderma affine.* Star with anchor. *5 Laetmophasma fecundum.* Wheel. *6 Leptosynapta inhaerens.* Plate with anchor. **B** Auricularia larva of *Labidoplax digitata* with uniformly placed ciliary band. Ventral view. Ventral parts of the band *heavily dotted*; sections overlapping dorsalwards *lightly dotted*. **C** Doliolaria larva of *Labidoplax digitata*. Ciliary band of the auricularia torn and put back together to form five separate ciliary rings running transversely across the body. **D–F** *Rhabdomolgus ruber*. Interstitial inhabitant of sandy bottoms with ten tentacles: 4–5 mm long. **D** Late developmental stage with ovary. Oral surface directed downwards for comparison with the Echinoida. **E** Phases of motion with one tentacle rubbing over the opening of the mouth. Stage with five to six tentacles; buds of further tentacles visible. **F** Tentacle, calcareous ring, ring canal and stone canal in optic section. *an* Anus; *cr* calcareous ring; *go* gonopore; *in* intestine; *mo* mouth; *nr* nerve ring; *ov* ovary; *po* polian vesicle; *pu* pulsatile peritoneal bladder; *ri* ring canal; *st* stone canal; *te* tentacle. (**A** Marinelli 1960; **B, C** Kaestner 1963; **D–F** Menker 1970)

Plesiomorphies in the ground pattern

The primitive opening of the stone canal in the body wall is only rarely realized in the sea cucumbers; it then lies in the same radius as the gonopore. Usually, the stone canal hangs freely in the body cavity with a button-like extension (madreporite; Fig. 49F). This apomorphous state possibly evolved more than once (ERBER 1983).

The planktotrophic auricularia belongs in the ground pattern of the Holothuroida. In addition, it can be assigned as the most primitive larva of all Echinodermata living today. The auricularia evolved from the developmental stage of a dipleurula through extension of the circumferential ciliary band into two rostral and caudal tips each (Figs. 30, 49B). The interpretation of the larva, however, does not mean that the Holothuroida must form the sister group of all other Echinodermata (SMILEY 1988). Rather, an auricularia was taken over in the ground patterns of the Eleutherozoa, Cryptosyringida as well as the Echinozoa and continued in the stem lineage of the Holothuroida. The planktotrophic larvae of other taxa of the Echinodermata can be explained by repeated, independent evolution from the state of a dipleurula/auricularia – the bipinnaria of the Asteroida by fusion of the two rostral tips of the ciliary band as well as the pluteus larvae through the development of arms with skeletal rods in the Ophiuroida and Echinoida (p. 106).

A phylogenetic systematization of the Holothuroida does not yet exist. I will mention the monophylum Apoda (*Labidoplax, Leptosynapta, Rhabdomolgus*) as an example of the strong evolutive change within the entity. The ambulacral system is limited to a ring canal to supply the ten tentacles on the oral surface. Radial canals and ambulacral tube feet are secondarily absent, as is also the respiratory tree (Figs. 47F, 49D).

Stomochordata[9]

We now come to the Stomochordata as the sister group of the Echinodermata in which all remaining Deuterostomia are combined. Within the Stomochordata the taxa Rhabdopleura and Pharyngotremata (Cephalodiscida + Cyrtotreta) can be validated as adelphotaxa. Accordingly, the traditional "Pterobranchia" (*Rhabdopleura* and *Cephalodiscus* + *Atubaria*) must be eliminated as a paraphyletic collection from the system of the Deuterostomia.

Autapomorphies (Fig. 33 → 3)

- Division of the body into three sections: (1) head (protosome) with shield-shaped expansion of the ventral side; (2) collar (mesosome); (3) trunk (metasome) with tail.
- Stomochord.
  Protrusion of a rostral buccal diverticulum into the protosome.
- Colonial organism in coenecia (tubes, cases).

The clear division of the body into head, collar and trunk belongs with desirable certainty in the ground pattern of the entity; we postulate its evolution in the stem lineage of the Stomochordata. This concept, however, is subject to some limitations because we cannot make any certain statements about the body form of the trimeric stem species of the Deuterostomia. Even so, a pair of arms with tentacles - realized in *Rhabdopleura* - must have been taken over from the ground pattern of the Deuterostomia into the stem lineage of the Stomochordata. The arguments for a possible evolution of the stomochord in this stem lineage have been presented earlier (p. 100).

Rhabdopleura and *Cephalodiscus* are colonial organisms in tubes. Accordingly, the evolution of this form of life can be postulated as an apomorphy in the stem lineage of an entity Stomochordata. The change to solitary animals without cases followed in the stem lineage of the Cyrtotreta.

Rhabdopleura - Pharyngotremata

Rhabdopleura

Small marine organisms with zooid lengths of 1–2 mm.

Few species that the usual classification places in a single supraspecific taxon *Rhabdopleura*. Thus, the establishment of a further superordinated taxon Rhabdopleurida is not applicable.

---

[9] tax. nov.

Autapomorphies (Fig. 33 → 4)

- Unpaired gonad.
- Vertical dwelling tube of regular rings.
- Zooids of the colony linked life-long.
- Lecithotrophic larva.

The state of a filter apparatus in the form of two arms with tentacles is a plesiomorphy (Fig. 50 B, C). The lack of gill slits is also primitive.

In contrast, the unpaired gonads in ♂ and ♀ can be unambiguously judged as an autapomorphy of the taxon *Rhabdopleura*. In comparison with the Cephalodiscida, the following peculiarities may probably also be considered as derived characteristics.

1. Tube system of the colony (Fig. 50 A). Main tube on the substrate divided into sections by transverse walls from which the vertical dwelling

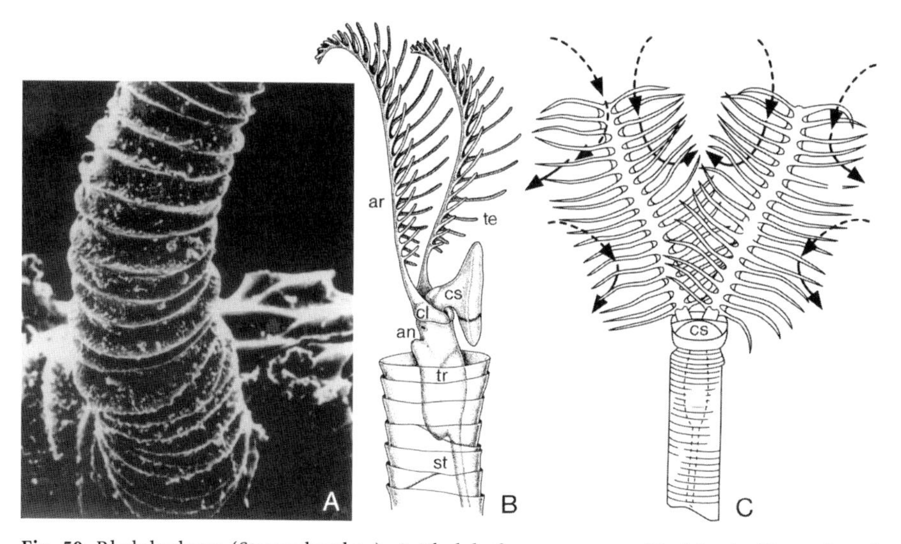

Fig. 50. Rhabdopleura (Stomochordata). **A** *Rhabdopleura compacta.* Upright dwelling tube of a zooid. Rings with regular protrusions on the outside. The dwelling tubes of a colony grow out of a horizontal main tube that is fixed to hard substrates. The main tube consists of diagonal pieces of rings (on the *right* of the picture). **B** Lateral view of a zooid of *Rhabdopleura* with unfolded arms. The two arms insert dorsally in the collar (mesosome); the two rows of tentacles of an arm are inclined to the ventral side, resulting in a V-shaped figure. In the biotope, the zooid anchors itself to the edge of the dwelling tube with its shield. **C** *Rhabdopleura normani.* Ventral view of a zooid in filter position. Lateral ciliary bands of the tentacles draw a stream of water through the filter apparatus; it enters frontal and is passed between the tentacles to the outside (abfrontal). *an* Anus; *ar* arm; *cl* collar; *cs* head shield; *st* stalk; *te* tentacle; *tr* trunk. (**A** Dilly 1976; **B** Benito & Pardos, in F.W. Harrison, E.E. Ruppert (eds.) (1997) Microscopic Anatomy of Invertebrates, Vol. 15: Hemichordata, Chaetognatha, and the Invertebrate Chordates, p. 17. Reprinted by permission of John Wiley & Sons, Inc.; **C** Halanych 1993)

tubes of the single zooids arise. Construction of the dwelling tubes from rings in regular sequence, the main tube of diagonally displaced pieces of rings.
2. The vegetatively formed zooids (budding) remain linked in the system of tubes by a tissue cord for their whole lives; this secretes a dark substance (black stolon).

Rhabdopleura has a biphasic life cycle. The simple lecithotrophic larvae with complete ciliation only live for a short time free in water (STEBBING 1970; DILLY 1973; LESTER 1988a,b). If a planktotrophic larva belongs in the ground pattern of the Deuterostomia (p. 94) and if the Cephalodiscida are the sister group of the Cyrtotreta (see below), then the lecithotrophic larvae of *Rhabdopleura* and *Cephalodiscus* must have developed convergently.

Pharyngotremata

I have taken the name Pharyngotremata over from SCHAEFFER (1987) for an assumed monophylum of the Deuterostomia with gill slits as a derived characteristic. With this measure the just described taxon *Rhabdopleura* must be excluded.

Autapomorphies (Fig. 33 → 5)

- Existence of gill slits.
- Filter apparatus consisting of several arms with tentacles[10].

The Cephalodiscida possess a pair of simple gill slits, the Cyrtotreta (Enteropneusta + Chordata) numerous U-shaped intestinal perforations. The hypothesis of a homology of the gill slits of both taxa and the evolution of a pair of simple slits in the stem lineage of an entity Pharyngotremata provides the most economical explanation of these facts.

The evolution of gill slits is possibly linked functionally with a reinforcement of the water stream in a complex filter basket consisting of several pairs of arms (GILMOUR 1978). According to this perspective, the evolutionary increase in the number of arms (realized in Cephalodiscida) from a primary state with two arms (*Rhabdopleura*) can also be postulated in the stem lineage of the Pharyngotremata. This means that the filter basket has been taken over by the Cephalodiscida as a plesiomorphy and completely reduced in the stem lineage of the Cyrtotreta.

[10] Male of *Cephalodiscus sibogae* with one pair of arms (HALANYCH 1996b).

With the interpretation of the Cephalodiscida and Cyrtotreta as sister groups the traditional taxon "Hemichordata" in the union of the "Pterobranchia" (p. 136) and the Enteropneusta disappears. The "Hemichordata" is to be deleted as a paraphylum.

Cephalodiscida - Cyrtotreta

Cephalodiscida

Autapomorphy (Fig. 33 → 6)

- Lecithotrophic larva.

Traditional classifications distinguish two taxa - the species-rich taxon *Cephalodiscus* and the species *Atubaria heterolopha*; our knowledge of the latter is based on studies of a single population in Japan[11].

With body lengths of up to 5 mm, the zooids of the Cephalodiscida are considerably larger than those of *Rhabdopleura* species.

The gill slits and a filter apparatus consisting of at most nine pairs of arms can, in comparison with *Rhabdopleura*, be evaluated without conflict as apomorphies (Fig. 51 C–E). Since we have presented grounds for their evolution in the stem lineage of the Pharyngotremata, they cannot be considered, however, as autapomorphies of the Cephalodiscida. On the other hand, the completely ciliated, lecithotrophic larva known for *Cephalodiscus* is possibly an autapomorphy of the Cephalodiscida (see above).

In *Cephalodiscus* the cases have many forms, from a simple (? primitive) coenecium with separate, parallel vertical tubes for individual zooids (*C. densus*) to irregular structures with coherent cavities for the zooids of the colony (*C. gracilis*, Fig. 51 A; VAN DER HORST 1939).

After extensive differentiation, zooids from vegetative reproduction (Fig. 51 B) detach themselves from the stalk of the mother animal and crawl independently about the colony. In comparison with *Rhabdopleura*, this seems to be a primitive behavior, to be set in the ground pattern of the Stomochordata.

Atubaria heterolopha is possibly of interest for phylogenetic considerations. No cases have been observed in the studied population; the animals are found as freely creeping individuals on hydroid polyps. Since we postulate the evolutive change to noncolonial organisms without tubes in the stem lineage of the Cyrtotreta, *Atubaria heterolopha* could well represent the adelphotaxon of the Cyrtotreta. However, the population structure and lifestyle of *A. heterolopha* must be studied in more detail before we can make further statements.

---

[11] "Atubaria is most likely a form of Cephalodiscus" (HALANYCH 1996b).

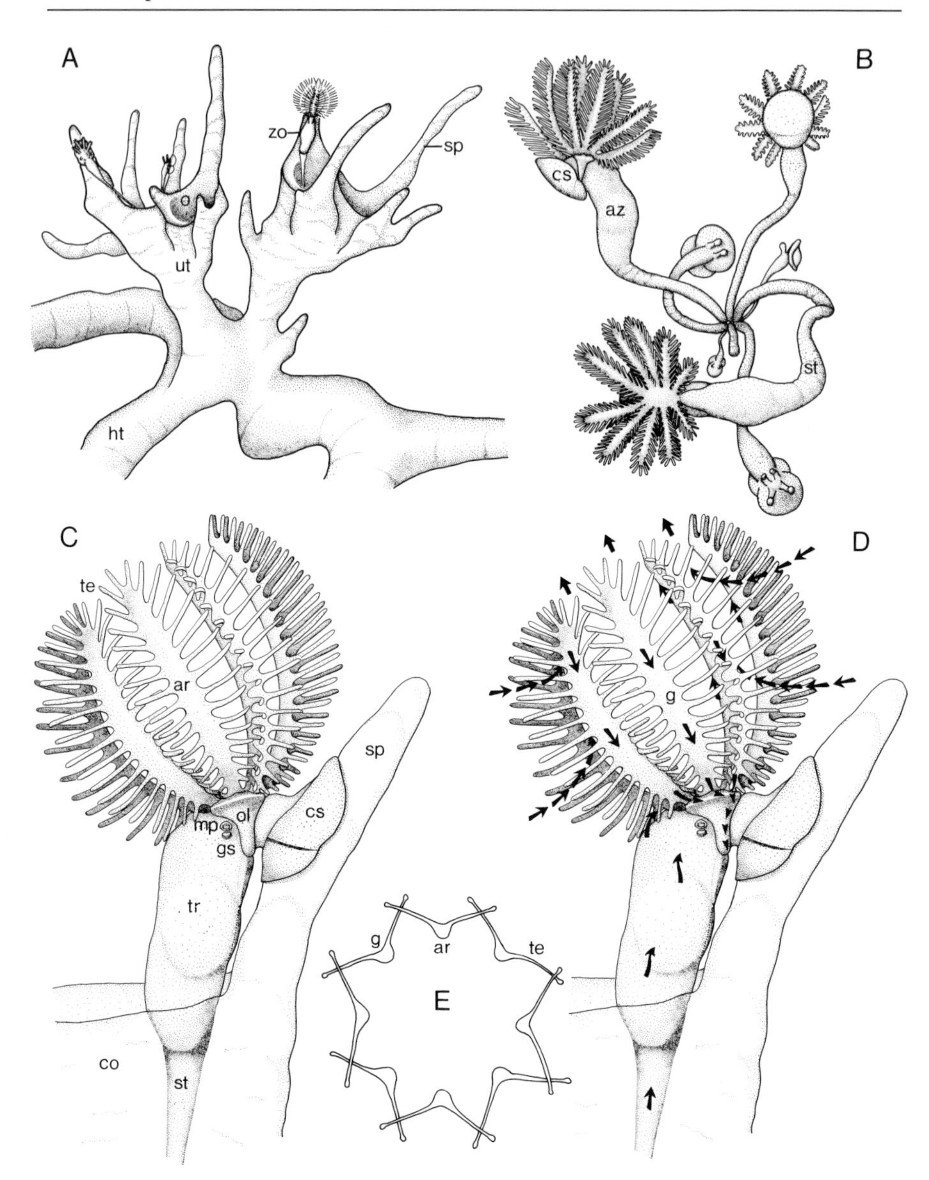
A
B
C
D
E
zo
sp
o
ut
ht
cs
az
st
te
ar
mp
ol
gs
tr
co
g

◀ **Fig. 51.** *Cephalodiscus gracilis* (Cephalodiscida). **A** Casing (coenecium) of horizontal tubes on the substrate and upright tubes with openings and spines. Three zooids are sitting on spines. **B** Colony of pedicled zooids in various stages of development. New zooids emerge near the end of the stalk of the colony founder that anchors the colony in the coenecium. **C** Zooid in filter position with the arms arranged to a basket. V-shaped figure of the individual arms with tentacles as in *Rhabdopleura*. Adhesion of zooids with the shield (protosome) to a spine of the shell. **D** Water flows. As a result of the evolutionary increase in number of arms in *Cephalodiscus* and their ring-like arrangement, the frontal sides of the arms (cf. *Rhabdopleura*) orient to the outer surface of the basket while the abfrontal sides become the inside covering. The lateral cilia of the tentacles therefore now draw water from the outside into the filter basket. The filtered water flows out over the tips of the arms. Retained particles are transported in the oral direction on the frontal sides of the arms as in *Rhabdopleura*. Cilia of the trunk and stalk (metasome) direct undesired particles and faeces into the backflow. **E** Cross section of the filter apparatus of an individual with eight arms. The tentacles mesh in the periphery; the transport grooves of the arms are on the outside. *ar* Arm; *az* adult zooid; *co* coenecium; *cs* head shield; *g* groove; *gs* gill slit; *ht* horizontal tube; *mp* pore of the mesocoel; *o* ostium; *ol* oral lamella; *sp* spine; *st* stalk; *te* tentacle; *tr* trunk; *ut* upright tube; *zo* zooid. (**A–D** Lester 1985; **E** Dilly 1985)

Cyrtotreta

Enteropneusta and Chordata were combined to a monophylum Cyrtotreta by NIELSEN (1995) on the basis of convincing arguments. The existence of a branchial gut with U-shaped gill slits forms a central synapomorphy.

In light of the hypothesis of an adelphotaxa relationship Enteropneusta – Chordata, the collection "Hemichordata" ("Pterobranchia" + Enteropneusta) must be deleted also from this side of the phylogenetic system of the Deuterostomia (p. 139).

Autapomorphies (Fig. 33 → 7)

- Branchial gut with numerous U-shaped slits.
  The construction of the branchial gut, or branchial region of the trunk, of the Enteropneusta is to be set in the ground pattern of the Cyrtotreta as the primary state. The essential elements are described below (Fig. 53).
  The gill slits develop at the beginning of the metasome from lateral intestinal pockets that perforate to the outside through small pores. In the stretched internal openings, tongues grow from above and transform them into U-shaped slits. Long cilia on the epithelia of the walls and tongues form the motor of a water flow directed outwards through which food particles covered in mucus by the acorn are sucked into the mouth and further to the intestine. The basal lamina of the epithelia thicken to a branchial skeleton with arches over the gill slits, unpaired rods in the septa between slits and paired braces in the tongues.
  With the evolutionary transformation to a filter apparatus, the branchial gut of the Chordata represents the apomorphous alternative. Here, a

ventral glandular organ (endostyle) produces flat mucous filters that are flicked up on the inner side of the branchial gut and filter fine particles from the inflowing water (p. 150).

- Vagile stem species.
  In the stem lineage of the Cyrtotreta, a change from colonial organism with cases (*Cephalodiscus*) to solitary inhabitants of soft sediments of the sea bottom occurred.
- Reduction of mesosomal tentacles.
  The filtration of fine particles out of free water disappeared with the abandonment of the tentacle apparatus of the sister group Cephalodiscida.
- Mucociliary mechanism for feeding.
  Covering of particulate food from the surface of the bottom or out of the sediment with mucous secretion of the protosome (acorn). Uptake with the water flow generated by the cilia of the branchial gut (see above).

Enteropneusta – Chordata

Enteropneusta

About 70 species of inhabitants of the sea bottom with uniform division of the body into three sections – acorn (proboscis), collar and trunk with large variations in length – *Saccoglossus pygmaeus* (North Sea) 2–3 cm, *Balanoglossus gigas* (Brazil) 2.5 m.

The Enteropneusta live in soft sediments and line their burrows with mucus. The irregularly wound tubes, as have been described in more detail for *Protoglossus koehleri* (BURDEN-JONES 1956), appear to be primitive. U-shaped burrows with entrance funnels and coiled faecal castings on the ground surface are highly differentiated; the colonization mode of *Balanoglossus clavigerus* is comparable to that of the lugworm *Arenicola marina* (Polychaeta).

The Enteropneusta are ciliary suspension feeders or deposit feeders (BARRINGTON 1965; BENITO & PARDOS 1997). The uptake of food via the mucociliary mechanism begins with movement of the acorn (protosome). Through alternating stretching and contraction, particles adhere to a mucus produced by dermal glands of the acorn. Epidermal cilia force the mucus-covered material backwards. The bands of mucus converge at the end of the acorn ventrally and pass through a ciliated groove with long cilia before the mouth – probably a chemoreceptor. A strong suction – generated by the cilia of the branchial gut – draws the mucus-covered food into the digesting section of the gut; the water itself flows from the phar-

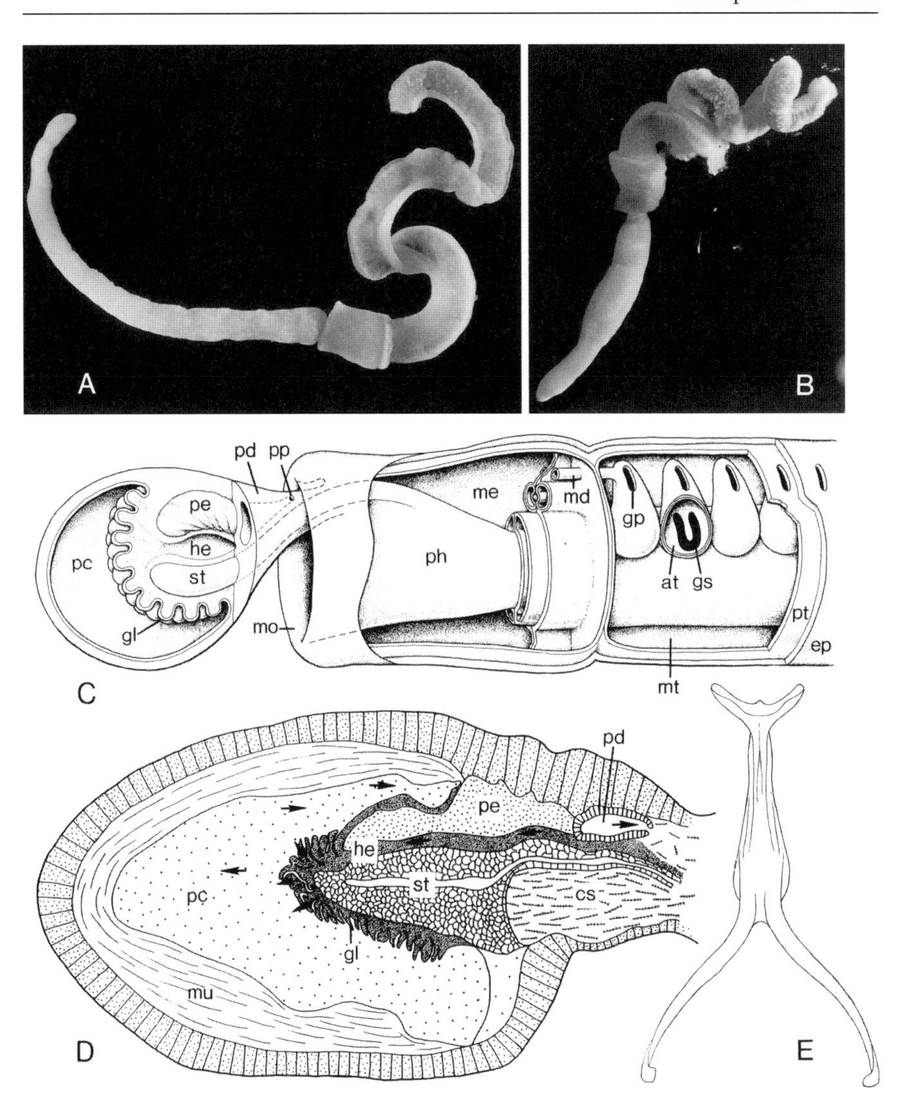

Fig. 52. Enteropneusta. **A, B** *Saccoglossus kowalevskii.* Photographs of a living individual with clear division in a long acorn (protostome), a short collar (mesosome) and the trunk (metasome) as main part of the body. **C** *Saccoglossus kowalevskii.* Diagram of the anterior end. Left body wall lifted up. **D** *Saccoglossus kowalevskii.* Sagittal section through the acorn. Terminally the excretory-active glomerulus lies on the stomochord. The fluid flow heart → glomerulus → protocoel → duct of the protocoel → exit through the pore of the protocoel is marked with *arrows.* **E** *Dolichoglossus caraibicus.* Acorn skeleton (ventral view). The unpaired anterior piece lies in the region of the neck-like tapering between acorn and collar under the stomochord; the two limbs enclose the pharynx (collar gut). *at* Atrium (gill pocket); *cs* acorn skeleton; *ep* epidermis; *gl* glomerulus; *gp* gill pore; *gs* gill slit; *he* heart; *md* duct of the mesocoel (opening in the atrium of the first gill organ); *me* mesocoel; *mt* metacoel; *mu* musculature; *pc* protocoel; *pd* duct of the protocoel; *pe* pericardium; *ph* pharynx; *pp* pore of the protocoel; *pt* peritoneum; *st* stomochord. (**A, B** Originals; E.E. Ruppert, Beaufort, North Carolina, USA. **C, D** Balser & Ruppert 1990; **E** van der Horst 1939)

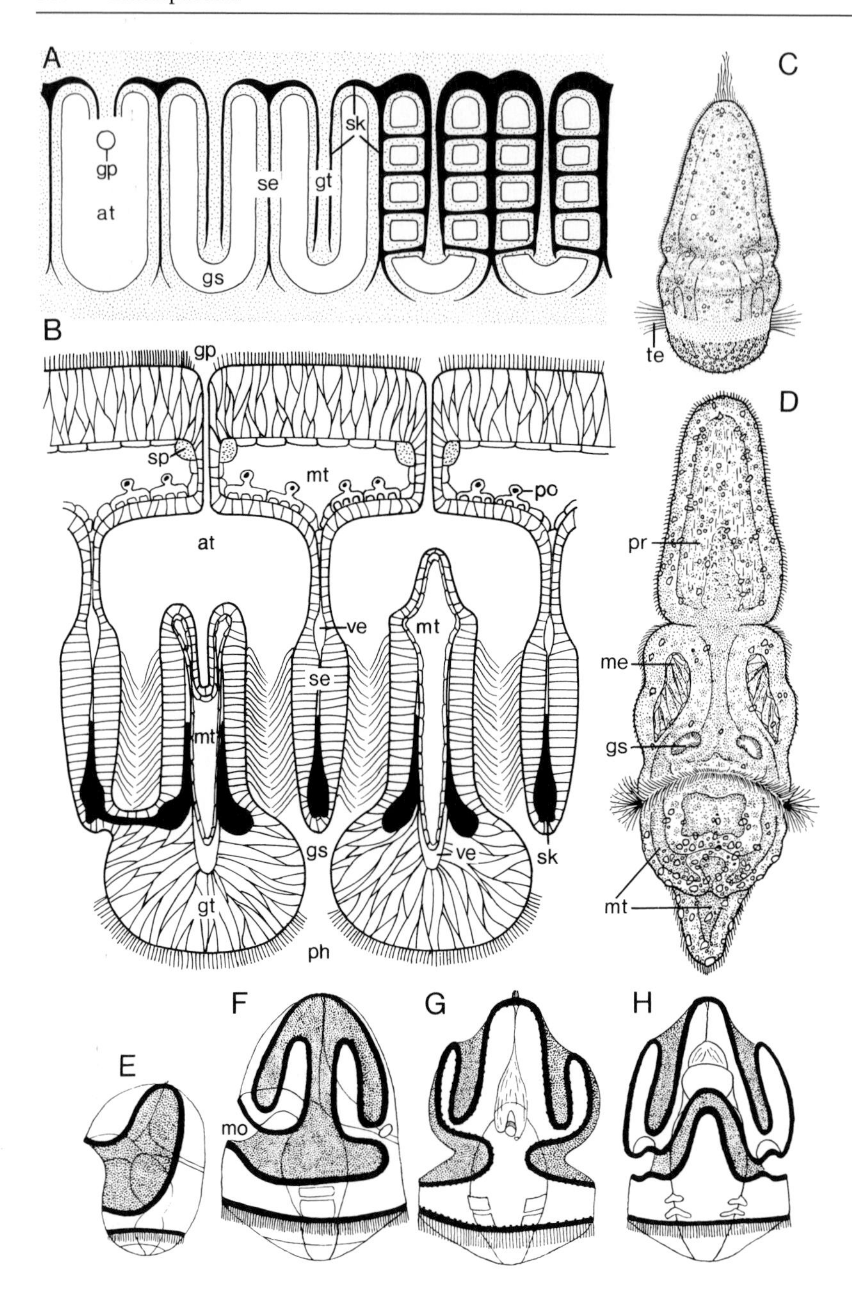
A
gp
at
gs
se
gt
sk
B
gp
sp
mt
po
at
ve
mt
se
mt
gs
ve
sk
gt
ph
C
te
D
pr
me
gs
mt
E
F
mo
G
H

◀ **Fig. 53.** Enteropneusta. **A** Scheme of the branchial gut, view of the inner side. The gill slits are separated by septa. In the middle two U-shaped slits as the result of dorsally ingrowing tongues. *Left* View in the gill pocket (atrium) and on the pore after removal of the tongue. *Right* Gill slits covered by transverse bars. Apomorphy within the Enteropneusta (Ptychoderidae). The gill skeleton as thickening of the basal lamina is drawn in *black*. **B** Parasagittal section through initial part of the metasome. The gill apparatus is cut transversely. Change from septum with one skeletal rod and tongue with two braces. Septum and tongue are ciliated in the region of the water passage to the atrium. **C, D** Lecithotrophic larvae of *Saccoglossus horsti*. **C** Young plankton larva (2 days). **D** Benthic, creeping larva at the age of 6 days. **E–H** Planktotrophic tornaria. Different developmental stages. **E** Young larva, lateral view. With simple ciliary band; telotroch just formed. **F–H** Later stages. Ciliary band with lobes and saddles. Side view (**F**), dorsal view (**G**), ventral view (**H**) with pre- and postorally running ciliary band. *at* Atrium (gill pocket); *gp* gill pore; *gs* gill slit; *gt* gill tongue; *me* mesocoel; *mo* mouth; *mt* metacoel; *ph* pharynx; *pr* protocoel; *se* septum; *sk* skeleton of the gill apparatus; *sp* sphincter; *te* telotroch; *ve* blood vessel. (**A** Gruner 1994; **B** Benito & Pardos, in F.W. Harrison, E.E. Ruppert (eds.) (1997) Microscopic Anatomy of Invertebrates, Vol. 15: Hemichordata, Chaetognatha, and the Invertebrate Chordates, p. 17. Reprinted by permission of John Wiley & Sons, Inc., modified; **C, D** Burden-Jones 1952; **E–H** Stiasny-Wijnhoff & Stiasny 1931)

yngeal lumen out through the gill slits. Otherwise, the branchial gut does not participate in the uptake of food or, at most, serves to trap finer particles (BARRINGTON 1965).

It is easy to delimit the Enteropneusta from *Rhabdopleura* and *Cephalodiscus* as well as from the Chordata by various "diagnostic" features. However, it is difficult to identify the Enteropneusta as a monophylum. The reason for this is simple. Much of what is discussed in traditional textbooks was evolved in earlier stages of the phylogenesis of Deuterostomia and is accordingly of a plesiomorphous nature for the Enteropneusta.

1. The division of the body into the three sections protosome, mesosome and metasome with three coelom sac sets, the basiepidermal nerve system, the excretory organ with glomerulus in the protocoel and the planktotrophic larva come from the ground pattern of the Deuterostomia.
2. The stomochord in the protosome is taken from the ground pattern of the Stomochordata. The long ongoing discussion of the pros and contras of a homology of the stomochord with the chorda dorsalis is not relevant in the present context. For the postulated adelphotaxa relationship between Enteropneusta and Chordata it is not important whether the stomochord entered the evolution of the chorda dorsalis as a "precursor" or was reduced in the stem lineage of the Chordata.
3. Ultimately, substantial features of the Enteropneusta originate from the ground pattern of the Cyrtotreta. The construction of the branchial gut and its function in the generation of the water flow to transport mucus-covered food has been discussed above (p. 142). The vagile stem species

of the Cyrtotreta could have been an organism with division of the body into acorn, collar and trunk – as is the case with the recent Enteropneusta.

Finally, there remains little that can be counted as the derived characteristics of the Enteropneusta with weak validation. The following features may possibly belong here.

Autapomorphies (Fig. 33 → 8)

- Subepidermal dorsal nerve cord (neurochord).
  In the collar (mesosome) the basiepidermal nerve system on the back side sinks to a subepidermal cord. The lack of comparable processes in *Rhabdopleura* and *Cephalodiscus* can be interpreted as plesiomorphy. On the other hand, the hypothesis of a homology between the neurochord of the Enteropneusta with the dorsal neural tube of the Chordata is rather unlikely. In Enteropneusta the sunken nerve plexus does not have a central nervous function; nerves neither enter nor exit from it (GOLDSCHMID 1996a). Thus, it is reasonable to hypothesize the evolution of the neurochord as a novelty in the stem lineage of the Enteropneusta.
- ? Supporting skeleton for the stomochord.
  The proboscis skeleton as an elastic plate appears to be a structure for supporting the stomochord (Fig. 52E). The anterior part lies underneath the buccal diverticulum; the arms encircle the intestine in the collar region. This supporting skeleton is absent in *Rhabdopleura* and *Cephalodiscus* and, with the lack of a stomochord, also naturally in the Chordata. The organ probably arose to support the peristaltic movements of the acorn on the change to a vagile lifestyle. This could have happened in the stem lineage of the Cyrtotreta, but possibly also later in the stem lineage of the Enteropneusta. An unequivocal evaluation is not possible.
- ? Tornaria with telotroch.
  In comparison with the dipleurula of the Echinodermata, a planktotrophic larva with telotroch is most certainly an apomorphous state (p. 96). The time point of its development is uncertain, however. The tornaria with telotroch was either taken over from the stem lineage of the Cyrtotreta (Enteropneusta + Chordata) or evolved first in the stem lineage of the Enteropneusta.

On the basis of molecular data, CAMERON et al. (2000) placed the "Pterobranchia" in the taxon Enteropneusta – namely, as the sister group of the subtaxon Harrimaniidae; the lifestyle as epibenthic microphages with tentacles should have arisen within the Enteropneusta. This notion is not compatible with the well-founded hypothesis of a homology between the tentacle apparatus of the "Pterobranchia" with that of the "Tentaculata" (p. 69).

Chordata

Three high-ranking entities of the Deuterostomia are combined in the taxon Chordata. These are the Tunicata, the Acrania and the Craniota - and they can all be validated as monophyla. We combine them on the basis of their supposed origin from a unique common stem species and mention three outstanding characteristics: (1) a dorsal neural tube of ectodermal origin, (2) a chorda dorsalis of entodermal genesis as well as (3) the differentiation of the branchial gut to a feeding system with "endless" mucous filter. We hypothesize the single evolution of these features and, in this way, justify the entity Chordata as a monophylum.

Furthermore, the pathway for the evolutionary genesis of the Chordata is even today the subject of extremely controversial discussions, as documented in recent publications (GEE 1996; SALVINI-PLAWEN 1998; NÜBLER-JUNG & ARENDT 1999; NIELSEN 1999). We stand somewhat apart since it is not the task of phylogenetic systematics to display evolutionary scenarios with the character of unprovable speculations. Rather, its job is to uncover kinship relationships and to define ground pattern features of monophyletic entities - in the present case, therefore, the Chordata, Tunicata, Acrania and Craniota. With this objective in mind, we will now concentrate on a few elemental considerations.

If the Enteropneusta with sedentary adult and pelagic larva can be discussed with good reasons as the sister group of the Chordata (p. 141), then it is reasonable to propose a comparable, biphasic life cycle in the ground pattern of the Chordata. Moreover, if it becomes possible, within the framework of this argument, to postulate a homology between the branchial gut of the Enteropneusta and the corresponding state of the foreguts in the Tunicata and Acrania, then the circumstances for the evolution of a new feeding apparatus for the latter can be specified. A branchial gut to trap particulate food from free water with a mucous filter can only have arisen in a organism with a fixed location. A sessile, epibenthic filterer with endostyle and peribranchial cavity evolved from a vagile particle feeder in sediment (Enteropneusta). In other words, we require an adult with a fixed location in the ground pattern of the Chordata, i.e., a life form as is realized in the "Ascidiacea" within the Tunicata (GARSTANG 1928; BERRILL 1955; HENNIG 1983).

When sessility belongs in the ground pattern of the Chordata, then the motility of the adult in free water is an apomorphous alternative; accordingly, it can form a synapomorphy of the Acrania and Craniota that are justifiable as adelphotaxa through a series of other characteristics (see below). This interpretation supports the current argumentation because it weakens the feasibility of the evolution of a branchial gut to a food filter in only a common stem lineage of the Tunicata and Acrania. This notion would demand the assumption that all apomorphous agreements between the Acrania and Craniota would have the character of convergencies – and this is most unlikely. The evolution of the feeding apparatus with an endless filter must have occurred in a stem lineage common to all Chordata.

For the genesis of the neural tube and the chorda dorsalis, we must again go back to the biphasic life cycle in the ground pattern of the Chordata. Naturally, a sessile adult does not provide the prerequisites for the evolution of neural tube and chorda. How one wants to create the connection to a specific larval form of the Enteropneusta is not relevant for our question. In every case only a larva can represent the phase of life in which the dorsal neural tube with a cerebral vesicle at the anterior end and the chorda dorsalis at the posterior end as stabilizer for the undulating tail evolved.

The evolution of the above-mentioned three essential autapomorphies in the stem lineage of the Chordata is spread over both phases of the life cycle – evolution of the new feeding apparatus in the benthic adult, evolution of chorda and neural tube in the pelagic larva.

Besides the sessile adult, the "Ascidiacea" have a freely moving larva with undulating tail, neural tube and chorda. However, it must be firmly emphasized at this point that they do not thus form the "basis group" of the Chordata. They are rather certainly Tunicata in which primitive organizational traits have been retained in both phases of the life cycle.

Nomenclature

In connection with the presented key statements, uniformity in the naming of the high-ranking subtaxa of the Chordata must be targeted. The synonyms for the Tunicata (Urochordata) and for the Acrania (Leptocardii, Cephalochordata) each refer respectively to one taxon and thus create only minor problems in communication.

However, in comparison with the Tunicata, we should also accept a well-known, unequivocal name for the sister group Acrania + Craniota. We can avoid new creations such as Holochordata (SALVINI-PLAWEN 1989, 1998) or Notochordata (NIELSEN 1995) when we turn back to the incorporation of the Acrania in the Vertebrata – as was done (LÖNNBERG et al. 1924) with the first description of the lancelet *Branchiostoma* (COSTA 1834) or *Amphioxus* (YARREL 1836) – and, together with HENNIG (1983) extend just this name to a monophylum of Acrania + Craniota. It is not relevant for this naming that the Acrania or the Myxinoida within the Craniota do not possess vertebrae.

However, there is a difference in the opinions held at that time. In contrast to the incorporation in the paraphylum "Pisces", today we interpret the Acrania as the adelphotaxon of all remaining Vertebrata – the entity Craniota.

Thus, in the phylogenetic system of the Chordata, we operate with the following names that will be used in our discussion.

Chordata
  Tunicata
  Vertebrata
    Acrania
    Craniota

Autapomorphies (Fig. 33 → 9)

- Biphasic life cycle with sessile adult and freely moving larva with an undulating tail.
  Adult.
- Branchial gut as feeding apparatus for taking up particulate food from free water. With endostyle for the production of an "endless" mucous filter. Inclusion in a peribranchial cavity or atrium.
  Larva.
- Ectodermal neural tube in its entire length under the dorsal side. With neuropore and neurenteric canal in the ontogenesis.
- Entodermal chorda dorsalis beneath the neural tube. Primarily in posterior body as stabilizer for movements of the tail.

The ontogenesis of the **tube-shaped central nervous system** is associated with two conspicuous phenomena at the anterior and posterior ends (Fig. 54). At the stage of the neurula a cell plate detaches from the ectoderm; it sinks down into the embryo. Two ectodermal neural folds form laterally and fuse with the sunken plate to give the neural tube. At the front the tube temporarily retains a dorsal pore to the outside; this is the neuropore. At the posterior end, the tube closes over the dorsally positioned blastopore. The peculiar, transient connection between the neural tube and the archenteron (primitive gut of embryo) is formed in this manner – the well-know neurenteric canal.

At the same time, the **chorda dorsalis**, as a fold protruding from the roof of the intestine, approaches the neural tube from below (Fig. 54). When we consider the genesis of the chorda in connection with the evolution of the larval undulating tail, its limitation to the posterior end of the

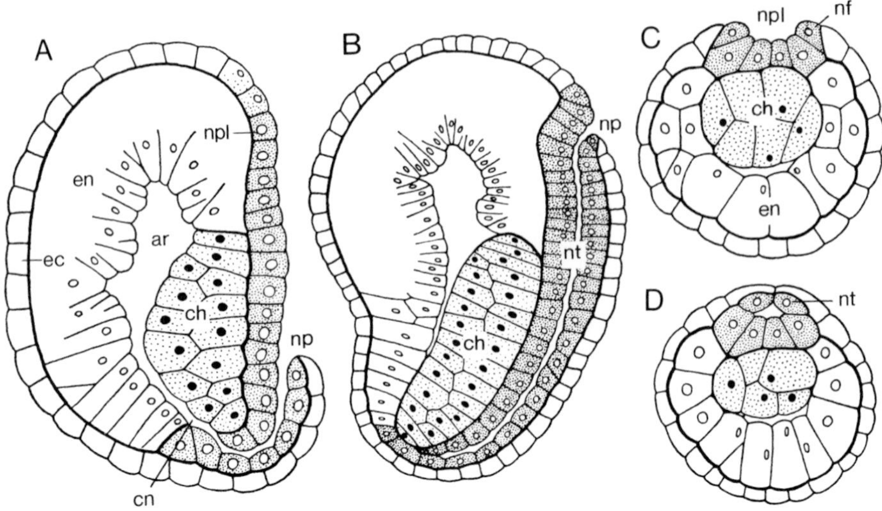

Fig. 54. Neural tube and chorda dorsalis as autapomorphies of the Chordata. Occurrence in the ontogenesis of *Clavelina lepadiformis* (Tunicata). **A** Early embryo (longitudinal section): Start of formation of the neural tube with neuropore and canalis neurentericus. **B** Later embryo (longitudinal section). Neural tube differentiated. Neuropore displaced to the anterior end; neurenteric canal closed. **C** Early embryo (cross section). Development of the neural tube with median neural plate and lateral neural folds. **D** Later embryo (cross section). Neural tube closed. Chorda dorsalis in both embryonal stages in the posterior body beneath the neural tube or its anlage. *ar* Archenteron; *ch* chorda dorsalis; *cn* canalis neurentericus; *nf* neural fold; *np* neural pore; *npl* neural plate; *nt* neural tube. (Siewing 1969)

larva is a plesiomorphous state in the Tunicata, its extension over the entire dorsal side (Acrania + Craniota) the apomorphous alternative.

We can now complete our discussion of the evolution of the **branchial gut** to a complex feeding apparatus. In the Enteropneusta, food particles covered with mucus on the outside of the body are sucked with the water stream of the branchial gut into the intestine. The "evolutive advance" of the Chordata consists in the use of suspended particles inside the branchial gut (Fig. 55). The ventral endostyle continuously produces two mucous filters that are transported to the left and to the right upwards on the inner sides of the gill slits. When water flows through these organic nets the suspended particles are filtered out, rolled up together with mucus on the roof of the branchial gut to a food cord and pushed into the oesophagus. The filtered water does not exit directly; it first passes into a peribranchial cavity or atrial chamber that surrounds the branchial gut as an ectodermal invagination inside the body.

The principle of the "endless mucous filter" (WERNER 1959) can only function when the fragile feeding apparatus is shielded against irritations from the surrounding environment. This is the task of the peribranchial cavity. For the above-described sessile stem species of the Chordata we ac-

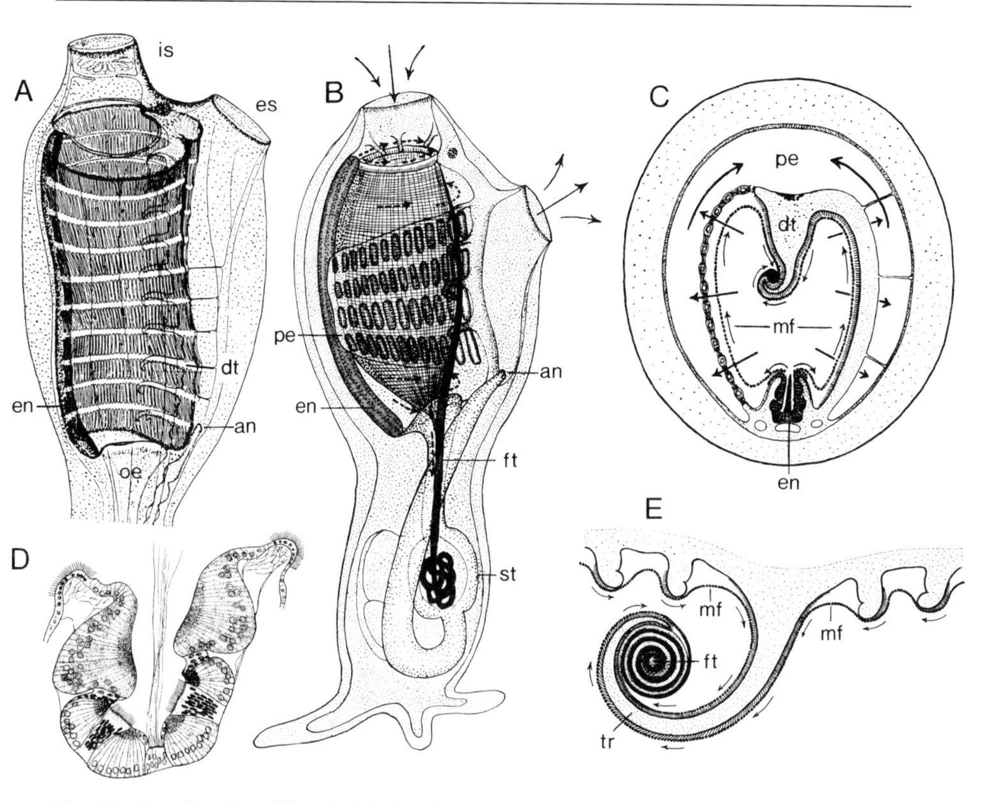

Fig. 55. *Clavelina lepadiformis* (Tunicata). Production of an endless mucous filter in the branchial gut. **A** Side view of anterior body. **B** Young animal with few gill slits, lateral view. **C** Cross section at the level of the branchial gut. **D** Endostyle (hypobranchial groove). Cross section. **E** Dorsal side with dorsal tongue. Cross section. Rolling up of the food cord. A flow of water is generated by the cilia on the gill slits. The water enters through an ingestion siphon, flows through the gill slits into the peribranchial cavity and is forced out again through an egestion siphon. Coarser particles are separated by the mouth tentacles and expelled again through the ingestion siphon by contractions of the branchial gut. The food particles entering the branchial gut are trapped by a mucous filter that covers the inside wall of the branchial gut and accordingly must be perfused by the water flow. The filter is created on the ventral side of the branchial gut from the endostyle (hypobranchial groove) – a stretched glandular organ that is divided into two mirror image halves by a median strip of long cilia. Ciliary rings of the branchial gut transport the mucous bands produced on both sides upwards. There they are taken up by the crescent-shaped tongues of the dorsal organ and rolled up to a thread. The food thread is then passed by ciliary beats into the oesophagus and is led further to the stomach. *an* Anus; *dt* dorsal tongue; *en* endostyle; *es* egestion siphon; *ft* food thread or cord; *is* ingestion siphon; *mf* mucous filter; *oe* oesophagus; *pe* peribranchial cavity; *st* stomach. (Werner & Werner 1954; Werner 1959)

cordingly postulate an atrium as protective device for the branchial gut. Continuation of the argumentation for the undisputed homology of the filter apparatus of the Tunicata and Acrania inevitably leads to the hypothesis of a single evolution of the functionally associated peribranchial cavity in the stem lineage of the Chordata. As BERRILL (1955, p. 196) stated "the tunicates and amphioxid atria represent two forms of one and the same basic structure." The differences in the ontogenesis with paired dorsolateral invaginations in the Tunica and that with a ventral folding in of the ectoderm in the Acrania must be interpreted as the result of evolutionary changes of a homologous organ.

Tunicata – Vertebrata

Tunicata (Urochordata)

The names Tunicata and Urochordata are commonly used as equivalents. I have chosen to use the name Tunicata because the tunica (mantle) as a coat outside the epidermis with a cellulose-like carbohydrate and mesenchymal cells possibly constitutes an autapomorphy of the entity. In contrast, the limitation of the chorda dorsalis to the posterior end of the larva is probably a ground pattern feature of the Chordata, i.e., a plesiomorphy in the character pattern of the Tunicata.

Thus, we come to the search for autapomorphies to validate the Tunicata as a monophylum. A sessile adult and a vagile larva with undulating tail already belong in the ground pattern of the Chordata and are thus not available as a plesiomorphy at this point. Arguments for possible, genuine features of the Tunicata are put together below.

Autapomorphies (Fig. 33 → 10)

- Tunica (mantle).
  Thick secretion of the epidermis from gel-like to cartilaginous consistence. With fibers of a cellulose-like carbohydrate (tunicin). Various free cells that migrate as connective tissue cells and blood cells from the body into the tunica.
  A tunica may already belong in the ground pattern of the Chordata, but there is no mandatory justification for such an assumption. On the other hand, the lack of tunicin and cells in the tegument of pelagic Doliolida and Appendicularia (see below) can be interpreted as a secondary state.
- Lack of a coelom.
  Since extensive coelom compartments exist in the Enteropneusta and the sister group Vertebrata, the lack of corresponding cavities in the Tunica-

ta must be an apomorphy. Only the pericardium ventral of the branchial gut is discussed as a derivative of the coelom.

- Reversal of heart beat.
  The heart of the Tunicata is formed by a folding in of the wall of the pericardium. The change in direction of blood transport by periodic reversal of the contraction waves of the musculature of the tubular heart is unique.
- Hermaphroditism.
  Gonochorism belongs in the ground pattern of the Enteropneusta as well as in the ground pattern of the sister group Vertebrata. The evaluation of hermaphroditism as an apomorphy seems to be unambiguous.

The Tunicata provide a new and impressive example for the deep gulf between traditional typological classifications and the objectives of phylogenetic systematics. In the usual classification of the tunicates, three taxa are placed side by side in an identical rank of classes – and this with clear differences and appropriate characteristics. (1) The **Ascidiacea** have a sessile adult and a pelagic larva with undulating tail; (2) the **Thaliacea** are freely swimming organisms at the level of an adult ascidian and (3) the **Appendicularia** present themselves as vagile, pelagic organisms in the form of sexually mature larvae of the Ascidiacea.

Why does phylogenetic systematics reject a division into these placatory, well circumscribed entities? On the one hand, a phylogenetic system of the Tunicata cannot allow three taxa with an identical rank. On the other hand, the conventional delineation of the taxa itself causes us problems – and this begins already with the sessile Ascidiacea.

"Ascidiacea"

"It may be assumed that the common ancestors of all Tunicata and even of all Chordata did not differ from the most primitive living members of the group 'Ascidiacea' in most features" (HENNIG 1983, p. 25).

The life cycle with sessile adult and vagile larva (Figs. 56, 57) is already a plesiomorphy in the ground pattern of the Tunicata and thus cannot be used for the justification of a system entity Ascidiacea. The evaluation of individual larval features such as the short lifetime of hours without feeding is controversial. There is nothing that can be identified unequivocally as an apomorphy of the sea squirts. In the interpretation of the Ascidiacea as a paraphylum there is even agreement between morphology and molecular findings. According to investigations on 18S rDNA, the ascidian taxon Phlebobranchiata (with *Ciona, Ascidia, Ascidiella*) is more closely related to the pelagic Thaliacea than to all the other "Ascidiacea" (SWALLA et al. 2000).

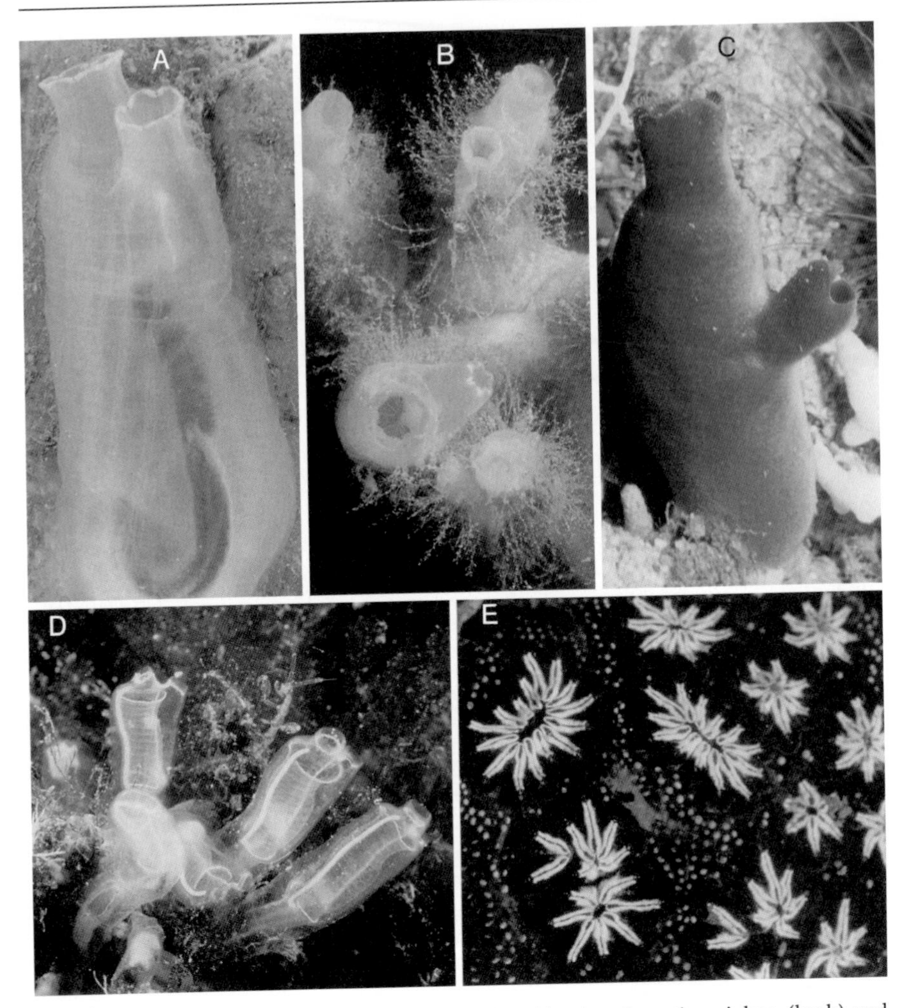

Fig. 56. Tunicata. "Ascidiacea". **A** *Ciona intestinalis*, side view. Ingestion siphon (back) and egestion siphon are close together. The U-shaped gut is full of food. **B** *Ciona intestinalis*. View looking down on several individuals growing close together. In one individual in the lower half of the picture mouth tentacles on the inside end of the ingestion siphon are visible. Thick growth of peritrich Ciliata. **C** *Halocynthia papillosa* with widely displaced egestion siphon. **D** *Clavelina lepadiformis*. Colonial group of glass-clear transparent individuals. The white lines are formed by pigment cells that scatter light. These are cells that are linked to blood canals of the branchial gut and the siphons (Burighel & Cloney 1997). **E** *Botryllus schlosseri*. Colonies touching each other in each of which several individuals are arranged star-shaped in a common, dark tunica. The egestion openings join to a central atrial pore. (**A, B, D, E** rocky tidal zone, Heligoland, North Sea; **C** Banyuls-sur-Mer, Mediterranean Sea. Originals)

Even so, we will concern ourselves with a few remarks about the primarily sessile sea squirts. With more than 1000 species, the spectrum of colonization areas ranges from rocks of the tidal zone (*Pyura*) to the deep sea; here, tiny organisms like *Dicarpa simplex* (Fig. 57 D) anchor themselves in soft sediment. Sea squirts have even taken up the task of locomotion in the substrate. Millimeter-small, vagile species from the taxa *Psammascidia, Dextrogaster, Psammostyela* or *Heterostigma* have conquered the mobile sand of the littoral zone (MONNIOT 1965). Colonies of *Diplosoma migrans* move themselves with tubular projections through the interstitial system of the "amphioxus sand" around Heligoland (Fig. 57 C). Perhaps they occasionally met *Branchiostoma lanceolatum* on their wanderings. However, even when the Tunicata in amphioxus sand certainly do not have anything in common with *Branchiostoma* except this habitat, the existence of vagile tunicates on the sea bottom rightly provokes the question about the routes of invasion of the pelagic zone by the "Ascidiacea".

Pelagotunicata

Did pelagic Tunicata develop only once or perhaps many times independently of sessile ancestors? The four monophyletic entities Pyrosomida, Salpida, Doliolida and Appendicularia are treated together without prejudice in "The biology of pelagic tunicates" (BONE 1998). At present, there are, however, diametrically opposite answers to our question. SWALLA et al. (2000) take up the traditional taxon Thaliacea for Pyrosomida, Salpida as well as Doliolida and interpret it as a monophylum within the "Ascidiacea". In contrast, GODEAUX (1998, p. 275) offers "evidence for thaliacean polyphyly."

With the postulate of a separate conquest of the pelagic zone, the Appendicularia are mostly widely separated. For SWALLA et al. (2000) they constitute the sister group of all other Tunicata. According to the interpretation of molecular findings by CHRISTEN & BRACONNOT (1998), they are not even tunicates. "Appendicularians and vertebrates form a very robust monophyletic unit, and the two *Oikopleura* spp. form a sister group to all vertebrates" (l.c., p. 268) – and this with exclusion of the Acrania (*Branchiostoma*).

HENNIG would turn in his grave; in 1983, he constituted a monophylum Pelagotunicata for the Pyrosomida, Salpida, Doliolida and Appendicularia, which implies the hypothesis of a single invasion of the pelagial in a stem lineage common to these four entities.

Of the following autapomorphies of the Pelagotunicata, at least the first three features form an evolutively coherent character complex.

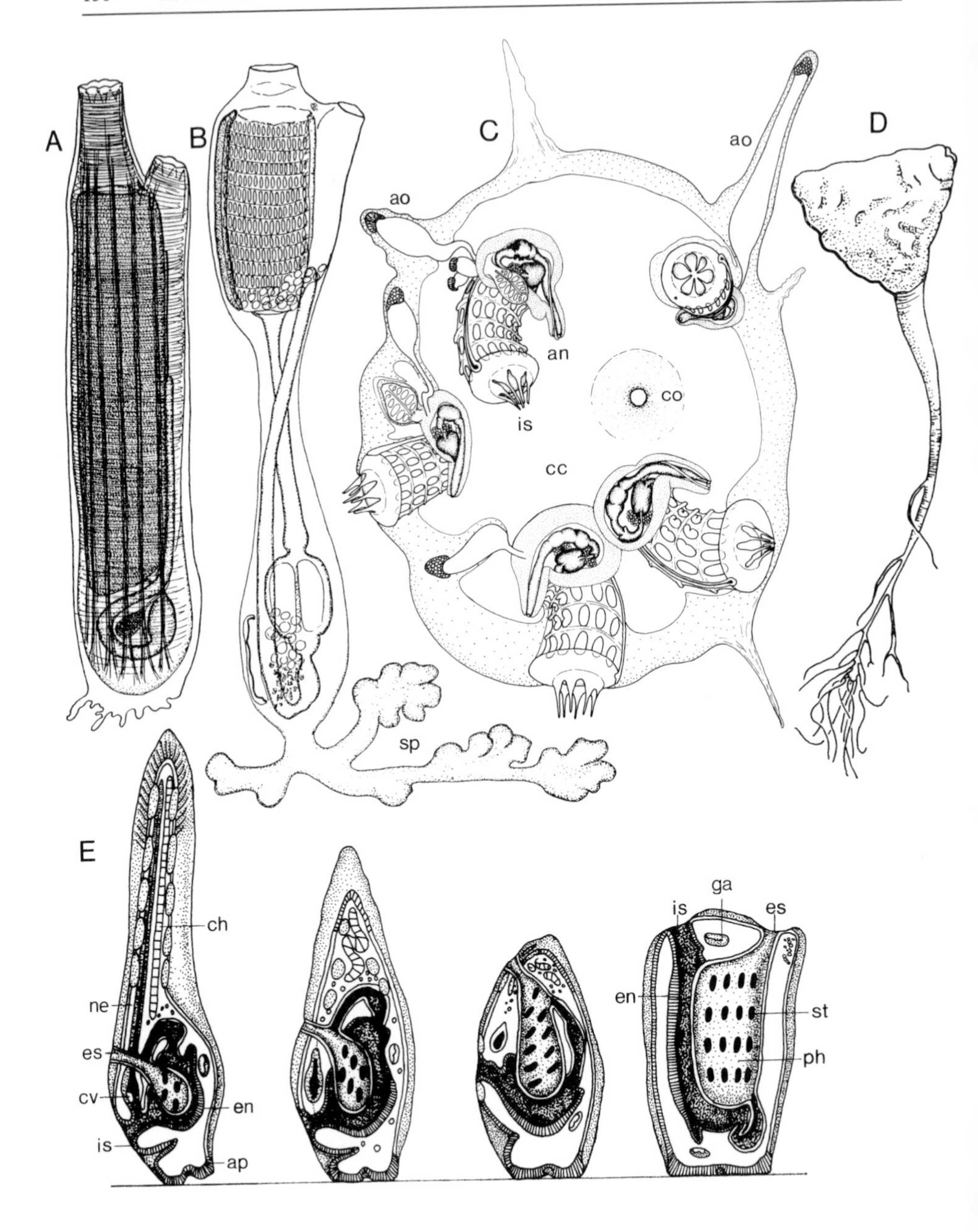
A
B
C
D
ao
ao
an
co
is
cc
sp
E
ch
ne
es
cv
en
is
ap
ga
is
es
en
st
ph

◀ **Fig. 57.** "Ascidiacea". **A** *Ciona intestinalis* with solitary living individuals. **B** *Clavelina lepadiformis* as example of a sessile colony. Zooid with extensive basal stolon from the buds of which new, freely protruding single animals emerge; they remain linked to the stolon. **C** *Diplosoma migrans*. Psammobiont, vagile colony of 1–3 mm diameter consisting of at most six individuals in a common tunica. Peripheral ingestion openings at arbitrary positions in the colony. Anal openings join to a common cloacal cavity that communicates with the surroundings through a pore. Three to five tube-like evaginations of the ectoderm per individual carry distal glandular caps; they form the locomotory and anchoring devices of the colony. **D** *Dicarpa simplex*. Deep-sea organism of 2 mm body length. Root appendix about 6 mm long. **E** *Clavelina*. Metamorphosis (from *left* to *right*). Anchoring of the swimming tail larva with adhesive papilla at the anterior end. Degeneration and absorption of the tail with nerve cord and chorda dorsalis. Rotation of the larval organs through 90° with orientation of the ingestion and egestion siphons to free water opposite the anchoring. *an* Anus; *ao* anchoring organ; *ap* adhesive papilla; *cc* common cloaca; *ch* chorda dorsalis; *co* cloacal opening; *cv* cerebral vesicle; *en* endostyle; *es* egestion siphon; *ga* ganglion of the adult; *is* ingestion siphon; *ne* nerve cord in tail; *ph* pharynx; *sp* stolon prolifer; *st* stigma (gill slit). (**A**, **B** Berril 1950; **C** Menker & Ax 1970; **D** Monniot 1965; **E** Gilbert & Raunio 1997)

Autapomorphies

- Existence in pelagic zone.
  At first, nothing more than a supposition for which, however, the next item will become a mandatory explanation.
- Displacement of the egestion opening to the posterior end.
  Ingestion opening (oral siphon) and egestion opening (atrial siphon) as outlet of the peribranchial cavity (atrial chamber) in opposition. The original function of the branchial gut as filter apparatus remains intact. The water flow is, however, used for a second, elemental task. The unidirectional flow through the body is now a motor for locomotion in free water. According to the principle of the most economical explanation, the single evolution of this structural change and functional extension of the peribranchial cavity is postulated.
- Lack of larval adhesive organs.
  With the abandonment of bonding to the substrate, the lack of adhesive devices at the anterior end of the larva can be interpreted as a secondary state.
- Metagenesis.
  Life cycle of two generations of individuals with differing reproduction. Obligate change of sexual reproduction and vegetative propagation. In this alternation of generations there are origin-related terms for the alternating individuals.
  Blastozooids: All asexually (vegetatively) created individuals. When they form gonads they become, as gonozooids, individuals of the sexual generation.
  Oozooids: All individuals arising sexually from egg cells of the gonozooids; in the ground pattern their development proceeds via a larva with tail.

The oozooids are individuals of the asexual generation. Vegetative buds that develop to blastozooids form on a stolon prolifer of the body.

Unfortunately, the kinship relations between the Pyrosomida, Salpida, Doliolida and Appendicularia have not yet been clarified satisfactorily. I will thus describe the four entities in succession and add two sections on possible phylogenetic connections.

1. Pyrosomida

Pyrosoma, Pyrosomella, Pyrostremma. In total, only eight species in warm waters, colonial organisms. Colonies from only a few centimeters in length (*Pyrosoma aherniosum, Pyrosoma ovatum*) to gigantic dimensions of more than 20 m (*Pyrostremma spinosum*; GODEAUX 1998; GODEAUX et al. 1988).

The cylindrical **colony** (Fig. 58 A) is closed in the oldest part; the common cloacal opening is at the other end. The blastozooids are embedded

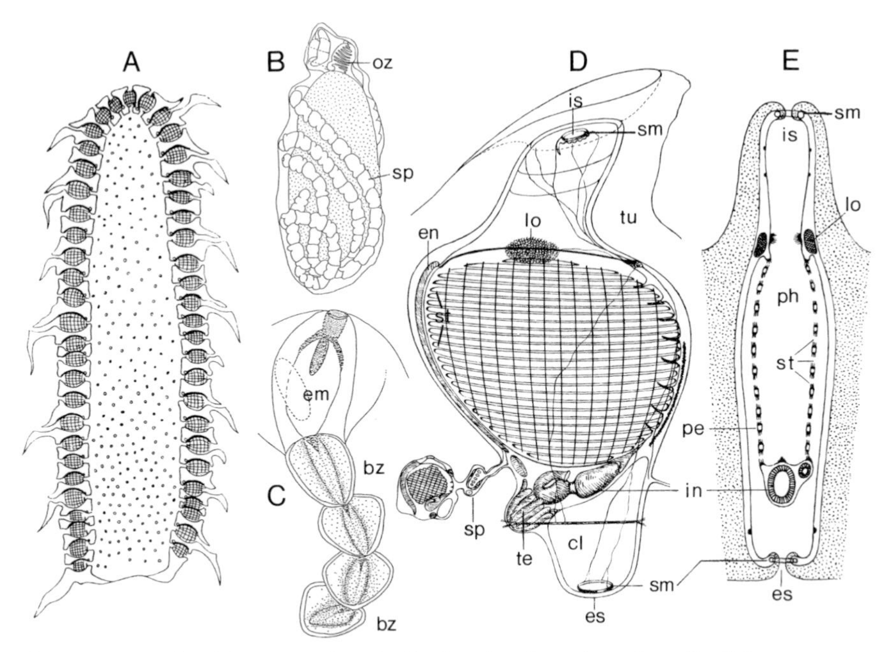

Fig. 58. Pyrosomida. **A** *Pyrosoma*. Longitudinal section through a colony. Parallel arrangement of the individuals in a common tunica. Cilia of the gill slits create a flow of water from the outside through the individual blastozooids into the common cloacal cavity. **B** *Pyrostremma agassizi*. Oozooid on a large drop of yolk that is entwined by the stolon prolifer. **C** *Pyrosoma atlanticum*. Stolon of the embryonal oozooid decomposes into four blastozooids. **D** *Pyrosoma triangulum*. Single blastozooid of the colony with stolon prolifer. **E** *Pyrosoma*. Sketch of a blastozooid in longitudinal section. *bz* Blastozooid; *cl* cloacal cavity; *em* embryo; *en* endostyle; *es* egestion siphon; *in* intestine; *is* ingestion siphon; *lo* light organ; *oz* oozooid; *pe* peribranchial cavity; *ph* pharynx; *sm* sphincter muscle; *sp* stolon prolifer; *st* stigma (gill slit); *te* testes; *tu* tunica. (**A** Brien 1948; **B, C** Godeaux et al. 1998; **D, E** Neumann 1933)

regularly side by side in the tunica – with their ingestion openings oriented towards free water, and egestion openings to a cloacal cavity inside the colony. Water flows through the individual blastozooids into the central cloacal cavity; on exiting the cloacal opening the flow drives the colony through the water.

In the **life cycle** of the Pyrosomida, the oozooid generation is only weakly developed; the larva is completely reduced.

In *Pyrostremma agassizi* (*P. vitjasi*), the oozooid sits on an enormous drop of yolk. The organs are only partially differentiated to functionality. A long stalk that entwines the yolk drop emerges from the posterior part of the endostyle (Fig. 58B). The stolon is degraded by constrictions into numerous buds from which the primary blastozooids of a future colony develop.

In comparison, the behavior of *Pyrosoma atlanticum* is more strongly derived. In the embryonal state, the oozooid already develops a stolon that divides itself into four primary blastozooids (Fig. 58C). While the oozooid degenerates, they fuse together as a "tetrazooid" colony to constitute the starting point for a new colony.

Through continuous vegetative propagation of the blastozooids, the colonies of the Pyrosomida grow to unions of countless individuals. The change to sexual reproduction with differentiation of the blastozooids to gonad-carrying gonozooids occurs only when the colony has reached a species-specific size.

Autapomorphies. Cylindrical colonies of blastozooids. Extensive regression of the oozooid generation. Lack of larva. Paired light organs lateral to the branchial gut.

Plesiomorphies. "Ascidian" construction of the branchial gut with numerous gill slits. A closing muscle on each of the ingestion and egestion openings. Water flow through blastozooids of the colony effected only by the cilia of the branchial gut.

On the basis of the primitive branchial gut the Pyrosomida may represent the sister group of all other Pelagotunicata.

? Taxon from Salpida and Doliolida + Appendicularia

In agreement, the Salpida and Doliolida possess several muscle circlets or bands in the periphery that encircle the body at regular intervals. Upon expansion they suck water into the ingestion opening; upon subsequent contraction the water is forced backwards. It is reasonable to hypothesize a homology of this musculature in the Doliolida and Salpida. They represent thereby two stages of the evolution.

The Doliolida continue to use a water flow generated by cilia of the gill slits for filtering food particles; the outflow from the egestion opening only drives the animal slowly forwards. Contacts to the mechanoreceptors of the flaps of the ingestion and egestion openings lead to abrupt contractions of the muscle bands that drive the individual rapidly forwards or backwards through the water.

In contrast, the Salpida have no cilia on their two large gill slits. Salps are active, agile swimmers with a continuous movement through contraction of the muscle circlets or bands. The water flow now serves a double function: transport of food particles as well as locomotion of the animal (BONE 1998b; MADIN & DEIBEL 1998). The simplest explanation for the outlined development of the musculature is the hypothesis of the single evolution in a common stem lineage for the Salpida and Doliolida + Appendicularia. This entity could be the sister group of the Pyrosomida. For the evaluation of the presumed neotenous Appendicularia, refer to p. 165.

2. Salpida (Desmomyaria)

Cyclosalpa, Helicosalpa, Thalia, Salpa, Ihlea. About 40 species. Worldwide distribution with emphasis in tropical and subtropical seas. The solitary oozooids usually measure a few centimeters. The oozooid of *Salpa maxima* reaches 15 cm; chains of blastozooids of this species may be several meters long (GODEAUX et al. 1998).

The most conspicuous feature of the barrel-shaped organism is certainly the **construction** of the **branchial gut.** A dorsal, oblique gill bar divides the body into two large cavities – the anterior region of the branchial gut and the adjacent cloacal cavity. The side walls dissolve into two enormous gill slits. The driving force of the water flow for feeding and locomotion is provided by the above-mentioned body musculature. There are rarely closed muscle rings, which may be a primary state; the muscles occur mostly in the form of ventrally open bands.

We now come to the **life cycle.** In the alternation of generations the oozooids and blastozooids are fully differentiated individuals with principally concordant organization. In comparison to the weakly developed oozooids of the Pyrosomida, this is an original state. However, the tail larva is secondarily absent also in the Salpida.

The Salpida are characterized by a strict change between solitary oozooids with vegetative propagation and colonial blastozooids with sexual reproduction. In the oozooid a stolon prolifer protrudes ventrally; this degrades into groups of buds from each of which a colony of blastozooids develops. The individuals have species-specific circular or linear arrangements and, accordingly, are differently linked together through stalks or adhesive plates.

The blastozooids start early, at millimeter sizes, with the production of sex cells; they change by definition to gonozooids. The eggs form first and are fertilized by sperm from older individuals. The growing embryos remain joined to the maternal gonozooid until they mature to young oozooids.

There are fundamental differences in shape and formation between the colonies of Salpida and Pyrosomida. The colonies of Pyrosomida grow through a continuous, asexual reproduction of the blastozooids; these arrange themselves around a cloacal cavity. In contrast, colonies of Salpida arise solely from buds of the oozooid (Fig. 59); in addition, there is no common cloacal cavity. This situation speaks in favor of an independent evolution of the colonial organization in Salpida and Pyrosomida; accordingly, a convergent reduction of the larva with tail must be assumed.

Autapomorphies. Only two large gill slits, separated by a medial bar from the wall of the branchial gut. Viviparous oozooid, with placental connection to the embryo. Colonial blastozooid generation. Lack of a larva.

Plesiomorphies. Oozooids and blastozooids each individuals without signs of regression.

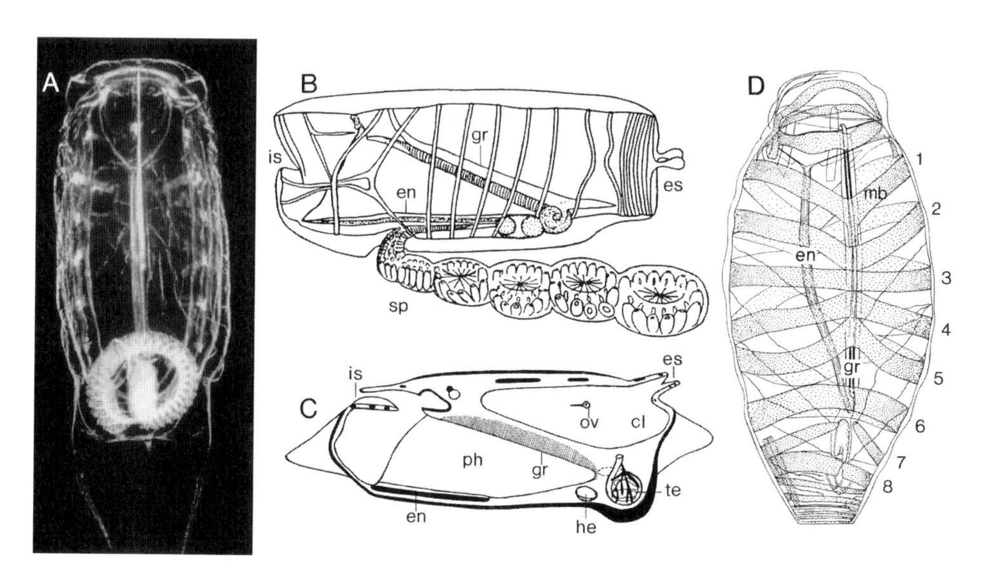

Fig. 59. Salpida. **A** *Thalia democratica.* Oozooid, Stolon prolifer emerges as a spiral from the body. **B** *Cyclosalpa affinis.* Oozooid. Stolon with arrangement of the blastozooids in wheels (*circles*); they detach themselves in this form as colonies. **C** Blastozooid of a salp colony in longitudinal section. **D** *Salpa racovitzai.* Oozooid in dorsal view. *cl* Cloacal cavity; *en* endostyle; *es* egestion siphon; *gr* gill rod; *he* heart; *is* ingestion siphon; *mb* muscle band; *ov* ovary; *ph* pharynx; *sp* stolon prolifer; *te* testes. (**A** Bone 1988a; **B** Berril 1950; **C** Neumann 1933; **D** Brien 1948)

? Taxon of Doliolida and Appendicularia

GARSTANG (1928) founded the hypothesis of a closer relationship of the Doliolida with the Appendicularia; in other words, in our terminology, the assumption of a sister group relationship between the two taxa.

In both the Doliolida and the Appendicularia, a mantle of cellulose-like tunicin with mesodermal cells is lacking. When judging the tunica as a ground pattern character of the Tunicata (p. 152), this is an apomorphy – and in the most economical interpretation – a synapomorphy of the Doliolida and Appendicularia. It would be important to know if the Doliolida share the composition of the cuticle from glycosylated proteins with the Appendicularia.

A further agreement is the periodic sloughing of the cuticle. In *Doliolum* and the Appendicularia, the cuticle is lifted up from the epidermis. In the former case, it is completely discarded; in the Appendicularia, it folds out to form the well-known filter house; but even this is abandoned after a few hours and replaced by a new cuticular house. The argumentation about a possible adelphotaxa relationship between the Doliolida and the Appendicularia has been recently reopened (HENNIG 1983; NIELSEN 1995); a new, precise analysis of GARSTANG's (1928) concepts seems to be urgently needed.

We must emphasize one consideration. If the Appendicularia should not be the sister group of the Doliolida and perhaps not even belong to the Pelagotunicata, then the Thaliacea would be a monophylum with the entities Pyrosomida, Salpida and Doliolida. In such a case, the name Pelagotunicata would be a synonym of Thaliacea; the autapomorphies of the Pelagotunicata listed above would accordingly be derived characteristics of the Thaliacea.

3. Doliolida (Cyclomyaria)

Doliolum, Dolioloides, Doliolina, Dolioletta. About 20 holoplanktonic species (GODEAUX 1998); predominately inhabitants of warm water.

The oozooids are on average between 5 and 15 mm long. The barrel-like body is tapered at the front and back, in the ground pattern encircled by nine muscle rings. A statocyst lies on the left side of the body; it is absent in other stages of the life cycle.

The body is divided into an anterior pharyngeal region and a posterior cloacal region – namely, by a thin, oblique septum that is perforated by two rows of four ciliated gill slits. The cilia generate the above-mentioned water flow in the pharynx (p. 157).

We come to the **alternation of generations** of the Doliolida – one of the adventure stories of zoology. "The complex cycle... is found in all species and sets doliolids apart from all other animals." (GODEAUX et al. 1988, p. 10).

Larva and oozooid

In free water, a larva with tail hatches from the fertilized egg cell. In the tail there is a chorda that is flanked on both sides by three rows of cross-striated muscles. In comparison with the ground pattern of tunicate larva, the Doliolida lack the dorsal neural tube. In the course of development, the anterior part of the larva gradually takes on the form of the adult zooid. The barrel is encircled by muscle bands; the tail with chorda degenerates.

The oozooid forms the stolon prolifer on the ventral side and at the posterior end a long dorsal appendage. Most of the inner organs disappear, i.e., the gills, the endostyle and the gut.

Blastozooids

In the life cycle, three generations of blastozooids now occur. According to the differing tasks, they are named as trophozooids (gastrozooids), phorozooids and gonozooids (Fig. 60 B, C).

The story begins with the production of a series of buds on the stolon prolifer of the oozooid. The buds migrate on the surface of the left body side to the base of the dorsal appendage. On the upper side of the appendage, they arrange themselves in three rows and are fixed. The buds of the lateral rows differentiate to the individuals of the first generation – the trophozooids. According to their destiny as suppliers of food for the colony, they have a highly enlarged gill apparatus; the process of transfer of digested food to the dorsal appendage has not been analyzed in detail. The buds of the central row grow to phorozooids as organisms of the second generation; they attach themselves with a short stalk. The phorozooids are the carriers of the third blastozooid generation. Their stalks carry a small "prebud"; from the division of which the buds that mature to the sexual gonozooids originate, they then detach themselves from the dorsal appendage. The eggs are fertilized with the sperm of younger gonozooids inside the body and are then released through the cloaca into the sea. A new life cycle begins with the development of larvae with tails.

Smaller differences between the individuals of various generations are mentioned in conclusion. The oozooids of all known species have nine muscle bands, all barrel-shaped blastozooid stages possess eight muscle bands. The oozooids each have four gill slits, the blastozooids five and more slits.

Autapomorphies. Oozooids with statocyst. Life cycle with three generations of blastozooids. Larva without neural tube.

Plesiomorphies. Existence of a larva with tail (lack of neural tubes as apomorphy). Life cycle without a generation of permanent colonial organisms. The individuals of all three blastozooid generations detach themselves from the dorsal appendage of the oozooid.

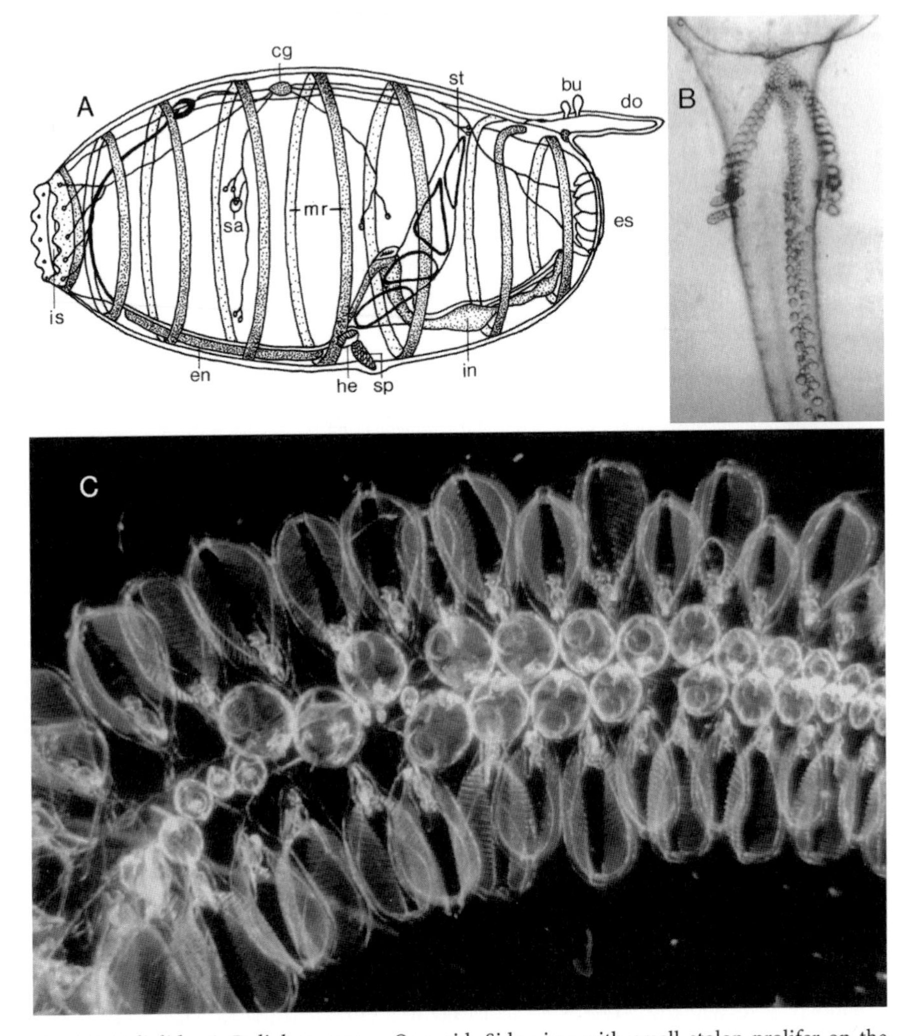

Fig. 60. Doliolida. **A** *Doliolum rarum*. Oozooid. Side view with small stolon prolifer on the ventral side and long dorsal appendage at the posterior end. **B** *Dolioletta gegenbaueri*. Base of the dorsal appendage from above. With three rows of buds. **C** *Dolioletta gegenbaueri*. Buds on the dorsal appendix have grown to phorozooids (median row) and trophozooids (two lateral rows). The trophozooids reach a length of 4–5 mm. *bu* Bud; *cg* cerebral ganglion; *do* dorsal appendage; *en* endostyle; *es* egestion siphon; *he* heart; *in* intestine; *is* ingestion siphon; *mr* muscle ring; *sa* statocyst; *sp* stolon prolifer; *st* stigma (gill slit). (**A** Brien 1948; **B** Godeaux et al. 1998; **C** Madin & Deibel 1998)

4. Appendicularia (Larvacea)

Oikopleura, Appendicularia, Fritillaria, Kowalewskia. With ca. 70 species in all oceans, especially in warm waters (FENAUD 1998a, c).

Division in an oval to stretched trunk of a few millimeters in length and a flat, muscular tail that can be several times the length of the trunk. Inhabitants of an extensive, gelatinous filter house that is secreted from cells of a specialized part of the epidermis, the oikoplast epithelium (Fig. 61).

We start with a cytological character – the cell constancy of the Appendicularia. The number of somatic cells is fixed in the development. Accord-

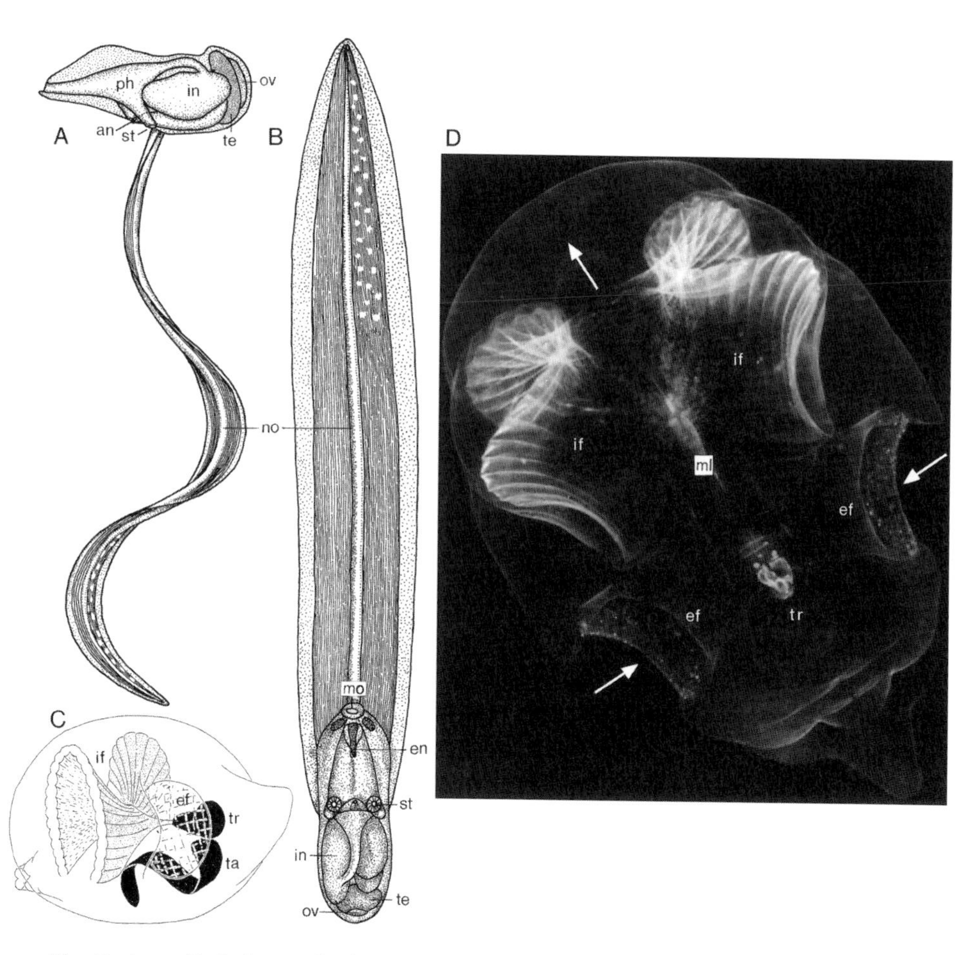

Fig. 61. Appendicularia. **A** *Oikopleura albicans.* Lateral view. Mouth opening *left.* Attachment of the tail ventral under the trunk. **B** *Oikopleura albicans.* Dorsal view. Mouth directed upwards; tail with orientation forwards under the body. **C** *Oikopleura labradoriensis.* House with occupant individual. One lateral inflow opening visible, the two wings of the food concentration trap and their median canal with connection to the animal's mouth. **D** *Oikopleura labradoriensis.* House after staining with carmine particles and ink from *Sepia officinalis. an* Anus; *ef* inflow filter; *en* endostyle; *if* inner filter of the food concentration trap; *in* intestine; *ml* midline between the wings of the trap; *mo* mouth opening; *no* notochord (chorda dorsalis); *ov* ovary; *ph* pharynx; *st* stigma (gill slit); *ta* tail; *te* testis; *tr* trunk. (**A, B** Alldredge 1976; **C** Flood & Deibel 1998; **D** Flood 1991)

ingly, growth proceeds only through an enlargement of the soma cells and multiplication of the germ cells.

The **trunk** has an extensive pharynx cavity with ventral endostyle (secondary lack in *Kowalewskia*) and a pair of lateral, ciliated gill slits. Similar to the other tunicates the endostyle produces a mucous filter. The cilia of the slits generate the water circulation in the pharynx. The anus opens ventrally between the gill slits.

The Appendicularia are protandrous hermaphrodites – with one strange exception. *Oikopleura dioica* that also lives off the coasts of the North and Baltic Seas is a dioecious organism, and here this is a secondary state. The gonads lie in the posterior part of the trunk. Sperm are released through sperm ducts of the testes. Egg cells can only become free through rupture of the ovary and the trunk wall; this leads to the death of the mature individual. The anterodorsal brain is swollen to a vesicle; it carries a statocyst.

The **tail** is flat like a knife blade; it inserts ventral to the trunk and lies with the right, broad side on the belly. The dorsal nerve cord lies left of the chorda which can be explained by a corresponding rotation through 90° in the development. The ten muscle cells developed above and below are lateral muscles morphologically, the outgrowths of the epithelium are morphological dorsal and ventral fins. In the recent review by FENAUD (1998 a, b), this process is surprisingly not mentioned, instead a dorsoventral flattening of the tail is discussed. The most conspicuous event in the development seems to be the rapid change in the orientation of the tail. From a position in the horizontal extension of the trunk, the tail suddenly turns under the body and then lies with its tip pointing towards the front.

The Tunicata are rich in unusual phenomena. The **house** of the Appendicularia is a unique structure among animals, as is the just discussed alternation of generations of the Doliolida. Oikoplasts as house-builders cover the anterior part of the trunk. The oikoplast epithelium secretes a house – a new external filter apparatus that is placed before the old, but still functioning mucous filter in the pharynx. The animal in the house pumps water from the surroundings through a complicated particle trap, sucks the water enriched with food in its pharynx, and filters it here a second time.

There exists not only one house shape, rather the houses of *Oikopleura*. *Fritillaria* and *Bathochordaeus* are constructed according to different architectural and functional principles (FLOOD & DEIBEL 1998). On the basis of the concordant secretion from oikoplast epithelium, a homology may be postulated and, therefore, a single evolution of the external filter apparatus in the stem lineage of the Appendicularia assumed.

The house of *Oikopleura* is "most likely composed of proteins heavily glycosylated by galactose, glucose and mannose" (FLOOD & DEIBEL 1998,

p. 119). It is the most thoroughly analyzed house and should thus provide the basis for the following survey (Fig. 61 C, D). Water flows into the house through two lateral openings; they are covered with cross-wise arranged fibrilla which hold back larger plankter that are unsuitable as food. Two large funnels lead from the inflow openings to the two median living chambers of the animal. The body is suspended in the trunk chamber; the forwards adjacent tail chamber is blindly closed at the front. The tail makes undulating movements in regular periods. As a tail pump it sucks water in through the inflow openings and forces it through two lateral passages (supply passages) to the edges of a complex food concentration trap. The massive structure lies transversely in the anterior part of the house, namely in the form of two bent back and multiply channeled wings that join in a median canal. The food concentration filter consists of three layers, an upper and a lower filter membrane as well as a middle mesh web of joining filaments. The water enters from outside between the two filter membranes in the trap and flows through the median canal further to an anterior outlet chamber. When the hydrostatic pressure here increases, a water jet is expelled outwards through a preformed opening and drives the house a short distance through the water in the opposite direction. In the food concentration trap the tangential flow of water over the filter leads to an accumulation of trapped particles in the median canal between the wings. From here, the occupant animal sucks the particles at short intervals into its pharynx – the cilia of the gill slits serve as motors. Finally, the mucous filter of the endostyle comes into operation and traps the particulate food once again; the filtered water passes through the two gill slits in the tail chamber.

We conclude with the last astonishing phenomenon. Although the house is highly complicated and certainly produced with a high expenditure of energy, it does not remain a single accomplishment in the animal's life. The animal can break through the back of its house and, after its flight into free water, can begin immediately with the secretion and spreading of a new house. In this way, a new, functional filter house is formed within a few hours. *Oikopleura* species build – depending on the temperature and the supply of food – 4–16 houses in 24 h (FENAUD 1986).

Autapomorphies. Abandonment of the primary adult phase in the biphasic life cycle of the Tunicata. Adoption of the gonads and the internal pharyngeal filter system in the pelagic larva that is now the adult form. Formation of a gelatinous filter house with a new filter apparatus in front of the existing system. The tail is the pump for generating water flow through the house; in connection with this new function a displacement of the flat tail under the trunk and a rotation of 90° to the left occur (? Convergent to comparable phenomena in the "Ascidiacea").

Plesiomorphy. In the case of an adelphotaxa relationship with the Doliolida, the existence of a nerve cord in the tail is primitive, the lack in the larva with tail of the Doliolida a derived state.

Vertebrata

Acrania and Craniota form adelphotaxa; they are combined under the name Vertebrata. Vertebrata and Tunicata are the sister groups of the Chordata at the next higher hierarchical level.

We repeat these hypotheses because we must come back to the biphasic life cycle in the ground pattern of the Chordata for the interpretation of the phylogenesis of the Vertebrata. With a sessile adult and vagile larva with undulating tail, this cycle is completely taken into the stem lineage of the Tunicata. In contrast, the Vertebrata lack a firmly attached adult. This is a situation that demands the following elemental statement. An essential process in the stem lineage of the Vertebrata consisted in evolutionary changes of the pelagic larva with tail under abandonment of an adult adhered to the ground. In the formulation of BERRILL (1955, p. 125), it has come to a combination of coexisting feature patterns of larva and adult in a novel swimming organism. Moreover, in this organism numerous evolutionary novelties developed as autapomorphies of the Vertebrata.

Autapomorphies (Fig. 33 → 11)

- Extension of the chorda dorsalis from the tail of the larva to the back of the entire body.
- Polymery under incorporation of the chorda dorsalis in a metameric musculature. Division of the body into numerous segments that, in their entirety, are probably homologous with the metacoel of the Enteropneusta. Ontogenetic formation of segments (metamers) through enterocoely as dorsolateral constrictions of intestinal diverticula. Breakdown of the single metamers to dorsal epimers with myocoel and ventral hypomers with splanchnocoel (Fig. 64). Development of musculature in the epimers; they change to myomers in which the original segmentation remains intact due to separation by myosepta. Fusion of hypomers with dissolution of dissepiments and ventral mesenteries to a uniform coelom cavity that encompasses the gut.
- Adoption of the gill apparatus with gill slits and endostyle that continues to produce an endless mucous filter, as well as the protective peribranchial cavity in the larval organization.
- Corresponding displacement of the gonads. The larva of the biphasic life cycle in the stem lineage of the Vertebrata is remodelled, in other words, to an adult.

- Formation of dorsal spinal nerves per segment. In the Acrania only dorsal nerves grow from the neural tube to supply the trunk musculature. Only these can be set in the ground pattern of the Vertebrata.
- Extensive agreements between the circulatory systems of the Acrania and Craniota. For the pattern of the common stem species, we emphasize the following major vessels that are realized in the Acrania (Fig. 63F).

  From the sinus venosus blood flows through the ventral aorta or endostyle artery under the branchial gut to the front. The two gill arch vessels rising up on both sides flow into two longitudinal stems at the top. These are the dorsal roots of the two aortae that join together behind the branchial gut to the aorta descendens (aorta caudalis). From the posterior body the subintestinal vein transports blood back to the sinus venosus. Furthermore, the hepatic vein and the paired cardinal veins of the body sides enter here. In the ground pattern of the Vertebrata, the vessels of the circulatory system did not have an epithelial cover; there was no central heart and no haemoglobin in the blood. This first evolutionary stage is documented in the organization of the Acrania.
- An unpaired fin on the back and tail as well as two ventrolateral metapleural folds (HENNIG 1983).

According to the presented hypothesis, the Vertebrata have a separate stem lineage besides the Tunicata. In other words, they have no phylogenetic relations to any subtaxon of the Tunicata.

This statement demands an exposition with the Appendicularia that are often taken for comparisons. Both here and there, a reduction of the primary adult and the adoption of gonads in the larva are postulated. On the other side, there are fundamental differences. In the Appendicularia, a pair of gill slits of the filter apparatus appear in the larva; numerous gill slits belong in the ground pattern of the Vertebrata as in *Branchiostoma* (Fig. 62). In the Appendicularia, the tail is rotated by 90°, lies flat beneath the trunk and serves as a pump for the water flow through the housing. Just this house is absent in the Vertebrata and nothing in the structure of the (primarily monociliated) epidermis or construction of the posterior body of the Acrania indicates that there could ever have been a house in the stem lineage of the Vertebrata.

The formal agreement in the lack of a sessile adult must be identified as the result of convergence in the stem lineages of the Vertebrata and Appendicularia.

Systematization of the Vertebrata · First Section

We will approach the phylogenetic system of the Vertebrata in two steps. The first section covers the primary aquatic vertebrates, the second section the Tetrapoda (p. 222). The hierarchical tabulation of the two sections will be supplemented in the running text by complementary diagrams of the phylogenetic relationships.

Vertebrata
- **Acrania**
- **Craniota**
  - **Cyclostomata**
    - **Petromyzonta**
    - **Myxinoida**
  - **Gnathostomata**
    - **Chondrichthyes**
      - **Elasmobranchii**
      - **Holocephali**
    - **Osteognathostomata**
      - **Actinopterygii**
        - **Cladistia**
        - **Actinopteri**
          - **Chondrostei**
          - **Neopterygii**
            - **Ginglymodi**
            - **Halecostomi**
              - **Halecomorphi**
              - **Teleostei**
      - **Sarcopterygii**
        - **Actinistia**
        - **Choanata**
          - **Dipnoi**
          - **Tetrapoda**

Acrania – Craniota

Acrania

About 30 species of a few centimeters in length. Conventional division into two supraspecific taxa – *Branchiostoma* (syn. *Amphioxus*) with paired gonads and *Epigonichthyes* (syn. *Asymmetron*) with gonads only on the right side of the body (RUPPERT 1997b).

Marine organisms with adults bound to soft bottoms and freely swimming young stages (the word larva is to be avoided, see below).

Sand and gravel bottoms near the coastline form the characteristic biotopes of the lancelets. The corresponding sediments around Heligoland are well known as "amphioxus sand". Lancelets usually stand obliquely in sediment, with the belly pointing upwards. The anterior end with a crown of cirri around the entrance to the mouth is lifted out in free water for the inflow of food particles. When the animal is washed out of the sand, it reinserts its body in the sediment by means of lateral, undulating movements. When colonizing muddy grounds, the lancelet lies with one side on the substrate (WEBB & HILL 1958).

The value of the Acrania for phylogenetic research can be illustrated by a methodologically oriented comparison with the well-known Onychophora. *Branchiostoma* provides for the phylogenetic systematization of the Vertebrata about the same information as *Peripatus* does for the Arthropoda (Vol. II, p. 77). In both cases, the character pattern presents three sets of characteristics or features with differing chronological origins. These are (1) the synapomorphies that were evolved in the stem lineage common with the sister group, (2) older plesiomorphies from stem lineages lying further back in the past and (3) the comparatively young autapomorphies developed first in their own stem lineage. In the following selection of the synapomorphies of the Acrania and Craniota, I make brief use of the already described autapomorphies of the Vertebrata (p. 168).

Synapomorphies of the Acrania and Craniota

- Chorda dorsalis with extension over the entire back.
- Polymery of the coelom with metamerically arranged musculature.
- Neural tube with dorsal spinal nerves in segmental arrangement.
- Circulatory system with sinus venosus as starting point of the arterial system and entrance for venous vessels.

  The synapomorphies of the Acrania and Craniota correspond to the derived agreements between the Onychophora and the Euarthropoda among the Arthropoda. These are, for example, the chitin cuticle with molting, the body cavity with a pericardial septum or the heart with ostia.

  The synapomorphies justify a monophylum each of the Vertebrata and the Arthropoda; they document the first stage in the phylogenesis of these entities.

Fig. 62. *Branchiostoma lanceolatum* (Acrania). Lancelet. **A** In biotope "amphioxus sand." **B** Anterior part of the transparent body. Cirri surround the entrance to the oral cavity. Dorsal → ventral: chambered dorsal fin – muscle segments – branchial gut – ventrolateral metapleural fold. (Banyuls-sur-Mer, Mediterranean Sea. Originals)

Fig. 63. *Branchiostoma lanceolatum* (Acrania). **A** View of the left side of the body. **B** Lateral view of anterior end. Chorda extends into the tip of the body. Cirri encircle the mouth. The wheel organ of ciliated cells in the oral cavity participates in the transport of water in the pharynx. The velum marks the boundary to the branchial gut. **C** Anterior end of the neural tube with expansion of the lumen, but without external swellings. Rostrally a group of cells with pigment granules. **D** Excretory organ in the subchordal coelom. Monociliated cells with podia on the coelom wall. Cell body with nucleus and cilium with circle of microvilli in the coelom cavity. **E** Aorta and intestinal plexus of the circulatory system in cross section. Without endothelial covering. **F** Diagram of the circulatory system from the left. The venous system runs together in the sinus venosus – from the tail the subintestinal vein, from the front and back the paired cardinal veins over the Cuvierian ducts as well as from the front additionally the hepatic vein ▶

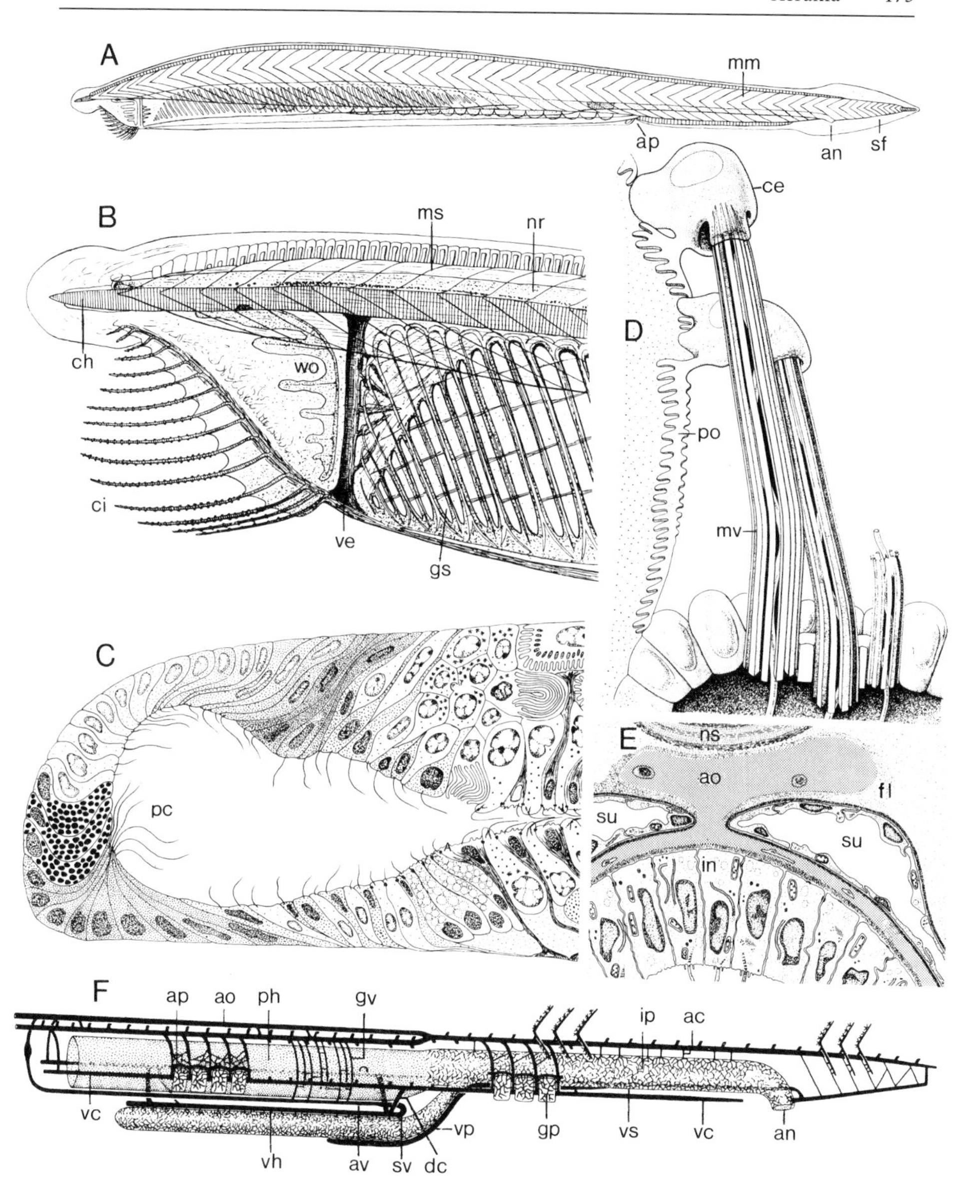

out of the midgut gland; this is supplied from the gut over the portal vein. The arterial system begins with the exit of the ventral aorta (endostylar artery) from the sinus venosus; it runs beneath the branchial gut forwards. Gill arch vessels course upwards to the paired roots of two aortae which fuse to the unpaired caudal aorta at the end of the branchial gut. *ac* Caudal aorta; *an* anus; *ao* root of aorta; *ap* atrial plexus; *at* atriopore; *av* ventral aorta; *ce* cell body; *ch* chorda dorsalis; *ci* cirri; *dc* Cuvierian duct; *fl* collagen fibril; *gp* gonadal plexus; *gs* gill slit; *gv* gill arch vessel; *in* intestinal epithelium; *ip* intestinal plexus; *mm* myomer; *ms* myoseptum; *mv* microvillus; *nr* neural tube; *ns* notochord sheath; *pc* pigment cell; *ph* pharynx; *po* podia (foot-like process); *sf* tail fin; *su* subchordal coelom; *sv* sinus venosus; *vc* cardinal vein; *ve* velum; *vh* hepatic vein; *vp* portal vein; *vs* subintestinal vein; *wo* wheel organ. (**A** Starck 1978; **B** Franz 1927; **C** Meves 1973; **D** Kümmel & Brandenburg 1961; **E** Rähr 1981; **F** Rähr 1979)

Plesiomorphies of the Acrania

- Single-layered epidermis which in the young stage consists throughout of monociliated cells. In the adult ciliated cells only occur scattered in the body wall (BEREITER-HAHN 1984).
- Lack of bone substance.
- Anterior end without head formation.
- No complex sensory organs such as lens eyes or static organs.
- Continuation of the branchial gut with numerous gill slits, endostyle and production of mucous filters.
- Simple hepatic cecum that forms digestion enzymes and stores glycogen and fat.
- Primitive features in the circulatory system: (1) blood vessels run in the space of the primary body cavity without epithelial covering; there is no endothelium. Coverage of the vessels by basal laminae of adjacent organs as well as connective tissue. (2) No contractile heart in the region of the sinus venosus. (3) Colorless blood cells as opposed to erythrocytes with haemoglobin in the Craniota.
  For the primary lack there is of course no genesis. We must exclude it when we draw the following conclusion. The time point of the evolution of material plesiomorphies lies in every case before the splitting of the Vertebrata into the stem lineages of the Acrania and Craniota. The apomorphous alternatives of all plesiomorphies evolved later in the stem lineage of the Craniota (p. 178) and document the second stage in the phylogenesis of the Vertebrata. Let us again take the Arthropoda for comparison. The body wall musculature and segmental nephridia can be interpreted as plesiomorphies of the Onychophora. The transformation of the body wall musculature into separate cords and the limitation to six pairs of nephridia constitute the apomorphous alternatives of the Euarthropoda from the second step in the phylogenesis of the Arthropoda.

Autapomorphies (Fig. 33 → 12)

- Endopsammic lifestyle.
  The larvae of the Tunicata are pelagic, as can be postulated without conflict for the stem species of the Vertebrata. A consequence of this statement is the hypothesis of an invasion of the Acrania from the pelagic to the protection-offering sediments of the sea bottom. It also possibly provides an explanation for the extensive "evolutionary standstill" of the Acrania with the retention of many primitive characters.
- Oral cirri (Fig. 63 B).
  The evolution of a crown of cirri at the entrance to the mouth near the sediment surface may be connected with the life in sandy bottoms. Functioning as a trap the cirri catch larger particles that are stirred up

from the sediment surface by the passage of the water flow into the branchial gut.

- Pelagic young stage.
  The migration into the sea bed led to the evolution of a new, freely swimming stage in the ontogenesis. This certainly has nothing to do with the larva in the biphasic life cycle of the Chordata (p. 147), but rather occurred first in the stem lineage of the Acrania. Thus, I avoid the use of the term larva in this case. Moreover, it would be difficult to justify in the absence of genuine larval characters.
- Asymmetry.
  At the beginning of ontogenesis, the primarily bilaterally symmetrical embryo develops a series of unusual asymmetries. They are particularly pronounced in the young stage. The opening of the mouth occurs on the left side of the body, the first gill slits on the right. This phenomenon remains without a convincing functional or evolutionary explanation.
- Pigment-cup ocelli.
  Simple photoreceptors consisting of a pigment cell and a photosensory cell are present in the neural tube; enormous numbers of them occur laterally and ventrally about the lumen of the canal. In comparison with the isolated distribution of corresponding pigment-cup ocelli in the Metazoa (SALVINI-PLAWEN & MAYR 1977), an independent development in the stem lineage of the Acrania is probable.
- Lack of a brain (Fig. 63 C).
  No differentiation of the neural tube at the anterior end; merely a slight expansion of the lumen. In comparison to the swellings of the neural tube in the larvae of the Tunicata and the brain of the Craniota, this situation can be identified as a secondary state. In the evolution of a hemisessile organism in sediment, the brain is certainly unnecessary.
- Subdivision of the splanchnocoel in the region of the branchial gut (Fig. 64).
  In the ontogenesis of the Acrania, the peribranchial cavity develops from an ectodermal fold, which protrudes inwards ventrally and spreads out around the branchial gut. This leads to a displacement of the original, uniform splanchnocoel. The longitudinal tubes of the paired, dorsal subchordal coelom as well as the unpaired endostylar coelom from beneath the branchial gut protrude markedly.
- Rostral expansion of the chorda dorsalis (notochord; Fig. 63 B).
  Underlying the entire neural tube through to the tips of the body – in contrast to the termination of the notochord behind the brain vesicle in larva of the Tunicata and beneath the brain in the Craniota.
  This conspicuous situation led to the widely used name Cephalochordata for the lancelets.

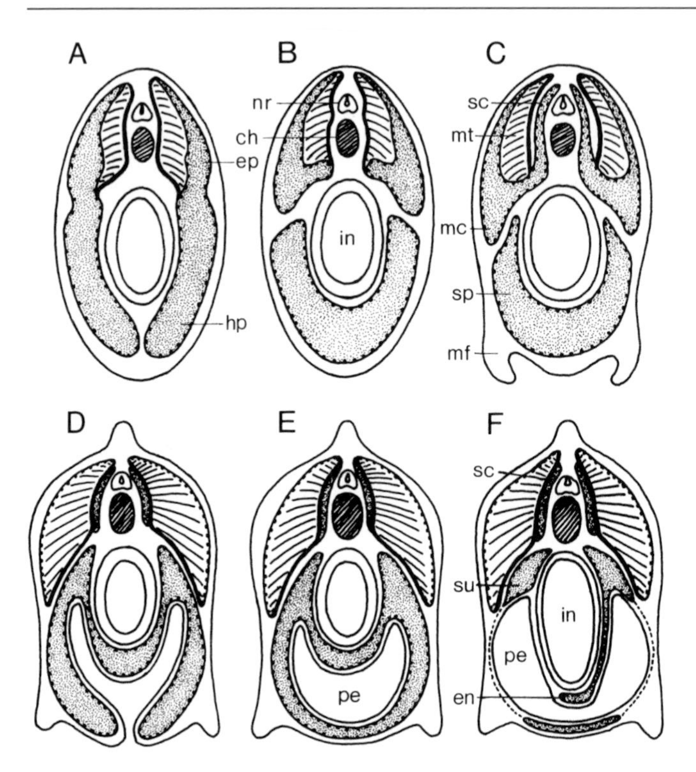

Fig. 64. Coelom of the Acrania. **A–C** Evolution to ground pattern in the segmental development of the polymeric coelom. **A** Start of the division into dorsal epimers and ventral hypomers. **B** Separation of epimers and hypomers. Differentiation of the epimers to myocoel with myotome on the inner surface. Fusion of the hypomers to splanchnocoel. **C** Development of sclerocoel from the myocoel. The sclerocoel pushes itself inwards of the myotome around the chorda and neural tube. **D, E** Decomposition of the splanchnocoel in the region of the branchial gut as a result of the development of the peribranchial cavity. **D** Ectodermal fold grows forwards from the ventral side in the region of the splanchnocoel. **E** Peribranchial cavity displaces the coelom. **F** Result of division of the splanchnocoel into longitudinal tubes. *ch* Chorda dorsalis; *en* endostylar coelom; *ep* epimer; *hp* hypomer; *in* intestine; *mc* myocoel; *mf* metapleural fold; *mt* myotome; *nr* neural tube; *pe* peribranchial cavity; *sc* sclerocoel; *sp* splanchnocoel; *su* subchordal coelom. (**A–C** Drach 1948; **D–F** Starck 1978)

- Chorda dorsalis with muscle plates.

  Densely placed plates or lamella of cross-striated musculature with gel-like filled intercellular spaces form the outstanding differentiation in the construction of the chorda dorsalis. This is unknown among the other Chordata. The evolution of a chorda with the ability for extreme stiffness is often placed in connection with the endopsammic lifestyle – with the occurrence of lateral undulations to penetrate into the ground and for locomotory activities in sediment.

- Excretory organs (Fig. 63 D).
  The paired, segmental excretory system of the Acrania is topologically linked with the subchordal coelom. In the region of the branchial gut glomerulus-like blood vessels penetrate on both sides in segmental sequence into the subchordal coelom. Groups of monociliated cells (cyrtopodocytes) with foot-like processes are here in the coelom wall; the podia or pedicles mesh together and form a thick cover on the ECM of the blood vessels. The long cilia of every cell are consistently surrounded by ten microvilli with triangular cross section. As rod-like elements they cross the subchordal coelom and stretch themselves into epithelial tube exits. The excretory canals open into the peribranchial cavity.
  From the combination of a protonephridial solenocyte with a metanephridial podocyte, an intermediate position of the excretory organs of the Acrania between "renal tubules" of invertebrates and vertebrates is derived (RUPPERT & SMITH 1988; RUPPERT 1997b). For the purposes of phylogenetic systematics, such an uncertain hypothesis is not much help. The question of a possible homology with the nephridial organs of any other entity of the Bilateria remains unanswered (BARTOLOMAEUS & AX 1992).
- Large number of gonads.
  Over 30 pairs of sac-like testes and ovaries are formed in the region of the branchial gut. They lie in echelons between the body sides inwards of the wall of the peribranchial cavity. Ripe gametes break out through the cavity.
  Gonads in comparable numbers and arrangement do not exist in other Chordata, this speaks for an autapomorphy in the ground pattern of the Acrania. In addition, the state of paired gonad rows in *Branchiostoma* must be considered as primitive, the limitation to the right side of the body in *Epigonichthyes* an evolutionary derivation within the Acrania.

The outlined autapomorphies occurred as evolutionary novelties first in the stem lineage of the Acrania – after splitting of the stem species of the Vertebrata into the stem lineages leading to the Acrania and Craniota.

For the sake of completeness, we will conclude the comparison with the Onychophora. To the autapomorphies of the Acrania correspond a series of characteristic features of the Onychophora such as tufts of long, unbranched tracheae in correlation with a terrestrial lifestyle, the oral papilla with mucus glands for trapping prey and defence on land or the jaws for cutting food objects. They all have contributed nothing to the rise of the Euarthropoda just as the autapomorphies of the Acrania have donated nothing to the evolution of the ground pattern of the Craniota as well as their later flowering diversification.

Craniota

If we proceed consequently with our reasoning, then the stem lineage of just this entity must now be discussed. From the manifold of evolutionary novelties that has occurred in the stem lineage through to the last common stem species of the recent Craniota, I have made the following selection.

Autapomorphies (Fig. 33 → 13; 68 → 1)

- Skin with multilayered epidermis.
  Several layers of ectodermal cells are an undisputed apomorphy in comparison with the monolayered epidermis of the Acrania. The formation of new cells to replace those used up in the periphery proceeds in the basal stratum germinativum. Beneath the ectodermal epidermis is the dermis of mesodermal origin as a stiff connective tissue. Epidermis and dermis form the skin (cutis) of the Craniota.
- Division of the body into head with brain and central sensory organs, trunk and tail.
- Bones.
  Calcified connective tissue with branched cells, an intercellular organic matrix, mostly collagen fibrils and intercellular minerals, especially calcium phosphate in the crystalline form of hydroxyapatite.
- Bony dermal skeleton.
  The paraphyletic † "Ostracodermata" (p. 184) from the Paleozoic period have led to the following important insight. An extensive peripheral bony skeleton as product of the dermis belongs in the ground pattern of the Craniota. Even the earliest stem lineage member from the Ordovician period possessed large bone plates on the head and bone scales on the rest of the body (Fig. 65).
  These facts and their interpretation have consequences. The bony dermal skeleton must have been reduced independently in the stem lineage of the Cyclostomata and within the Gnathostomata.
- Cartilaginous endoskeleton.
  The endoskeleton of cartilage is present in all Craniota and occurs as first solid substance in the ontogenesis. The stem species of the Craniota probably had an endoskeleton made exclusively of cartilage (JANVIER 1998). It is, however, still not clear what in detail belongs in the ground pattern. The well-developed endoskeleton of the Gnathostomata is fundamentally different from the sparse cartilage elements of the Myxinoida and Petromyzonta. Possible cranial homologies of the three entities are the ear and nose capsules as well as the embryonal trabecula in the form of cartilage rods at the anterior end of the chorda. In the axial skeleton the segmental, dorsal arcualia of the Petromyzonta and Gnatho-

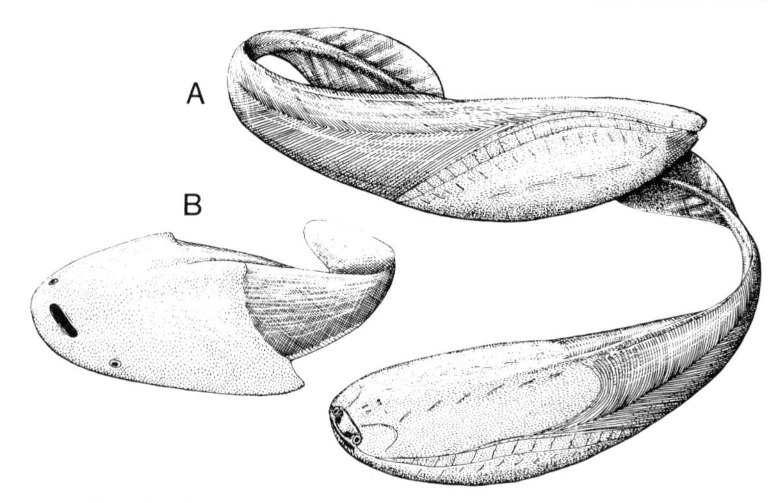

Fig. 65. Reconstruction of early, primarily jaw-less Craniota with extensive dermal skeleton. **A** † *Scacabambaspis* belong to the first fossilized Craniota from the Ordovician period (Bolivia, 450 million years old). The head covering consists of two large bone plates on the dorsal and ventral sides. Trunk and tail with rod-like scales. **B** † *Bannhuanaspis* from the Devonian period (Vietnam, 400 million years old). With wide, flat head shield and button-like scales. (Janvier 1998)

stomata are probably homologous (JANVIER 1998). Arcualia are absent in the Myxinoida – presumably a secondary deficit.

- Brain of five sections.

  Ontogenetic development out of two primary vesicles.

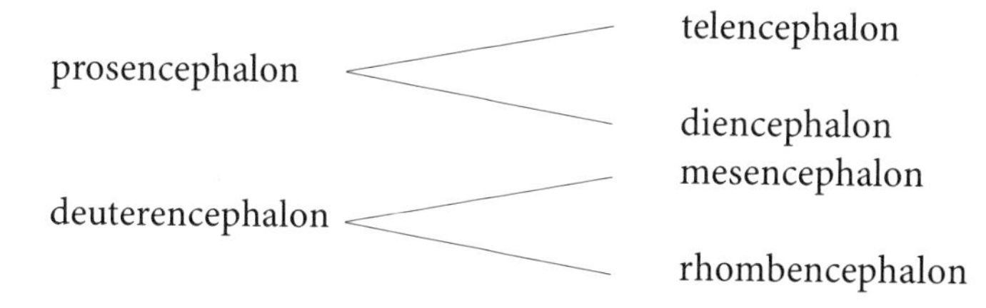

The rhombencephalon, in turn, divides into the metencephalon with cerebellum and the myelencephalon. The latter links the brain with the spinal cord.

- Dorsal spinal ganglia and ventral spinal nerves.

  The ganglion cells of the dorsal spinal nerves lie beside the neural tube and join together here to spinal ganglia. Ventrally new spinal nerves emerge from the spinal cord.

  Dorsal and ventral spinal nerves are separated from each other in the ground pattern of the Craniota.

- Paired lateral eyes.
  The optic cup occurs in the ontogenesis as a protrusion of the wall of the diencephalon. The optic cup induces the formation of the lens; it folds itself from the skin ectoderm inwards.
- Unpaired dorsal ocelli.
  Two evaginations directed dorsally of the diencephalon that are arranged medially one behind the other. In the Petromyzonta these are the rostral parietal eye (parapineal organ) and the caudal pineal organ.
  Within the Craniota the epiphysis develops from the pineal organ.
- Nasal organ.
  Cyclostomata have a single median "nasal opening" – in the Myxinoida at the tip of the body, in the Petromyzonta on the roof of the head. "Nasal opening" is in quotation marks because the adjoining unpaired passage enters in the formation of the adenohypophysis. In the Myxionoida, an unpaired nasal sac follows; in the Petromyzonta, it is divided into two halves by a median septum. On the other hand, the Gnathostomata possess paired lateral nasal sacs with separate nasal openings. The homologous hypophyseal duct is separated from the olfactory organ. The evaluation of the alternative is disputed; however, for the systematization of the Craniota, we do not necessarily need it.
- Lateralis system and labyrinth.
  The peripheral line system with neuromasts evolved in the stem lineage of the Craniota to detect liquid flows. The widespread development of closed canals with local pores is possibly the primitive state in the Craniota primarily living in water. The open lateral lines of the Cyclostomata would then be apomorphous, which could in this case be connected with the reduction of the bony dermal skeleton. The labyrinth is a specialized section of the lateralis system for orientation in space. The existence of only two vertical semicircular canals in the Cyclostomata is probably plesiomorphous, the existence of three semicircular canals with one horizontal canal an apomorphy in the ground pattern of the Gnathostomata (STARCK 1982; HENNIG 1983).
- Hypophysis.
  Composed from the adenohypophysis of the roof of the mouth with primary connection to the outside (see nasal organ) and the neurohypophysis of the floor of the diencephalon.
- Circulatory system.
  Blood vessels with epithelial linings (endothelium), a muscular heart as well as erythrocytes with haemoglobin constitute evolutionary novelties in the circulatory system of the Craniota.
  In the ground pattern the heart is a purely venous organ of four sections. The sinus venosus is already present in the Acrania; it was taken

over from the ground pattern of the Vertebrata. Forwards it is followed by the atrium, ventricle and the conus arteriosus.

- Thyroidea.
  Evolutionary derivation from the endostyle in branchial gut of the Chordata. Ontogenetic transformation of the endostyle into the thyroid gland in the ammocoete larva of the Petromyzonta.
- Pancreas.
  In comparison to the uniform midgut gland of the Acrania, in the stem lineage of the Craniota, the pancreas, alongside the liver, has differentiated as a separate source of digestive enzymes.
- Gill apparatus as a "pure" respiratory organ.
  From the primary double function, the task of a filter of particulate food has been eliminated. In the Craniota, there is no mucous filter as product of the endostyle. The water flow through the gills is no longer generated by cilia, but by muscle pumps. The number of gill slits is highly reduced in comparison with the Acrania. In the Myxinoida a maximum of 15 slits are formed; 7 pairs of gill slits belong in the ground pattern of the Gnathostomata.
- Spiral valve of the intestine.
  Evolutionary novelty to increase the surface of the absorbing intestinal epithelium.
- Opisthonephros from nephrons with Malphigian corpuscules.
  Irrespective of temporary nephridial systems in the ontogenesis, uniform, paired kidneys in the state of opisthonephros can be postulated for the adult of the stem species of the Craniota. Structural and functional unit of the opisthonephros is the nephron of the renal corpuscule (Malphigian corpuscule) and efferent renal tubule. In the renal corpuscule, a capillary blood vessel glomerulus is surrounded by the double-layered Bowman's capsule of the nephron. Primary urine is excreted through the inner layer in the cavity between the two layers of the capsule.
- Neural crest (Fig. 66).
  Embryonal cells of the ectoderm detach themselves from the epithelium of the neural fold upon formation of the neural tube and arrange themselves over the tube to a crest. From here, the cells of the neural crest migrate ventrally – to form many different structures such as gill arches, spinal ganglia, pigment cells or dermal skeleton.
  The neural crest is known for the Petromyzonta and Gnathostomata, but not for the Myxinoida. This does not tell us much because there is lack of information about young embryos of this entity.
  We emphasize this feature at the end of the autapomorphies. Neural crests are absent in the Tunicata and Acrania. The neural crest is dis-

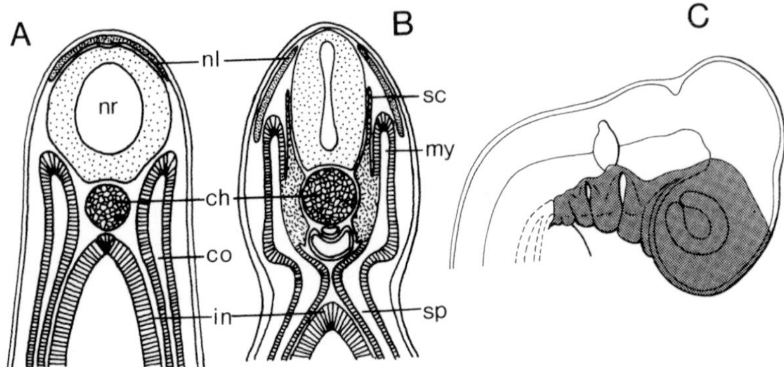

Fig. 66. Division of metameric coelom cavities into myocoel and splanchnocoel (autapomorphy of the Vertebrata) as well as the neural crest as autapomorphy of the Craniota. **A** *Squalus* (Chondrichthyes). Young embryo in cross section. Coelom still uniform; dorsal neural crest. **B** *Squalus*. Later embryo. Division of the coelom into dorsal myocoel (with sclerocoel around neural tube and chorda) and ventral splanchnocoel around the gut; cell accumulations of the neural crest migrate downwards laterally. **C** Chicken embryo. Neural crest tissue in definitive position on the ventral side (*shaded*). *ch* Chorda dorsalis; *co* coelom; *in* intestine; *my* myocoel; *nl* cells of the neural crest; *nr* neural tube; *sc* sclerocoel; *sp* splanchnocoel. (Janvier 1998)

cussed as the "most peculiar character" of the Craniota (JANVIER 1998; KARDONG 1998).

Systematization

In the first step, we are concerned with the recognition of the phylogenetic relationships between the three high-ranking taxa Petromyzonta, Myxinoida and Gnathostomata as well as the representation of the hypothesized kinship relationships at the upper hierarchical level of the system of the Craniota.

The interpretation of the Petromyzonta and Myxinoida as primarily jawless Craniota is undisputed. The hypothesis of a single evolution of the jaw apparatus with palatoquadrate cartilage and mandible in the stem lineage of the Gnathostomata is also unquestioned. A subject of much controversy today, however, is the question of whether the Petromyzonta and Myxinoida together form a monophylum or whether the taxon Petromyzonta represents the sister group of the Craniota (Fig. 67). The competing relationship hypotheses are shown below.

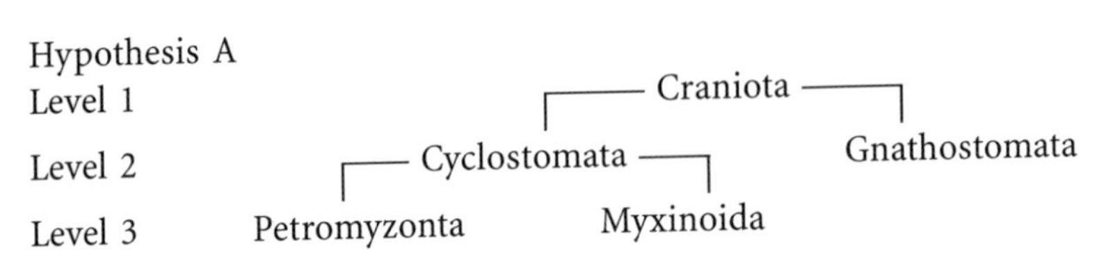

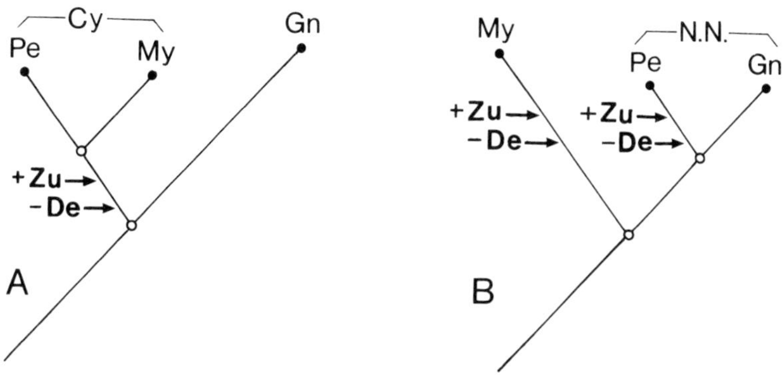

Fig. 67. Diagram of the discussed kinship relationships between the Petromyzonta (*Pe*), Myxinoida (*My*) and Gnathostomata (*Gn*). In hypothesis **A**, the Petromyzonta and Myxinoida form the adelphotaxa of the monophylum Cyclostomata (*Cy*). The tongue apparatus (+*Zu*) evolved once in the stem lineage of the Cyclostomata, the dermal skeleton (–*De*) was reduced once. In hypothesis **B**, the Petromyzonta and Gnathostomata are the sister groups of an unnamed entity *N.N.* This hypothesis demands the assumption of a twofold, independent evolution of the tongue apparatus and the double reduction of the dermal skeleton in the stem lineages of the Petromyzonta and Myxinoida

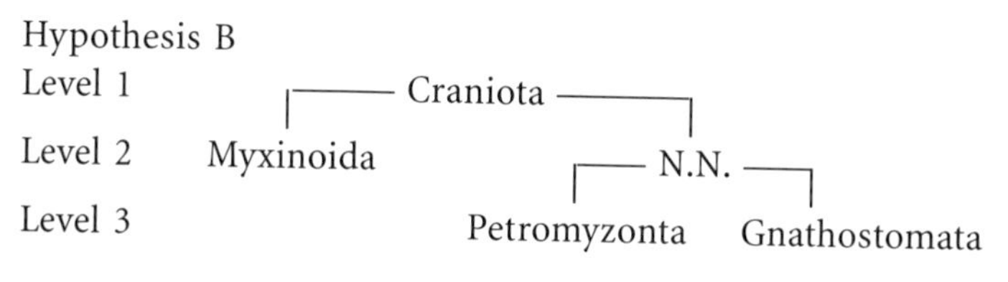

The traditional Cyclostomata hypothesis A was convincingly confirmed by a detailed analysis of the tongue apparatus in *Lampetra* and *Myxine* (YALDEN 1985). In Fig. 69, eight subtle agreements in the construction of the tongue apparatus from cartilaginous elements, protractors and retractors are demonstrated that are interpreted point for point as homologies. I can see no justifiable alternative to this interpretation that has, as a consequence, the homologization of the entire tongue apparatus in the Petromyzonta and Myxinoida. Moreover, this is with desirable certainty not an original homology – not a symplesiomorphy. The uptake of solid food and blood in the Cyclostomata undisputedly represents an apomorphy in comparison to the microphagy in the ground pattern of the Craniota. The tongue apparatus with rasping teeth must, in other words, form a synapomorphy between the Petromyzonta and the Myxinoida. Molecular findings also support the monophyly of the Cyclostomata (MALLAT & SULLIVAN 1998).

We come to a second important statement. The presented evaluation of the tongue apparatus is compatible with the simplest explanation for the evolution of the naked, eel-like body of the Cyclostomata. First of all, it is undisputed that this represents a complete reduction of the dermal skeleton of primarily jawless Craniota. The principle of the most economical explanation then requires the assumption of a single reduction in a common stem lineage of the Petromyzonta and Myxinoida, i.e., in the stem lineage of an entity Cyclostomata in which the reduction of the dermal skeleton and the evolution of a new tongue apparatus occurred concurrently.

Hypothesis B of an adelphotaxa relation between the Petromyzonta and the Gnathostomata must be discussed in the light of this.

LØVTRUP (1977) presented a list of "characters common to Hyperoartii (Petromyzonta) and Gnathostomata". When evaluating them, we must take into account that, in comparison with the Petromyzonta, the Myxinoida exhibit a series of undisputed, conspicuous signs of reduction, for example, the lack of dorsal fins or the highly reduced eyes (p. 189). A corresponding interpretation for the lack of "synaptic ribbons in retinal receptors" in the eyes seems reasonable, as well as for the lack of arcualia in the axial skeleton or the spiral valve in the gut.

The labyrinth organ of *Myxine* with only one vertical canal, but two ampulla at the ends, could have arisen from the fusion of two vertical canals, as has been realized in the Petromyzonta (STARCK 1982). This state is plesiomorphous compared to the three arches in the Gnathostomata (p. 192).

Correspondingly differentiated evaluations are necessary for the physiological and biochemical agreements between the Petromyzonta and the Gnathostomata.

As far as I can see, the second hypothesis has no convincing arguments to counter the problem of a twofold, independent development of the tongue apparatus and a twofold, independent reduction of the dermal skeleton in the Petromyzonta and Myxinoida.

Accordingly, we follow the basal systematization shown below, whereby the division in the adelphotaxa Chondrichthyes and Osteognathostomata has been taken in advance of the relevant discussion (Fig. 68).

Craniota
- **Cyclostomata**
  - **Petromyzonta**
  - **Myxinoida**
- **Gnathostomata**
  - **Chondrichthyes**
  - **Osteognathostomata**

At this point, we must take a methodologically oriented glance at fossil remains. The name † "Ostracodermata" for the armoured Craniota of the Paleozoic era (Fig. 65) is increasingly disappearing from the systematic literature of the fish-like vertebrates – and thus shares the fate of the name

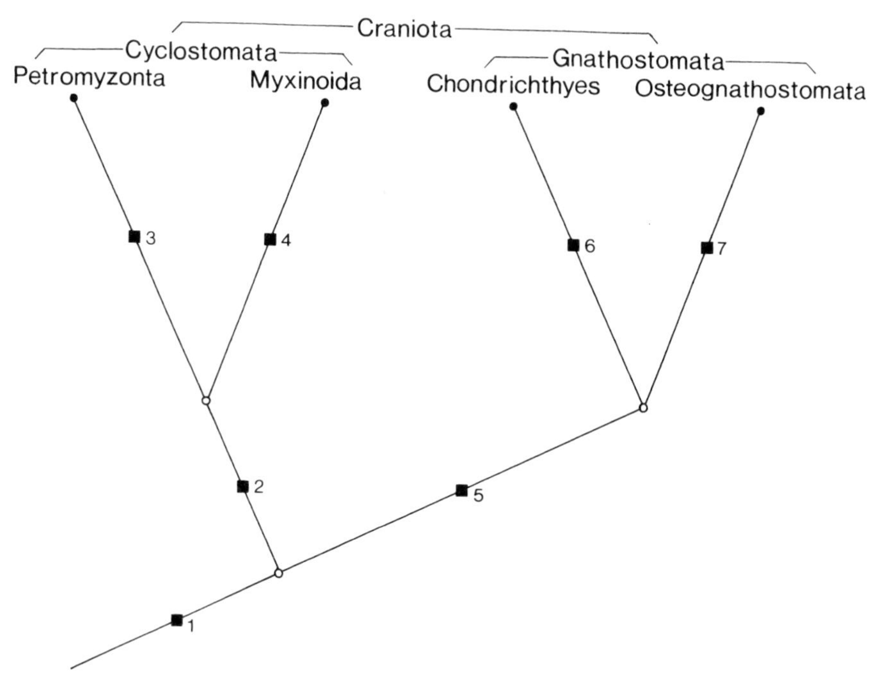

Fig. 68. Diagram of the phylogenetic relationships between the highest-ranking subtaxa of the Craniota

"Pisces" itself. In the "Early Vertebrates" (JANVIER 1998), the "ostracoderms" are only mentioned in passing; "Pisces" are completely ignored. In both cases, they are paraphyletic collections of early Craniota.

Taking the presented systematization of the recent Craniota as a base, the "Ostracodermata" with a bony dermal skeleton could, in principle, belong in three different stem lineages – the lineage of the Craniota, the Cyclostomata or the Gnathostomata. Phylogenetic systematics treat fossils as members of stem lineages. It must attempt to find the incorporation in the correct stem lineage by means of the identification of autapomorphies from the ground patterns of the recent entities, in our case of the Craniota, Cyclostomata or Gnathostomata.

Of course, the name "Agnatha" for a paraphyletic collection of recent and fossil Craniota with a primary lack of jaws is also obsolete – and the collection itself is to be dissolved.

As justification, I must remark that the stem lineage of the Gnathostomata probably began with agnathous Craniota – then it is hardly imaginable that the first stem species of the Gnathostomata from the Paleozoic era already possessed a jaw apparatus. Fossil Gnathostomata without jaws would then have to be incorporated in the stem lineage of the Gnathostomata before fossil stem lineage members with jaws, such as † Placodermi.

Cyclostomata – Gnathostomata

Cyclostomata

In the context of the above discussion, it may be added that recently – and curiously – differences between the Petromyzonta and Myxinoida have also been put forward to oppose the monophyly of an entity Cyclostomata. However, these are of no value because they could have occurred in arbitrary numbers in the separate stem lineages of the Petromyzonta and Myxinoida. Decisive are solely the agreements for which the evolution in a stem lineage common to these two entities can be postulated. We thus come to the autapomorphies of the Cyclostomata. The unprotected skin and the discussed tongue apparatus must be supplemented with two further characters.

Autapomorphies (Fig. 68 → 2)

- Naked, eel-like body.
  Result of a complete reduction of the dermal skeleton of primarily agnathous Craniota.
- Tongue apparatus with longitudinal rows of teeth.
  Organ for the uptake of bigger pieces of food.
- Gill pouch.
  The gill slits expand in the middle part to small pouches in which the gill lamella lie. Since this is only realized in the Petromyzonta and Myxinoida, the most economical explanation is a single evolution of the apomorphy.
- Unpaired gonad.
  From the paired anlage in ontogenesis, a dorsally placed, unpaired organ arises.

Petromyzonta – Myxinoida

The sister groups are species-poor entities with only few supraspecific taxa above the species level. The frequent life cycle of the Petromyzonta with the adult in the sea and ammocoete larva in fresh water is to be placed in the ground pattern of the Cyclostomata. This demands the assumption that the purely marine lifestyle of the Myxinoida with direct development represents a derived behavior.

Petromyzonta

Entity with a disturbingly large number of names. Practically all of them turn around the lamprey *Petromyzon* – Petromyzonta, Petromyzones, Petromyzoniformes, Petromyzontiformes, Petromyzontida or Petromyzoidea. The name Hyperoartia is also in use; it refers to the closed roof of the mouth without connection to the unpaired nasal sac.

What is the origin of the German trivial name "Neunauge" for the lamprey? Nine eyes can be counted when one does not look too closely (Fig. 69A). Starting from the back, there are the seven round pores of the gill slits, followed by the actual eye as number 8 and on top of the head the unpaired nasal opening as number 9.

The ground pattern of the biphasic life cycle of the Cyclostomata is documented in Europe by the sea lamprey *Petromyzon marinus* (length 1 m) and the river lamprey *Lampetra fluviatilis* (length up to 40 cm). These are anadromous migratory fish. The adult individuals live in the sea. For the uptake of food they attach themselves by suction with their round mouth to fish, scrape the musculature of the prey, and suck in blood and pieces of tissue. For reproduction they migrate upstream to the limnetic spawning region. Here, the ammocoete larvae develop as filterers of particulate food. After metamorphosis the young animals return to the sea; they take up food for the first time in the sea.

The larvae and adults of the small brook lamprey *Lampetra planeri* (length ca. 16 cm) both live in fresh water. The marine phase in the life cycle is completely suppressed. Adult individuals do not take up food; the gut degenerates already during metamorphosis.

In comparison with the adelphotaxon Myxinoida, practically all features of the Petromyzonta seem to be more primitive. I mention at this point only the two dorsal fins (Fig. 69A). As an autapomorphy, I interpret the differentiation of the mouth to a suction apparatus with round outline and funnel-like invagination.

Autapomorphies (Fig. 68 → 3)

- Suction mouth.

  The ventrally oriented mouth forms a large disk that is supported by a cartilage ring at the edge. The inwardly directed mouth funnel is fitted with numerous, pointed horny teeth.

  In comparison with the lack of a suction mouth in the Myxinoida and merely one palatine tooth in the oral cavity, the situation in the Petromyzonta should be understood as an optimization of the anchoring to prey objects.

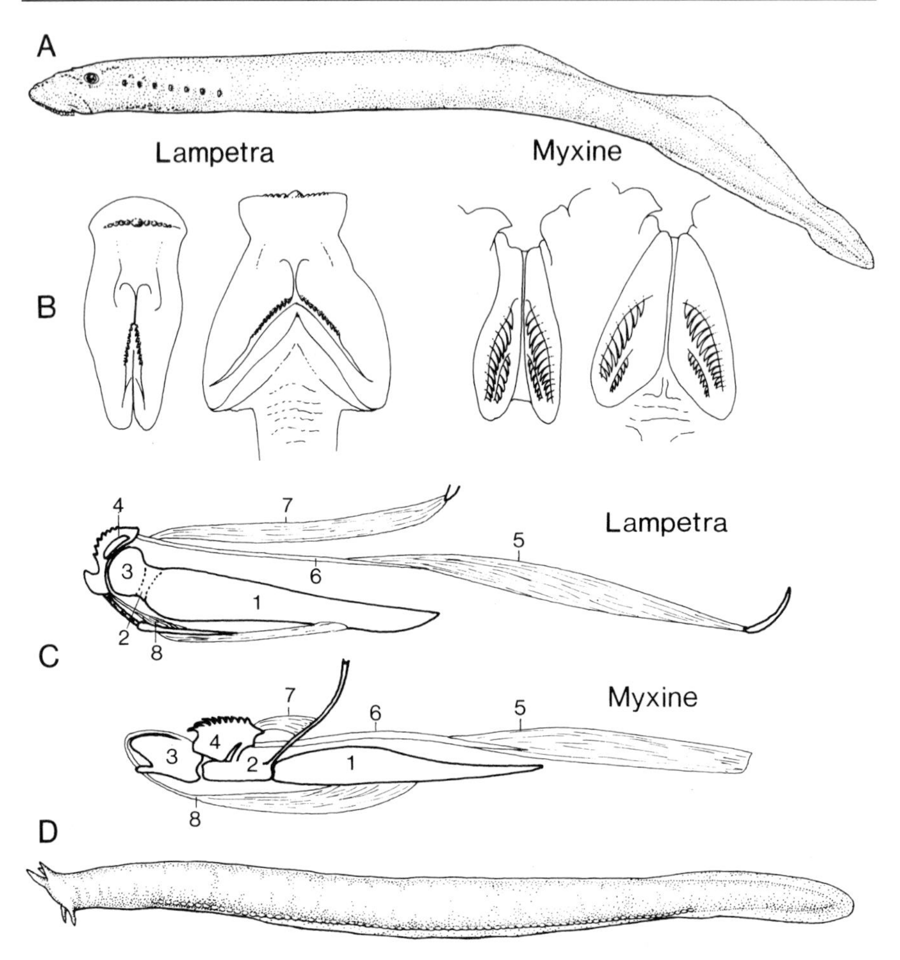

Fig. 69. Cyclostomata. **A** *Lampetra fluviatilis* (Petromyzonta). Habitus. **B** Arrangement of the horny teeth of the tongue apparatus in rows in *Lampetra* and *Myxine* (dorsal views after dissection). In each case on the *left* in closed position, on the *right* opened, from behind. *Lampetra* had one anterior transverse and two posterior longitudinal rows of small teeth. In *Myxine* there are two posterior, longitudinal rows of large, sharp teeth on each side. **C** Diagram of the tongue apparatus of *Lampetra* and *Myxine* in side view. *1–8* Marked similarities in specific elements of the cartilage and musculature that can be interpreted as homologous agreements. *1* Median ventral cartilage as central supporting skeleton; *2* rostrally adjacent soft cartilage (*Lampetra*) or rigid cartilaginous element (*Myxine*); *3* roll-like or U-shaped cartilage; *4* supporting cartilage for the rows of teeth. Paired in *Lampetra*; an unpaired, complex cartilage in *Myxine*; *5* central retractor is a median muscle dorsal of the main cartilage; *6* long tendon of the retractor runs forwards and inserts at cartilage 4 (supporting cartilage); *7* a pair of accessory retractors that arise from cranial cartilage, insert in the tendon of the main retractor (*Lampetra*) or at cartilage 4 (*Myxine*). Five pairs of central protractors arise from the median main cartilage in *Lampetra* and *Myxine*; homology uncertain. *8* Paired accessory protractors are joined with cartilage 4. **D** *Myxine glutinosa* (Myxinoida). Habitus. (**A, D** Strenger 1963; **B, C** Yalden 1985)

Myxinoida

The marine hagfish are the only Craniota without osmoregulation, as is frequently emphasized. The blood is isoosmotic with seawater – as is usual in invertebrates. The interpretation is a subject of much controversy. If a life cycle without a limnetic, ammocoete larva is derived (p. 186), then also the lack of osmoregulation in the Myxinoidea is possibly a secondary state.

Myxinoida, Myxini and Myxiniformes are used as names for the hagfish with their enormous production of mucous secretion from rows of dermal glands. The name Hyperotreta is the opposite of Hyperoartia (Petromyzonta); it refers to the open connection of the nose and roof of the mouth. However, the two names are very similar and are thus not particularly appreciated.

In *Bdellostoma*, there is a varying number of 5–15 gill slits; they open separately from each other – and this is a primitive state. In *Myxine* – with the species *Myxine glutinosa* in the North Atlantic – the ducts leading from the six gill pouches fuse on each side to a common end piece; there is then only one gill pore on each side.

Autapomorphies (Fig. 68 → 4)

- Direct development without ammocoete larva.
- Lack of arcualia in the axial skeleton.
  If the arcualia of the Petromyzontes are the remnants of a reduced axial skeleton then their complete lack in the Myxinoida constitutes the apomorphous end state of the reduction.
- Tentacles on the head (Fig. 69 D).
  Two pairs of tentacles around the terminal nasal opening as well as two further pairs around the subterminal mouth. Since this is otherwise unknown in the ambient of the Myxinoida, the assumption of its evolution in the stem lineage of the Myxinoida is the most economical explanation.
- Dermal glands with mucus production.
  Evaluation corresponding to the above character.
- Two pairs of longitudinal rows of strong horny teeth in the tongue apparatus.
  In comparison with a transverse row and two unpaired longitudinal rows of small teeth in the Petromyzonta (Fig. 69B), YALDEN (1985) interpreted the state of the horny teeth in the Myxinoida as an apomorphy.
- Lack of dorsal fins in comparison to two fins in the Petromyzonta.
- Highly reduced eyes without cornea, lens and eye muscles.

Gnathostomata

The name-giving element of the monophylum is the jaw apparatus with vertically biting upper and lower jaws – the palatoquadrate and the mandible (Fig. 70 A). In comparison with the sister group Cyclostomata, there is no doubt about the apomorphous character of the apparatus, nor about

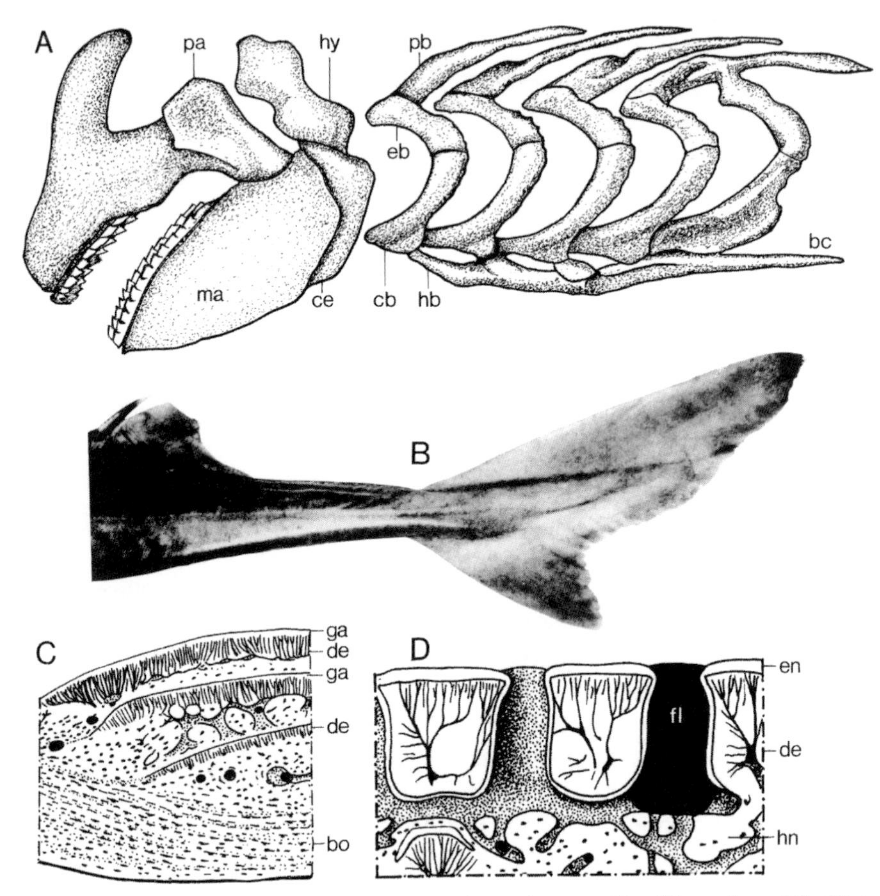

Fig. 70. Gnathostomata. **A** Splanchnocranium of *Squalus acanthias* (Elasmobranchii). Division into mandibular arch (palatoquadrate and Meckelian cartilage), hyoid arch (hyomandibula + ceratohyal) and five branchial arches. **B** Heterocercal tail fin of *Squalus acanthias* with entry of chorda dorsalis and vertebral column in the upper flap. **C–D** Scales of Osteognathostomata. **C** Actinopterygii with ganoid scales. † *Andreolepis*. Vertical section through a scale with alternating layers of ganoine and dentine. **D** Sarcopterygii with cosmoid scales. † *Porolepis*. Vertical section. Invagination of cosmine (enamel + dentine) with formation of small, bottle-shaped cavities. *bc* Basibranchial copula; *bo* bone; *cb* ceratobranchial; *ce* ceratohyal (hyoid); *de* dentine; *ep* epibranchial; *en* enamel; *fl* bottle-shaped cavity in the scale; *ga* ganoine; *hb* hypobranchial; *hn* horizontal canal network; *hy* hyomandible; *ma* Meckelian cartilage; *pa* palaoquadrate; *pb* pharyngobranchial. (**A** Marinelli & Strenger 1954–1973; **B** Jarvik 1980; **C, D** Janvier 1998)

the most economical explanation of a single evolution of the jaw in the stem lineage of the Gnathostomata.

This statement is completely independent of the scenarios that have been put forward about the possible routes of evolution of the jaw apparatus and the remaining visceral skeleton. The classical interpretation of the jaw as a transformation product of one branchial arch of agnathous predecessors is today opposed by the hypothesis of a homology of the jaw with cartilaginous parts in the velum of ammocoete larva (Petromyzonta) – and this leads to the statement "...a more parsimonious explanation could be that jaws have always been jaws" (JANVIER 1998, p. 258).

Autapomorphies (Fig. 68 → 5)

- Endoskeleton with central cartilage and perichondral bone.
  The stem species of the Craniota had an endoskeleton of pure cartilage (p. 178) – and this state was taken over in the stem lineage of the Cyclostomata. An alternative exists in the sister group Gnathostomata. In the usual formulation the Chondrichthyes possesses a cartilaginous endoskeleton, the Osteognathostomata ("Osteichthyes" and the remaining Gnathostomata) an internal skeleton of bone. As a refinement of this difference JANVIER (1998) put forward the following differentiated picture. In the stem species of the Gnathostomata, cartilage and bone were present in the endoskeleton. The extensive central cartilage was surrounded in the periphery by a bone mantle. Globular calcification occurred inside the cartilage. In the stem lineages of the adelphotaxa Chondrichthyes and Osteognathostomata, divergent changes occurred. In the Chondrichthyes, the perichondral bone was degraded and replaced by prismatic, calcified cartilage. Endochondral bone evolved in the stem lineage of the Osteognathostomata, resulting in the complete ossification of cartilage.
  The endoskeleton encompasses the skull skeleton with neurocranium and splanchnocranium, the axial skeleton as well as the accessory skeleton with paired and unpaired fins.
- Neurocranium.
  The braincase encloses the brain, the nasal capsules, the eyes and the labyrinth.
- Splanchnocranium.
  The visceral skeleton is divided into the mandibular arch, the hyoid arch and the gill arches (branchial apparatus; Fig. 70 A).
  1. Mandibular arch of palatoquadrate and mandible (Meckelian cartilage, mandibular cartilage). Upper and lower jaws are linked with the neurocranium by joints and bands in the ground pattern.
  2. Hyoid arch of hyomandibula and ceratohyal (hyoid). The hyomandibula articulates with the neurocranium. Connections between the palatoquadrate and the hyomandibula probably also belong in the ground pattern.

3. Five pairs of gill arches (visceral arches) that each consist of four skeletal rods (pharyngo-, epi-, cerato- and hypobranchials).

- The first gill slits lie in front of the first gill arch. Four further pairs of gill slits open outwards separately between the gill arches.
- On each side there is a spiracle between the mandibular and hyoid arches. Nothing is blown out of these "blow holes", rather water is sucked in. Their existence is an autapomorphy in the ground pattern of the Gnathostomata, independent of hypotheses of their evolution.
- Arcualia in the axial skeleton.
  On each side there are four arcualia per segment near the chorda dorsalis – dorsally the basidorsal and interdorsal as well as ventrally the basiventral and interventral.
  The dorsal arcualia probably already belong in the ground pattern of the Craniota (p. 178). Only the ventral elements may be considered with certainty as evolutionary novelties of the Gnathostomata.
- Paired fins.
  Paired pectoral fins and pelvic fins (ventral fins) evolved in the stem lineage of the Gnathostomata. The fins insert on girdles of the endoskeleton, the pectoral and pelvic girdles.
  I do not know of any convincing statements on the structure of the paired fins in the last common stem species of the recent Gnathostomata.
- Unpaired fins.
  Two dorsal fins, a ventral anal fin and the tail fin belong in the ground pattern of the Gnathostomata. They must be differentiated in their evaluation.
  A tail fin already exists in the Acrania and Cyclostomata; it may originate from the ground pattern of the Vertebrata. The Petromyzonta have two dorsal fins, one behind the other. The agreement with the Gnathostomata may go back to the ground pattern of the Craniota. Only the anal fin seems to be a certain autapomorphy of the Gnathostomata. Furthermore, there is the shaping of the posterior end to a heterocercal (or epicercal) tail fin with axis of chorda and vertebral column directed upwards. The heterocercal fin (Fig. 70 B) occurred as an evolutionary novelty in the stem lineage of the Gnathostomata; it experienced manifold changes within the entity.
- Horizontal septum in muscle segments.
  Division of myomers by a septum beneath the forwards-directed tip.
- Axons of the nerve cells surrounded by a myelin sheath.
- The dorsal and ventral roots of the spinal nerves join to uniform nerve cords.
- Labyrinth with horizontal semicircular canal as third element.
- Testes and kidneys linked. The exit of sperm proceeds through the outlet canal of the mesonephros.

Chondrichthyes – Osteognathostomata

Chondrichthyes

Two derived features unambiguously characterize the cartilaginous fish as a monophylum of sharks, rays and chimaeras.

Autapomorphies (Fig. 68 → 6; 71 → 1)

- Loss of perichondral bone – acquisition of prismatic, calcified cartilage.
  The peripheral stiffening of cartilaginous skeletal elements by a layer of crystallized calcium phosphate in the form of tiny platelets (Fig. 72 D, E) is an evolutionary novelty from the stem lineage of the Chondrichthyes. It is apparently connected with the reduction of perichondral bone that was postulated for the ground pattern of the Gnathostomata. Thin layers of perichondral bone in some sharks are explained as remnants of this reduction process.
- Copulatory organ (clasper) from pelvic fins.
  The males of all cartilaginous fish have a conspicuous copulatory organ for the direct transfer of sperm (Figs. 72 G, 73 E). The gonopod (pterygopod, mixipterygium) is formed by modification of the metapterygium – a basal cartilage of the pelvic fin.

Possibly, the large basal cartilaginous plates that carry the radials or rays in the fins form a further autapomorphous feature of the Chondrichthyes.

The exoskeleton of placoid scales in the skin, teeth on the edges of the jaws and spines on the dorsal fins is sharply separated from the endoskele-

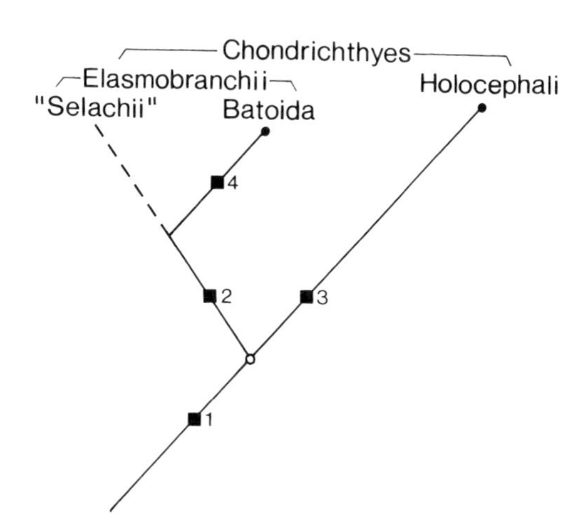

Fig. 71. Diagram of the phylogenetic relationships of the Chondrichthyes with the Elasmobranchii and Holocephali as sister groups

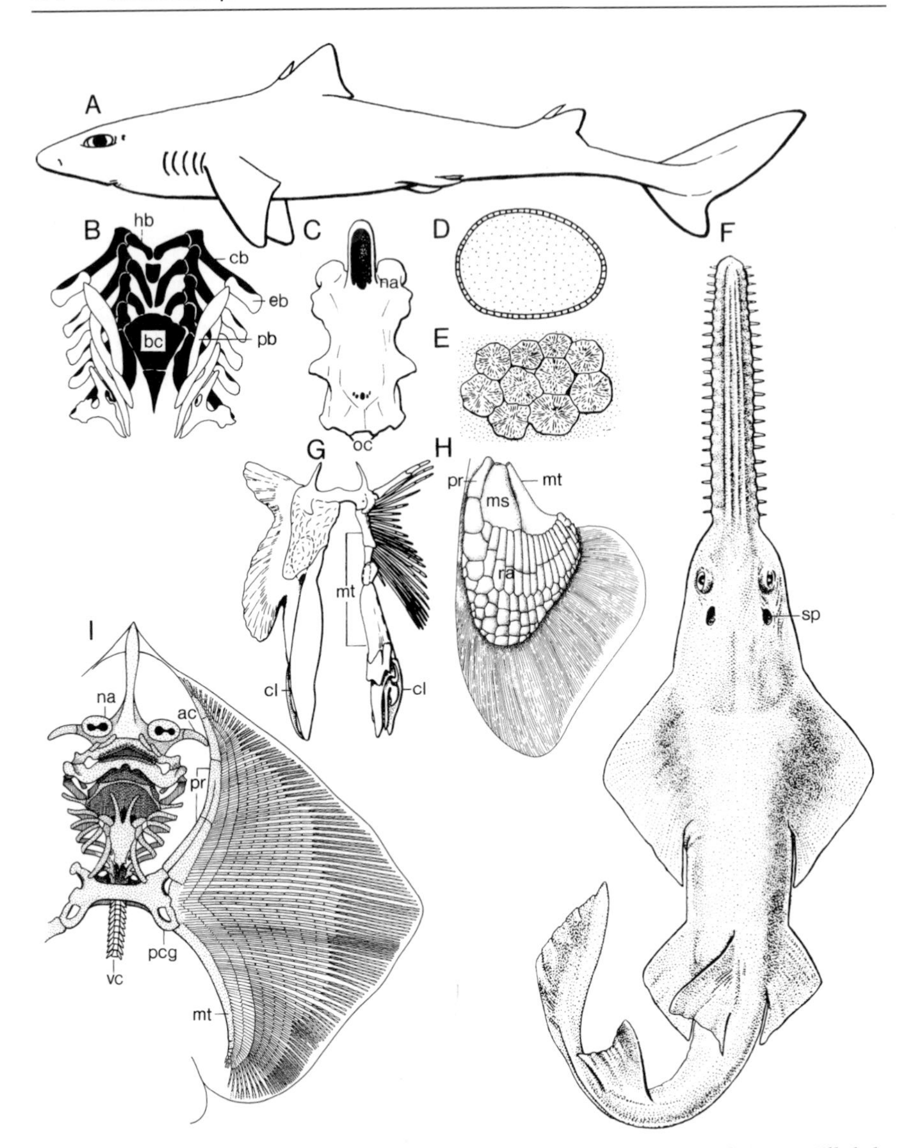

Fig. 72. Elasmobranchii (Chondrichthyes). **A–C** *Squalus* ("Selachii"). **A** Lateral view. **B** Gill skeleton in dorsal view. Ventral elements *black*. **C** Skull capsule in dorsal view. At the back with paired occipital condyles. **D** Cross section through a radial (fin) with prismatic calcified cartilage on the surface of the hyaline cartilage. **E** Detail of the surface of a radial; calcification in the form of plates. **F** *Pristis clavata* (Batoida). Dorsal view. **G** Pelvic fins on pelvic girdle of a male ray with copulation system (clasper). Right soft parts removed to demonstrate the endoskeleton. **H** *Heterodontus francisci* ("Selachii"). Pectoral fin with three large basalia and distally adjacent radials. **I** *Raja* (Batoida). Skull, shoulder girdle and skeleton of the left pectoral fin. Ventral view. *ac* Antorbital cartilage (connects pectoral fin with skull capsule); *bc* basibranchial copula; *cb* ceratobranchial; *cl* copulatory organ (clasper); *ep* epibranchial; *hb* hypobranchial; *ms* mesopterygium; *mt* metapterygium; *na* nasal capsule; *oc* occipital condyle; *pb* pharyngobranchial; *pcg* shoulder girdle; *pr* propterygium; *ra* radial; *sp* spiraculum (blow hole); *vc* vertebral column. (**A–E, G** Janvier 1998; **F** Arambourg & Bertin 1958; **H, I** Starck 1979)

ton of cartilage in the Chondrichthyes. Placoid scales have a crown of dentine and a base of acellular bone with pulp. Placoid scales and teeth as derivatives of scales were probably taken over from the ground pattern of the Gnathostomata and would accordingly be plesiomorphies in the organization of the Chondrichthyes.

Five pairs of gill slits probably belong in the ground pattern of the Chondrichthyes. Six or seven pairs of gill slits in some sharks (*Hexanchus, Heptranchias*) are considered as apomorphous states.

Elasmobranchii (sharks and rays) and Holocephali (chimaeras) form the adelphotaxa of the Chondrichthyes. In a comparison of the sister groups, we recognize once more the weak justification of the far-reaching plesiomorphous entity Elasmobranchii as monophylum in contrast to the highly derived Holocephali with a whole series of conspicuous autapomorphies.

Elasmobranchii – Holocephali

Elasmobranchii

Autapomorphies (Fig. 71 → 2)

- Ventral cartilaginous elements of the gill arches (hypobranchials) directed backwards (Fig. 72B).
- Paired condyles on the skull.
  The skull capsule "displays a more-or-less distinctly paired occipital condyle" (JANVIER 1998, p. 60; Fig. 72C).
- Enameloid tissue of the teeth.
  Enameloid is an enamel-like tissue formed by the epidermis and ectomesenchyme (cells from the neural crest), whereas enamel is solely the product of epidermal cells. The thin layer of enameloid on the teeth of Elasmobranchii consists of layers with differing microstructures. At least one haphazardly fibered layer of enameloid occurs in sharks and rays (JANVIER 1998).

The common classification of the Elasmobranchii into the Selachii (sharks) and Batoida (rays) is discarded by phylogenetic systematics. The Selachii cannot be justified as a monophylum and, accordingly, the sharks and rays cannot be interpreted as adelphotaxa.

"Selachii" ("Plagiotremata")

There are no features of the sharks that can unequivocally be described as autapomorphies. They form a paraphyletic collection of relatively primitive cartilaginous fish.

Plesiomorphies (in comparison with the Batoida): Spindle-shaped body. Lateral position of spiracle and gill slits.

Scyliorhinus (cat shark), *Squalus* (spiny dogfish), *Mustelus* (hound shark), *Squatina* (monkfish), *Sphyrna* (hammerhead shark).

Batoida (Hypotremata)

Autapomorphies (Fig. 71 → 4)

Dorsoventral plated body with shifting of the gill slits to the ventral side. The enlarged pectoral fins are joined with the braincase in front of the eyes by particular, antorbital cartilage (Fig. 72 I). Anterior vertebrae fused.

Raja clavata (thornback ray), *Raja batis* (common skate), *Myliobatis* (eagle ray), *Torpedo* (electric ray), *Pristis* (sawfish).

Holocephali

Species-poor group of marine Chondrichthyes. *Chimaera monstrosa* distributed in the North Atlantic Ocean and Mediterranean Sea.

Autapomorphies (Fig. 71 → 3)

- Holostyli.
  The complete fusion of the palatoquadrate to the cranium may be mentioned in the first place as a marked character of the skull construction (Fig. 73 B).
- Jaw with tooth plates.
  In place of the individual teeth of the sharks and rays there are two pairs of strong tooth plates in the upper jaw and one pair in the lower jaw (Fig. 73 C).
- Lack of spiracle – only four gill slits.
  The spiracle is formed, but disappears in the ontogenesis. One gill slit less than in the ground pattern of the sister group Elasmobranchii.
- Gill cover from a skin fold.
  Coverage of the gill slits by a fleshy fold originating from the hyoid arch. Convergence to bony operculum in the Osteognathostomata.
- Large spine of the dorsal fin with articulation in a thick cartilaginous plate (Fig. 73 A).
- Calcified rings surrounding the canals of the lateralis system (Fig. 73 D).
- Male with additional anchoring system for the copula.
  Besides the copulatory organ on the pelvic fin from the ground pattern of the Chondrichthyes, the Holocephali have retractable claspers in front of the pelvic fins and an unpaired tenaculum on the head (Fig. 73 A, E).

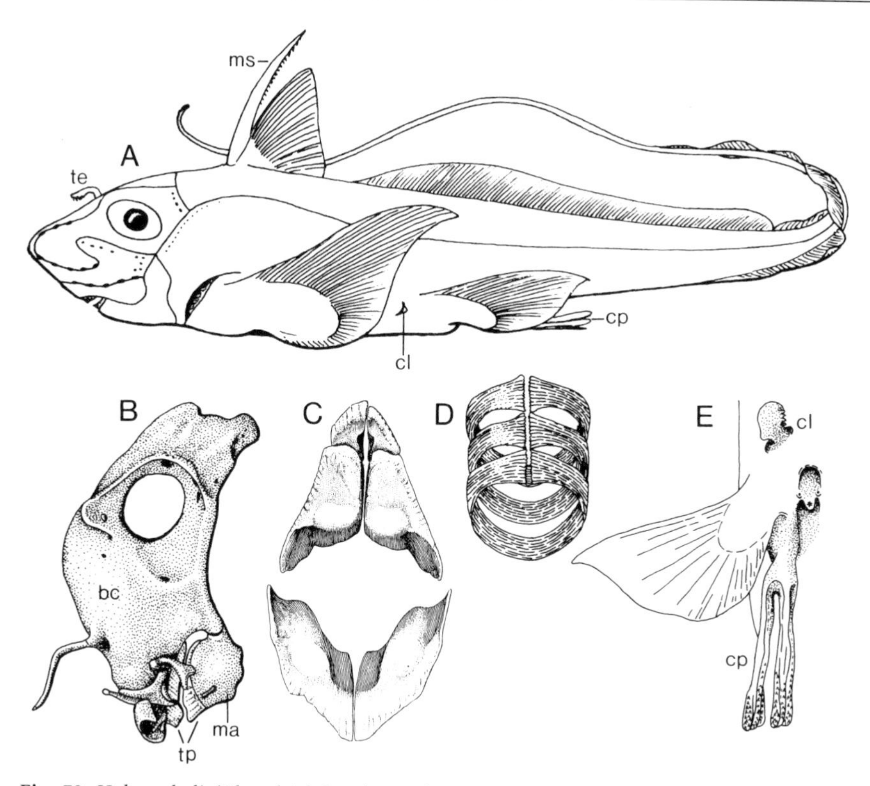

Fig. 73. Holocephali (Chondrichthyes). **A** *Chimaera*. Male. **B** *Hydrolagus*. Skull capsule with fused palatoquadrate and free Meckelian cartilage (lower jaw). Side view. **C** *Chimaera cubana*. Tooth plates. Two pairs in upper jaw, one pair in lower jaw. Inside view. **D** *Chimaera*. Calcified rings in lateral line system. **E** *Chimaera*. Male. Pelvic fin with large copulatory organ and small clasper in front of the fin. *bc* Brain capsule; *cl* clasper in front of pelvic fin; *cp* copulatory organ of the pelvic fin; *ma* Meckelian cartilage; *ms* mobile spine of the dorsal fin; *te* tenaculum; *tp* tooth plates. (**A, B, D, E** Janvier 1998; **C** Klausewitz 1963)

- Reduction of the placoid scales.
  In principle, the skin of the Holocephali is naked. Only a few hook-shaped placoid scales on the anchoring organs of the male and a few scutes at the posterior end (JANVIER 1998).

Osteognathostomata

HENNIG (1983) placed the Chondrichthyes as sister group of the entire remaining Gnathostomata under the name Osteognathostomata. In this way, the traditional division of the fish-shaped Gnathostomata into the Chondrichthyes (cartilaginous fish) and Osteichthyes (bony fish) became obsolete. Under exclusion of the Tetrapoda, the "Osteichthyes" form a paraphylum and have no place in the phylogenetic system of the Gnathostomata. I have no sympathy for the widespread attempts to rescue the Osteichthyes by inclusion of just the Tetrapoda (JANVIER 1998). Phylogenetic systematics will certainly not gain any friends by classifying the Tetrapoda in the bony fish.

The most satisfying solution is a new name for the adelphotaxon of the Chondrichthyes. In the light of the currently held hypothesis of an evolution of endochondral bone in the stem lineage of the Osteognathostomata, we must agree that the name was even then well chosen.

Autapomorphies (Fig. 68 → 7; 74 → 1)

- Ossified endoskeleton.
  The Osteognathostomata are characterized by endochondral bones that originate in ontogenesis through internal ossification of preformed, cartilaginous skeletal elements (substitute bones). In addition, the perichondral ossification was taken over from the ground pattern of the Gnathostomata as a plesiomorphy.

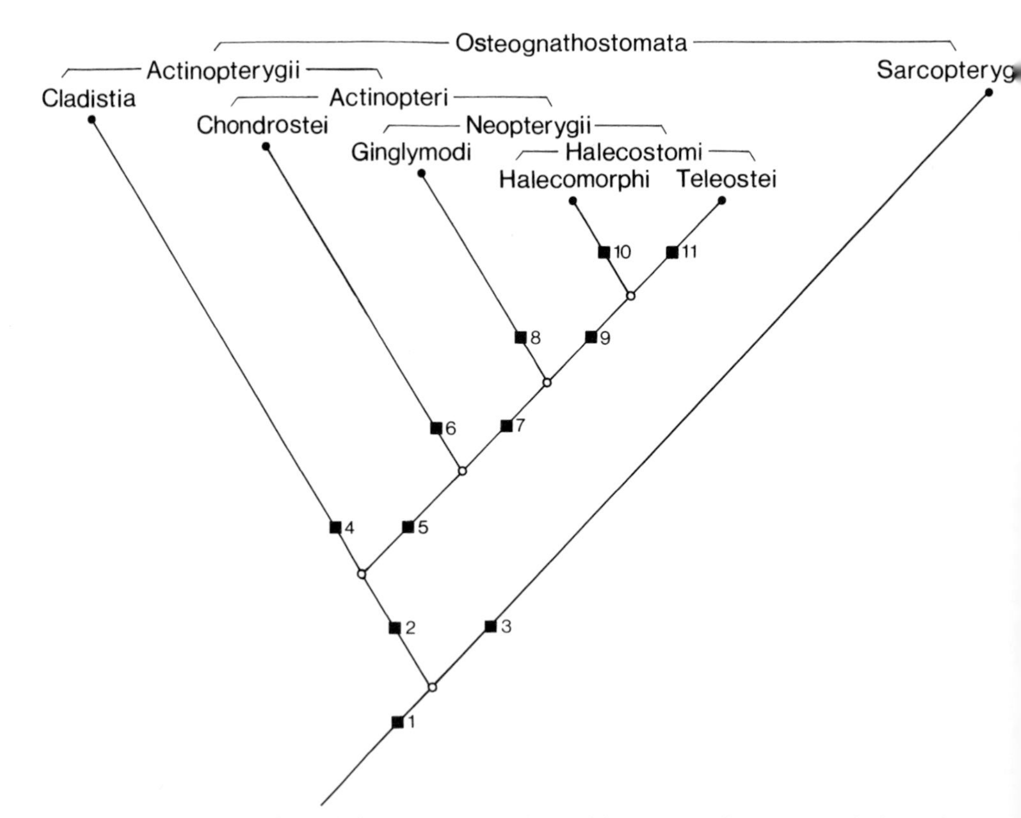

Fig. 74. Diagram of the phylogenetic relationships of the Osteognathostomata with the Actinopterygii and Sarcopterygii as adelphotaxa

- Exoskeleton of two components without preceding cartilaginous formation in the development.
  1. Large dermal bones cover head and shoulder girdle. Dermal bones also form the attachment sites for teeth (premaxilla, maxilla, dentary).
  2. Large scales on the body and the fins.
- Lepidotrichia.
  Lengthened scales that are formed to rows of small plates in the fins.
- Operculum.
  A gill cover of several bones covers the gill slits; it originates from the hyoid arch.
- Separation of the nasal openings.
  A skin bridge crosses the nasal pit and leads to a complete separation of an anterior and a posterior nasal opening. Only unfused skin flaps are present in the adelphotaxon Chondrichthyes.
- Labyrinth with large otoliths (statoliths).
- Pharynx with air-filled diverticula.
  A lung/air bladder organ exists in the Actinopterygii and the Sarcopterygii. This possibly goes back to intestinal diverticula that were filled with air in a common stem species.

Two forms of the dermal scales play a part in the characterization of the highest-ranking adelphotaxa of the Osteognathostomata – the ganoid scales of the Actinopterygii and the cosmoid scales of the fossil Sarcopterygii (SCHULTZE 1977).

Ganoid scales (Fig. 70 C)

Ganoine is a shiny enamel layer that covers the dentine of scales and dermal bones. Ganoine is, like enamel, secreted from the epidermis. Unique for the ganoid scales of the Actinopterygii is the construction from alternating layers of ganoine and dentine (JANVIER 1998). With this structure, their rhomboid shape and the articulated connection (see below), the ganoid scales, are interpreted as an autapomorphy of the Actinopterygii.

Cosmoid scales (Fig. 70 D)

Within the Sarcopterygii cosmine only occurs in certain † Porolepiformes, the mostly Devonian Dipnoi as well as some † Osteolepiformes. Cosmine is a double layer of enamel and dentine that invaginates inwards in scales or bones and covers small, bottle-shaped cavities there; they open through tiny pores to the outside.

According to SCHULTZE (1977), ganoid and cosmoid scales have arisen as alternative apomorphies from a common ground form. In this case, cosmoid scales could therefore be an autapomorphy of the Sarcopterygii.

However, they could also represent the original scales in the ground pattern of the Osteognathostomata.

A very similar pore-canal system with bottle-shaped cavities existed in † *Tremataspis* (Osteostraci), i.e., in jawless relatives of the Gnathostomata; it is discussed as a convergence (JANVIER 1998).

For the changes in the fins to be discussed below, the plesiomorphies in the ground pattern of the Osteognathostomata must be emphasized. As in the Chondrichthyes, there are two dorsal fins, a ventral anal fin and a heterocercal tail fin.

Actinopterygii – Sarcopterygii

Actinopterygii

The Actinopterygii or ray-finned bony fish constitute the majority of the primarily aquatic Gnathostomata; among them belong the Teleostei as bony fish in the stricter sense with well over 20,000 species. Of particular interest are some species-poor taxa with differing, primitive characters – the Cladistia, Chondrostei, Ginglymodi and Halecomorphi. Their mutual relationships have been well analyzed (LAUDER & LIEM 1983). Like the Insecta that constitute the very first taxon of the Metazoa with a consequent phylogenetic systematization (Vol. II, p. 259), the Actinopterygii are "the first major vertebrate group whose interrelationships and systematics have been worked out in detail by means of the cladistic method" (JANVIER 1998, p. 68).

The ray-finned bony fish owe their name to the dermal, radial supports (lepidotrichia) in the fins.

Autapomorphies (Fig. 74 → 2)

- A single dorsal fin (Fig. 75).
  Since two dorsal fins are present in the Chondrichthyes and in the sister group Sarcopterygii, the reduction of one fin in the stem lineage of the Actinopterygii can be postulated with good justification. The median fin radii insert directly into the body without the intervention of a basal muscular flap.
- Rhomboid ganoid scales with articulated connection (Fig. 76 A).
  Armour of rhomboid scales that are arranged in oblique rows on the trunk. There is an articulated connection in each row in that the pointed peg of one scale meshes into an acetabulum of the preceding scale.

Acrodine as a transparent tooth cap of hypermineralized hard substance (Fig. 76 B) and the limitation of enamel to the base of the teeth are empha-

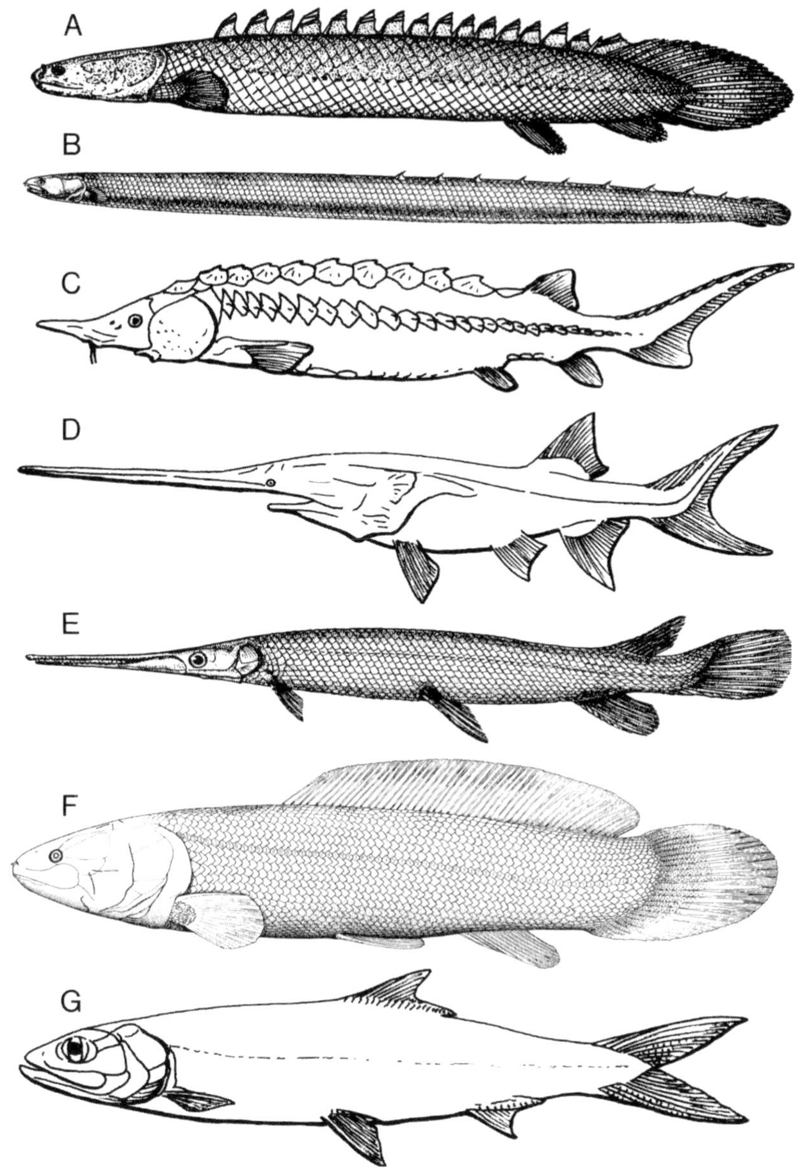

Fig. 75. Representatives of high-ranking subtaxa of the Actinopterygii. **A** *Polypterus*. Reedfish (Cladistia). **B** *Erpetoichthyes* (*Calamoichthyes*) *calabaricus* (Cladistia). **C** *Acipenser*. Sturgeon (Chondrostei). **D** *Polyodon*. Shovelnose sturgeon (Chondrostei). **E** *Lepisosteus osseus*. Longnose gar (Ginglymodi). **F** *Amia calva*. Bowfin (Halecomorphi). **G** *Elops*. Bonefish (Teleostei). (**A** Romer & Parsons 1977; **B** Daget 1958; **C, D, G** Janvier 1998; **E** Arambourg & Bertin 1958; **F** Jarvik 1980)

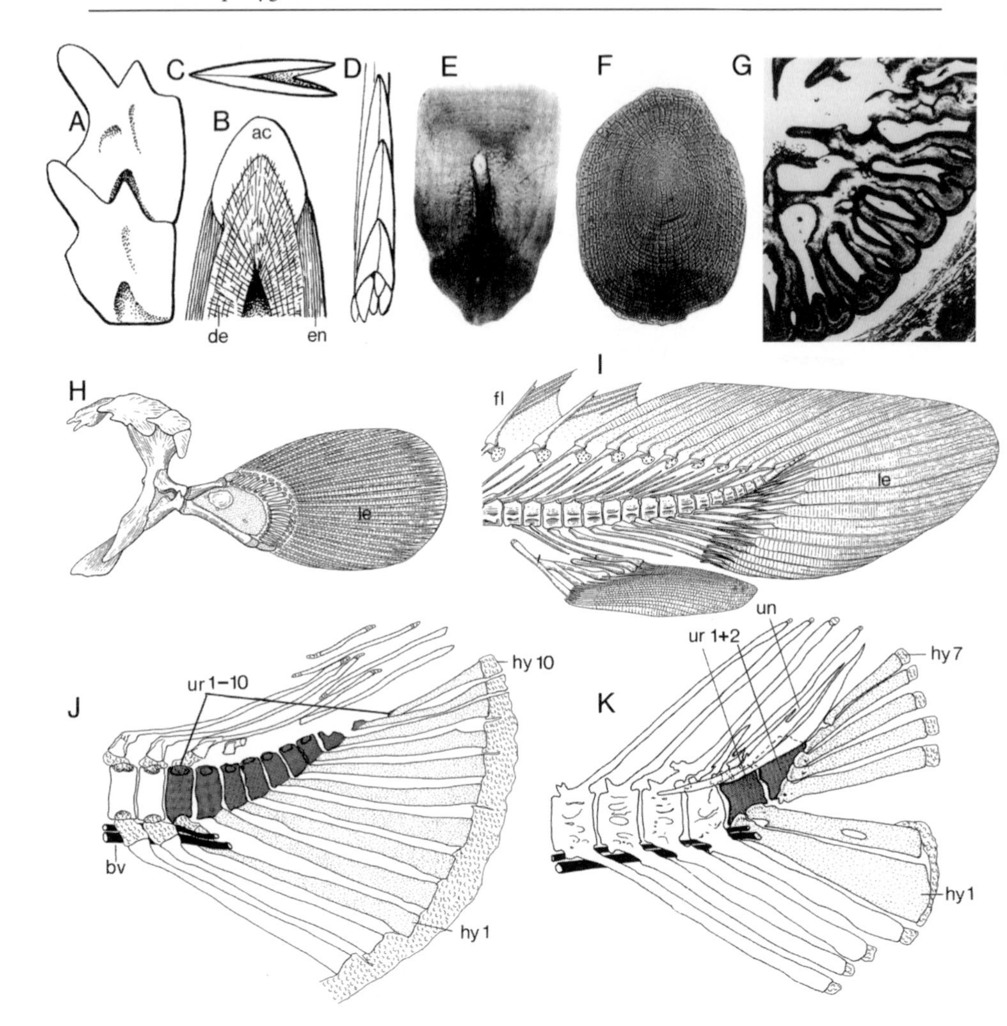

Fig. 76. Autapomorphies of the Actinopterygii and diverse subtaxa. **A** *Polypterus* (Cladistia). Ganoid scales with joint condyles and pits. Autapomorphy of the Actinopterygii. **B** Tooth of the Actinopterygii with acrodine cap (? Plesiomorphy or apomorphy). **C, D** *Lepisosteus* (Ginglymodi). Fulcra. Double scales of the pectoral fin from the front (**C**) and in natural position (**D**). Autapomorphy of the Actinopterygii. **E** *Amia calva* (Halecomorphi). Scale. **F** *Gadus morrhua* (Teleostei). Cycloid scale. Autapomorphy of the Teleostei. **G** *Lepisosteus* (Ginglymodi). Horizontal section through a folded tooth. Autapomorphy of the Ginglymodi. **H** *Polypterus* (Cladistia). Shoulder girdle and skeleton of the pectoral fin. Articulation through two basal elements. Autapomorphy of the Cladistia. **I** *Polypterus bichir* (Cladistia). Skeleton of the posterior end. Finlets of the dorsal fin each of a spine and a few delicate lepidotrichia. Autapomorphy of the Cladistia. **J** *Amia calva* (Halecomorphi). Tail skeleton. Primitive characters in comparison with the Teleostei are ten uralia and ten hypuralia. An autapomorphy is the fan-like arrangement of the hypuralia. **K** *Hiodon alosoides* (Teleostei). Tail skeleton. Derived characters are two uralia and seven hypuralia, whereby the latter are ordered in two groups. *ac* Acrodine; *bv* blood vessel; *de* dentine; *en* enamel; *fl* finlet; *hy* hypuralia, *le* lepidotrichia; *un* uroneuralia; *ur* uralia. (**A–D** Janvier 1998; **E, I** Jarvik 1980; **F** Starck 1982; **G** Schultze 1969; **H** Starck 1979; **J, K** Schultze & Arratia 1989)

sized as a further character of the Actinopterygii (acrodine is formed from ectoderm and ectomesenchyme, enamel solely from ectoderm).

However, if the extension of enamel over the entire tooth is interpreted as an autapomorphy of the Sarcopterygii (p. 209), the described state of the Actinopterygii with acrodine caps must represent a plesiomorphy.

Cladistia – Actinopteri

Cladistia

Polypteriformes, Polypterini, Brachiopterygii.

In fresh waters of Africa. *Polypterus* – bichirs with several species; length up to 1 m (Fig. 75 A). *Erpetoichthyes* (*Calamoichthyes*) *calabaricus* – African reedfish; pelvic fin reduced (Fig. 75 B).

Autapomorphies (Fig. 74 → 4)

- Division of the dorsal fin into numerous small fins (finlets; Fig. 76 I).
- Secondary diphycercal tail fin.
  Dorsal and ventral parts of the fin are about equal in size. The vertebral column, however, is slightly bent upwards and thus documents the remnant of the primitive heterocercal state of the Gnathostomata (Fig. 76 I).
- Articulation of the paired fins with the shoulder girdle over two large basal elements (Fig. 76 H).

The Cladistia are more primitive than the Actinopteri in the following two characters.

1. The paired fins have a muscular basal section covered with scales.
2. Paired lungs arise from the ventral side of the intestine. Homologous correspondences for both states are found in the Sarcopterygii.
   Plesiomorphies such as the spiracle and spiral intestine are continued in the ground pattern of the Actinopteri.

Actinopteri

Autapomorphies (Fig. 74 → 5)

- Unpaired lung/air bladder organ dorsal of the intestine with mainly hydrostatic function.
- Paired fins without muscular basal section.
  The fins supported by numerous thin rays arise directly from the body wall without the intervention of fleshy tissue.

- Fulcra (Fig. 76 C, D).
  Arrow-like double scales on the anterior edge of the fin. The fringing fulcra arise from the fusion of two neighboring, extended rhomboid scales (JANVIER 1998).

Chondrostei – Neopterygii

Chondrostei

Palaeopterygii, Acipenseriformes.

The two taxa Acipenseridae and Polyodontidae of conventional classifications are possibly sister groups.

Acipenseridae

Body with five longitudinal rows of large bone scales (plesiomorphy). Short rostrum (plesiomorphy; Fig. 75 C). Jaw of adult without teeth (apomorphy).

Acipenser and *Huso* with spiracles (plesiomorphy) in Eurasia and North America. *Acipenser sturio* (common sturgeon) and *Huso huso* (hausen, beluga).

Scaphirhynchus (shovelnose sturgeon) and *Pseudoscaphirhynchus* (pseudo-shovelnose sturgeon) in Asia. Without spiracle (apomorphy).

Polyodontidae

Body without bone scales (apomorphy). Very long rostrum (apomorphy; Fig. 75 D). Jaw with small teeth (plesiomorphy).

Only two species. *Polyodon spatula* (Mississippi paddlefish) in North America. With oar-like rostrum; long gill trap for filtration of plankton. *Psephurus gladius* (Chinese paddlefish) in China. With sword-shaped rostrum.

Sturgeons represent giants among the fish-like Osteognathostomata – *Psephurus gladius* reaches 7 m, *Huso huso* can be up to 9 m in length.

Primitive characters are the pronounced heterocercal tail fin and the spiral valve in the hindgut. Spiracles were also taken over into the organization of the Chondrostei from the ground pattern of the Gnathostomata, but have been lost within the entity (see above).

Sturgeons are bottom dwellers and this explains various characters – especially for obtaining food from the bottom. The following survey includes, in addition, special skull features that have been presented in the literature on ichthyology.

Autapomorphies (Fig. 74 → 6)

- Skeleton mainly of cartilage with persisting chorda.
  Extensive reduction of bones from the primitive state of the Actinopterygii.
- Five longitudinal rows of large scales on the body that are modified to proper bone plates (Fig. 75 C).
  The following difference must be considered in the evaluation. The existence of rows of bone plates is an autapomorphy in the ground pattern of the Chondrostei; we postulate their evolution in the stem lineage of the Chondrostei and their presence in the last common stem species of the recent members. When we now enter into the Chondrostei the described state will be a plesiomorphy of the Acipenseridae – namely because the comparable system of bone plates has been reduced in the Polyodontidae.
- Limitation of the ganoid scales to the root of the tail.
  The original covering of the body with ganoid scales is reduced to a residual area at the tail, apparently in connection with the evolution of large bone scales.
- Formation of a rostrum with low-set mouth on the ventral side.
- Evolution of barbells (tactile threads) in front of the mouth to locate bottom organisms as objects of food.
- The palatoquadrates are joined at the front by a symphysis.
- Reduction of the posterior myodome.
  The myodome is a canal of the skull base in which the eye muscles insert in most Actinopterygii.

Neopterygii

Autapomorphies (Fig. 74 → 7)

- Homocercal tail fin.
  Secondary externally symmetrical fins with end part of the vertebral column bent up dorsally.
- Spiracle closed.
- Identical numbers of radii and lepidotrichia in the dorsal fin and tail fin.
  Originally the number of lepidotrichia exceeds the number of bony radii supporting them.
- Lack of clavicle.

Ginglymodi – Halecostomi

Ginglymodi

Inhabitants of larger rivers and lakes of North America from southern Canada to Costa Rica. Nine recent species in two supraspecific monophyla – *Lepisosteus* and *Atractosteus* (WILEY 1976). Lepisosteiformes is a synonymous group name for the meter-long gars that are often compared in body shape and lifestyle with the pike *Esox lucius*. Both here and there, dorsal fins and anal fins are displaced far backwards to the base of the tail (Fig. 75E). In this way, the power for lightning quick attacks from a still ambush arises – for catching prey as a predator. In *Lepisosteus* there is also a long, forceps-like snout.

Of the primitive characters, the strong skin armour of ganoid scales is to be emphasized. As in *Polypterus* (Cladistia), the scales are arranged in oblique rows and articulated with each other. Among the recent Actinopterygii, only the Ginglymodi and the Cladistia have the plesiomorphous state of a closed scale covering of rhomboid scales.

Autapomorphies (Fig. 74 → 8)

- Highly lengthened snout.
  Maxilla reduced. Teeth of the upper jaw emerge from a series of infraorbital bones.
- Position of dorsal and anal fins close in front of the tail.
- Opisthocoelous vertebra.
  The vertebrae are convex at the front and concave at the back. With articulated joints. Complete displacement of the chorda dorsalis by the vertebrae.
- Folded tooth.
  Dentine penetrates increasingly into the twisted fold of the pulp towards the base of the tooth (Fig. 76G). The structure, designated as plicidentine, arose independently of comparable structures within the Sarcopterygii (SCHULTZE 1969; PREUSCHOFT et al. 1991).

Halecostomi

The Halecostomi encompass the last adelphotaxa pair of the Actinopterygii to be discussed in this survey. With one recent species, *Amia calva*, the Halecomorphi stands against an enormous cohort of over 20,000 Teleostei. These are the bony fish as such that are often separated as Teleostei sensu stricto from the other fish-like Osteognathostomata with endochondral bone.

Autapomorphies (Fig. 74 → 9)

- Roof tiling-like coverage of the scales with overlap of the posterior edges (Figs. 75 F, 76 E). Apomorphous state in comparison with the rhomboid scales that abut another.
- Vertebra with median neural spines.
- Freely motile maxilla and an interopercular bone. Structural basis for a suction mechanism for the uptake of food; objects of prey are sucked into the mouth with a flow of water (LAUDER & LIEM 1983).
- Quadratojugal reduced or fused with the quadratum.

Halecomorphi – Teleostei

Halecomorphi

Synonyms: Cycloganoida, Amiiformes.

Amia calva – Bowfin in fresh waters of North America (Mississippi, Lake Huron, Lake Erie). Length up to 1 m.

The most conspicuous character is the long dorsal fin extending far over the back (Fig. 75 F). It makes undulating movements on slow locomotion of the animal.

As the only recent species, *Amia calva* forms a monophyletic supra-specific taxon together with a series of fossil taxa.

Autapomorphies (Fig. 74 → 10)

- Unique articulation of the jaw.
  The bones quadrate and symplecticum participate in the joint.
- Highly lengthened dorsal fin.
- Diplospondyly.
  Two biconcave vertebrae per segment in the posterior body section.
- Fan-like arrangement of hypuralia (Fig. 76 J) at the caudal end of the vertebral column.
  The evolution of the fan is seen in connection with the accentuated dorsal flexure of the last tail vertebra (uralia).

Teleostei

DE PINNA (1996) characterized the Teleostei as a "highly corroborated monophyletic group" and pointed out 27 autapomorphies.

The emphasis of the justification as a monophylum lies in the structure of the caudal skeleton. We compare the tail fin of *Amia calva* (Halecomorphi) with that of *Hiodon alosoides* (Teleostei) (SCHULTZE & ARRATIA

1989; Fig. 76J, K). Reference point for the terminology is the exit of the caudal arteries and veins (bv) from the haemal canal and their branching. The vertebrae lying behind this are called uralia, the adjacent vertebra in the anterior direction, the preuralia. Hypuralia are then ventral bone pieces coming from the uralia; morphologically these are haemal spines without haemal canal.

Autapomorphies (Fig. 74 → 11)

- Diural caudal skeleton.
  In principle, the Teleostei have only two uralia that are formed by fusion of vertebral anlagen. *Amia calva* as adelphotaxon exhibits the plesiomorphous alternative with ten uralia.
- Hypuralia.
- (a) Seven or less bone pieces.
  Apomorphous in comparison with eight and more hypuralia in the other Actinopterygii. *Amia calva* has ten hypuralia.
- (b) Division into two groups.
  In all Teleostei, the series of hypuralia is clearly divided into a dorsal and a ventral group. The caudal fin rays are similarly divided (Fig. 75G). There is no comparable division in *Amia* or in the other Actinopterygii.
- (c) Hypuralia 1 and 2 are carried by one vertebra.
- (d) Enlargement to broad bone plates (H1 and H2 in the example of *Hiodon alosoides*).
- Uroneuralia.
  Several paired bones that are arranged as long rods beside the tail skeleton. They arise through modification of the neural arch of the ural vertebrae.
- Movable premaxilla.
- Unpaired basibranchial tooth plate.
- Cycloid scales (Fig. 76F).
  Round scales with concentric strips of bone without ganoine certainly belong in the ground pattern of the Teleostei. Ganoine is probably reduced convergently to the situation in the Halecomorphi. In any case, there are fossil taxa with and without ganoine. The concentric rings in the scales of Teleostei are surely a derived state. The scales of *Amia*, in contrast, show a subparallel arrangement of superficial stripes. This pattern also occurs in various other Neopterygii, but can as yet not be satisfactorily explained because of the disjunctive distribution (GARDINER et al. 1996).

Sarcopterygii

The Sarcopterygii (fleshy fins) encompass three high-ranking entities with recent members. These are the Actinistia with fossil remains from the Devonian era and the well-known recent *Latimeria chalumnae*, the Dipnoi with six currently existing species and the Tetrapoda that have evolved to a species-rich entity after their conquest of the terrestrial environment.

At present, there are two hypotheses, each with prominent supporters, for the mutual relationship of the monophyla: (1) Actinistia and Tetrapoda are sister groups (SCHULTZE 1986, 1991, 1994) and (2) Dipnoi and Tetrapoda form adelphotaxa (AHLBERG 1991; JANVIER 1998). I adopt the second hypothesis – especially because the as yet hardly considered agreement in the existence of a lymphatic system appears to be a convincing synapomorphy of the Dipnoi and Tetrapoda.

At first, however, we are concerned with the characterization of the Sarcopterygii in their entirety as a monophylum. What can we postulate with good justification as evolutionary novelties for the last common stem species of the recent Actinistia, Dipnoi and Tetrapoda?

Autapomorphies (Fig. 74 → 3; 78 → 1)

- Monobasally paired fins (Fig. 77).
  Only the metapterygium remains in the endoskeleton of the pectoral and pelvic fins; all other fin rays are reduced. The fins each insert with a single skeletal element at the shoulder or pelvic girdle. This first mesomer of the aquatic Sarcopterygii corresponds to the humerus and femur of the terrestrial Tetrapoda.
  The musculature of the fins forms a club-like mass around the axis of the skeleton (? Synapomorphy with the muscular basal section of the fins in the Cladistia and then a plesiomorphy for the Sarcopterygii).
- Genuine enamel covers the entire tooth. Apomorphous state in comparison with the adelphotaxon Actinopterygii in which the prismatic enamel of ectodermal origin only covers the base of the tooth.
- Sclerotic rings of more than five bone plates around the eyes.
- Intercranial joint.
  Latimeria has a "kinetic skull". A joint-like cleft in the skull roof divides the braincase into an anterior ethmosphenoid and a posterior otooccipital. *Latimeria* is the sole recent member of the Sarcopterygii with an intracranial joint. Among the proof in various fossil Sarcopterygii the † Osteolepiformes from the stem lineage of the Tetrapoda are of particular interest. This situation supports the interpretation of the intracranial cleft as an apomorphous ground pattern character of the Sarcopterygii

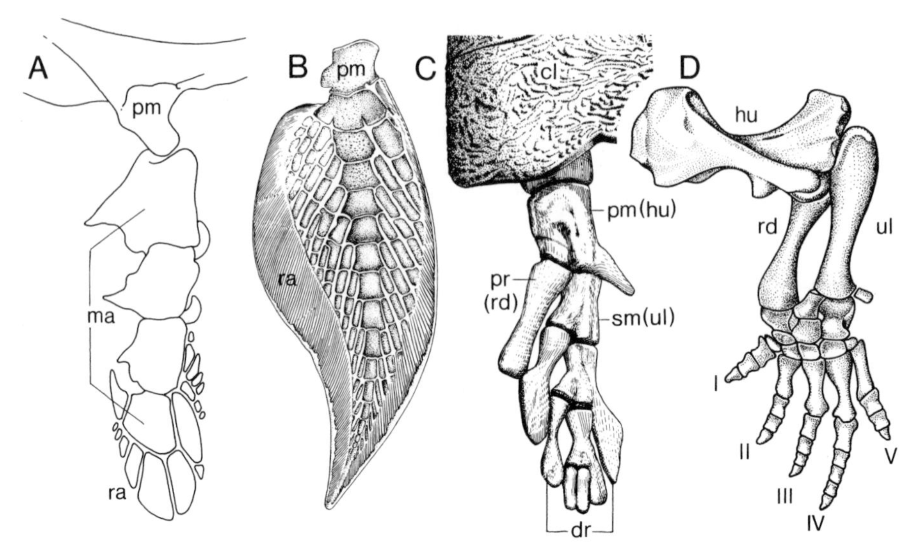

Fig. 77. Skeleton of paired extremities in the Sarcopterygii. Anterior extremities with monobasic articulation at the shoulder girdle. **A** *Latimeria chalumnae* (Actinistia). Left pectoral fin. Skeleton with a row of mesomers that are interpreted as elements of the metapterygial axis. **B** *Neoceratodus forsteri* (Dipnoi). Right pectoral fin. Metapterygial skeletal axis with two rows of regularly arranged radials. **C** † *Eusthenopteron foordi* (Osteolepiformes). Stem lineage member of the Tetrapoda. Left pectoral fin from the outside. Bones are partly homologized with elements of the Tetrapoda extremities. **D** † *Ophiacodon* ("Pelycosauria"). Stem lineage member of the Mammalia. Left arm. Primitive five-rayed extremity of the Tetrapoda. *cl* Cleithrum; *dr* distal radials; *hu* humerus; *ma* metapterygial axis; *pm* proximal mesomer; *pr* proximal radial; *ra* radials; *rd* radius; *sm* second mesomer; *ul* ulna. (**A** Rosen et al. 1981; **B, D** Starck 1979; **C** Jarvik 1980)

that has been lost in the Dipnoi and in the lineage to the recent Tetrapoda.

- Vena cava posterior (vena cava caudalis).
  The lower vena cava is a new vessel that carries all blood from the posterior body to the heart. The original cardinal veins disappear.

Cosmoid scales with a peripheral pore-canal system (p. 199) could form a further autapomorphy, but possibly are a plesiomorphy in the ground pattern of the Sarcopterygii. In any case, cosmoid scales must have been lost independently in the stem lineages of the Actinistia, Dipnoi and Tetrapoda.

Actinistia – Choanata

Actinistia

Latimeria chalumnae (length up to 1.8 m) is the sole recent species of the Actinistia with a well-known population in the region of the Comoro Archipelago. Rocky precipices of the islands at a depth of a few hundred meters are the habitat.

Latimeria chalumnae is a member of the supraspecific taxon Actinistia with a series of fossil taxa from the Devonian († *Miguashaia*, † *Diplocercides*) and Mesozoic eras († *Undina*, † *Macropoma*).

In comparison to the sister group Choanata, various **primitive features** from the ground pattern of the Gnathostomata are retained. *Latimeria* has a plesiomorphous fish nose with anterior and posterior nasal openings on the surface of the head (Fig. 80 A, B). *Latimeria* has a longitudinally extended heart with a linear arrangement of sinus venosus, atrium, ventricle and conus arteriosus. As in the fish-like Gnathostomata there is a primary and a secondary circulatory system (p. 215). *Latimeria* possesses plesiomorphous conical teeth without folds.

As the next step, we consider the characters of *Latimeria* the consecutive evolution of which in the lineage leading to *Latimeria chalumnae* is partially documented by the sequence in the character patterns of fossil stem lineage members (AHLBERG 1991).

Autapomorphies of *Latimeria chalumnae* (Fig. 78 → 2)

- Large rostral organ in the snout. It opens to the outside through several pores (Fig. 80 D). ? Electrosensory organ.
- Vertebral column highly reduced. Vertebrae only represented by thin ventral and dorsal arcualia (Fig. 80 E).
- Chorda dorsalis as an unconstricted, thick tube with fluid (Fig. 80 E).
- Bone skeleton extensively replaced with cartilage.
- Diphycercal tail fin with equally sized dorsal and ventral parts; these are separated by a small fin part in the extension of the chorda dorsalis (Fig. 80 A, B). († *Miguashaia* has a primitive, heterocercal tail fin).
- Basal muscular lobes in the second dorsal fin and in the anal fin. The skeleton of these two fins consists of a series of small elements in the lobes; they articulate with a single basal plate (Fig. 80 A, B).
- First dorsal fin with hollow spines (= fused lepidotrichia). The spines sit directly on the back.
- Unpaired sac of the intestine that is filled with fatty substances. It is interpreted as a homologue of the lung/air-bladder organ.
- Ovoviviparity. Large eggs remain in the body. Development of young animals in the oviducts.

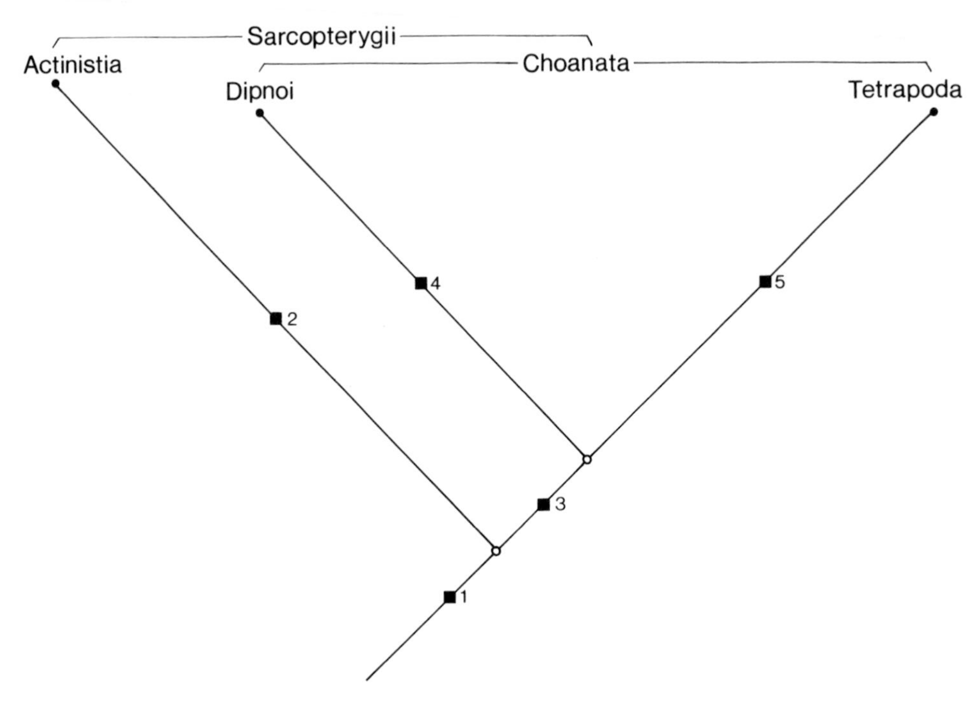

Fig. 78. Diagram of phylogenetic relationships of the Sarcopterygii with the Actinistia and Choanata as the highest-ranking adelphotaxa

Choanata

The established name Choanata for a monophylum from the Dipnoi and Tetrapoda has recently been questioned because the choana of these two entities are apparently not homologous. The name Choanata need not be discarded for this reason. Its retention seems to be decisively better than the lately suggested substitution by the name "Rhipidistia", i.e., a confusing extension of a name created for fossil taxa († Porolepiformes + † Osteolepiformes) to the recent Dipnoi and Tetrapoda (AHLBERG 1991; JANVIER 1998).

For the justification of the monophyly of the Choanata, there is one outstanding character from fossil remains and one from the organization of the recent representatives. We will discuss these within the framework of the following autapomorphies.

Autapomorphies (Fig. 78 → 3)

- Folded teeth with plicidentine (Fig. 81 A, B).
  Plicidentine is an orthodentine that is positioned towards the pulp of the teeth in folds with an almost vertically standing "folding axis"

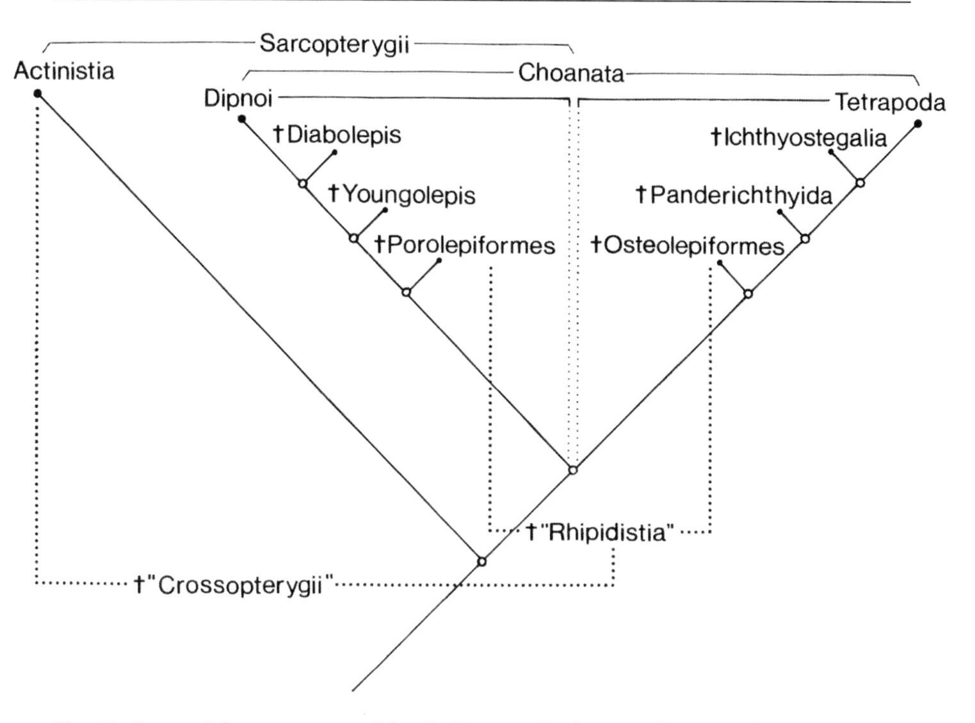

Fig. 79. Sequential arrangement of fossil Choanata in the stem lineages of the Dipnoi and Tetrapoda. Justification in text. Two measures follow from the position of the † Porolepiformes as a stem lineage member of the Dipnoi and of the † Osteolepiformes as a stem lineage member of the Tetrapoda. (1) In the first step, the taxon "Rhipidistia" encompassing them is abandoned as a paraphyletic, unnatural group of fossil Sarcopterygii; (2) thereafter, the † "Crossopterygii" as a union of the just eliminated † "Rhipidistia" and the Actinistia disappears from the phylogenetic system of the Sarcopterygii. (Original with use of data from Ahlberg 1991; Janvier 1998)

(SCHULTZE 1969). While only primitive, simple conical teeth occur in the Actinistia, folded teeth with plicidentine are realized in all fossil stem lineage members of the Tetrapoda and in practically all stem lineage members of the Dipnoi (exception † *Diabolepis*). This situation can lead to only one rational conclusion. Folded teeth with plicidentine evolved in the stem lineage of the Choanata. It is an irony of fate that folded teeth are absent in the recent Dipnoi and in the Amphibia among the Tetrapoda, which then inevitably requires the interpretation of a convergent secondary loss. Those who find it hard to accept this argumentation may be convinced by the supposed synapomorphous existence of a lymphatic system in the Dipnoi and Tetrapoda.

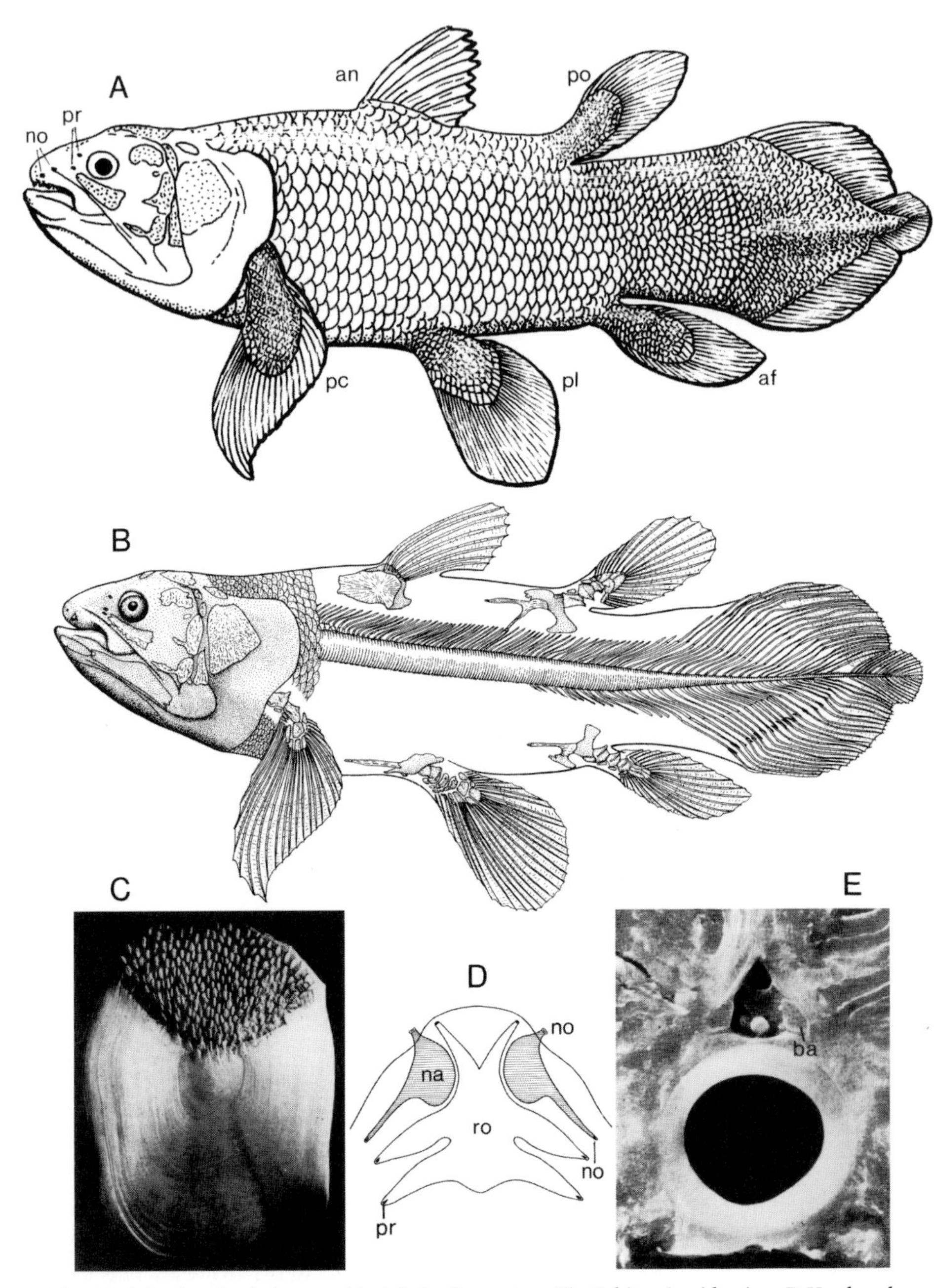

Fig. 80. A *Latimeria chalumnae* (Actinistia, Sarcopterygii). Habitus in side view. **B** Head and trunk skeleton. **C** Isolated scale of the skin. **D** Nose and unpaired rostral organ. Nose with anterior and posterior openings on the surface of the head. Rostral organ with three pairs of pores. **E** Cross section through chorda dorsalis and neural tube; lateral from this the basidorsalia. *af* Anal fin; *an* anterior dorsal fin; *ba* basidorsalia; *na* nose; *no* external nasal openings; *pc* pectoral fin; *pl* pelvic fin; *po* posterior dorsal fin; *pr* pores of the rostral organ; *ro* rostral organ. (**A** Janvier 1998; **B** Jarvik 1980; **C, E** Millot & Anthony 1958; **D** Starck 1982)

- Lymphatic system (Fig. 81 C, D).
  Cyclostomata, Chondrichthyes and Actinopterygii have a primary and a secondary circulatory system. The primary blood vessels supply the nerve system, the musculature and the visceral tract. The secondary system serves primarily to perfuse the inner and outer surfaces of organs (VOGEL 1985).
  Coming to the Sarcopterygii, *Latimeria* exhibits exactly this pattern with two circulatory systems. In contrast, Dipnoi and Tetrapoda have a single circulatory system, which is, however, supplemented by a new lymphatic vessel system. The tiny lymphatic vessels in connective tissue have no influx from the blood circulation; instead, they take up excess interstitial tissue liquids and transport them by micropumps to the venous part of the circulatory system (VOGEL & MATTHEUS 1998; VOGEL et al. 1998).
  In comparison with the two circulatory systems in the above-discussed "fish" including *Latimeria*, a uniform circulatory system with an additional lymphatic system is without doubt an apomorphy – in the most economical explanation, a synapomorphy of the Dipnoi and Tetrapoda.
- Agreements in the circulatory system.
  Lung arteries transport oxygen-depleted blood to the lungs. Lung veins combine to an unpaired trunk that leads oxygen-enriched blood to the atrium of the heart. In the hearts of Dipnoi and Tetrapoda the conus arteriosus has an S-shape.

Returning to the choana, according to the definition of the term as posterior nostrils into the oral cavity or palate, the Dipnoi and Tetrapoda do indeed both have choana. However, this alone does not tell us much. The agreement gave rise to a long discussion in regard to homology or convergence that obviously has now be decided in favor of the second alternative.

The taxon † *Diabolepis* from the stem lineage of the Dipnoi had the characteristic, apomorphous tooth plates of the recent lungfish (see below) in combination with the two external openings of a primitive fish nose (AHLBERG 1991; SCHULTZE 1991). The choana of the Dipnoi can therefore only have occurred in the stem lineage at a time after *Diabolepis* – presumably by migration of the posterior nostril inwards. Irrespective of how this may have happened, the simple facts demand the assumption of an independent occurrence of choana in the Dipnoi and Tetrapoda.

We thus come to the Tetrapoda. Here, there is a controversial discussion about the evolutionary pathway to the choana because a new element enters the game, the tear duct or nasolacrimal duct that joins the nasal cavity with the orbit. The duct as such is undisputedly a derived element of the Tetrapoda. Its existence has led, however, to two possibilities for the interpretation of choana in the Tetrapoda. (1) The posterior nasal openings have transformed to choana as in the Dipnoi; the tear duct is a new forma-

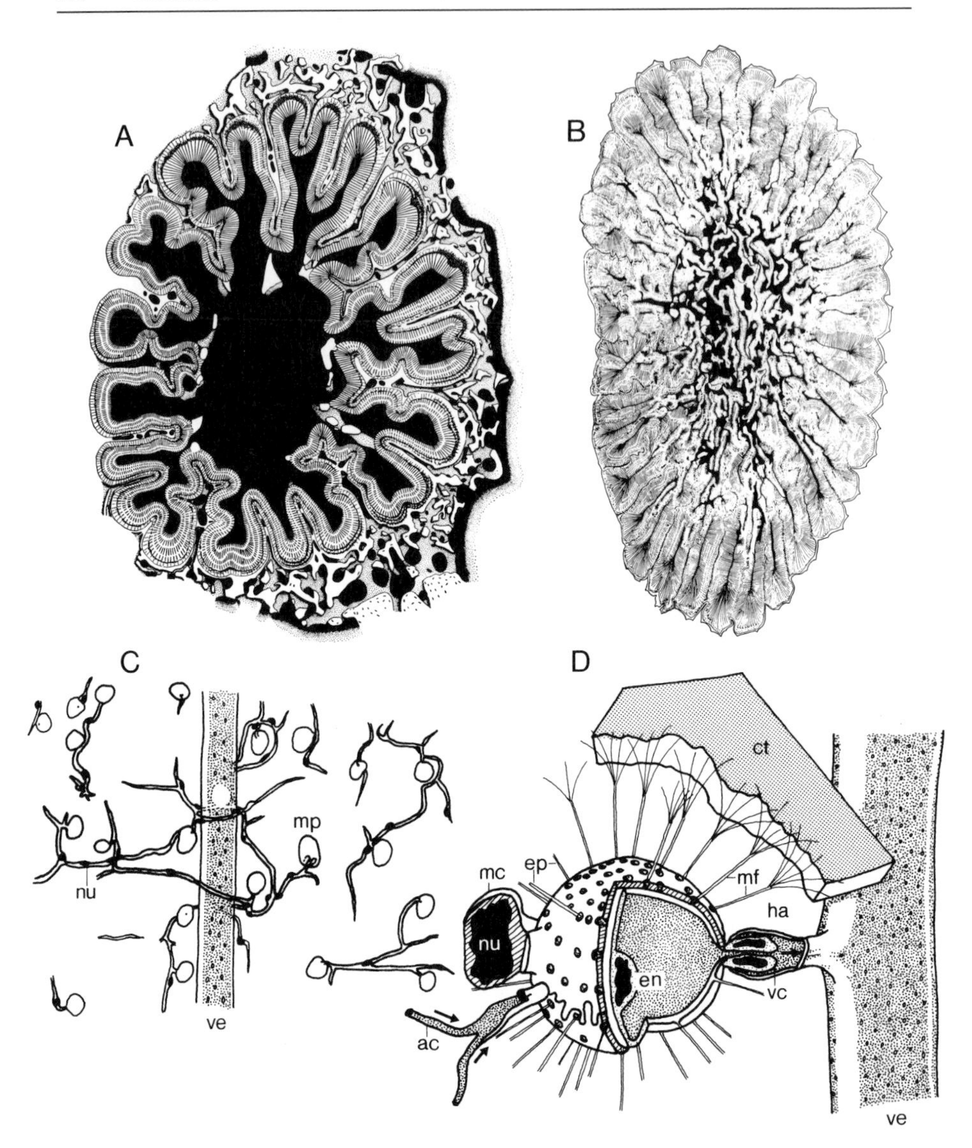

Fig. 81. Autapomorphies of the Choanata. **A, B** Folded teeth with plicidentine. **A** † *Eusthenopteron foordi* (Osteolepiformes). Horizontal cut through tooth that is surrounded by bone substance. **B** † *Porolepis* (Porolepiformes). Horizontal cut through fang of lower jaw. **C, D** Lymphatic vessel system of *Lepidosiren paradoxa* (Dipnoi). **C** Lymphatic capillary network with micropumps from the tail fin. **D** Diagram of a lymphatic micropump with connection to a vein. The pump is lined by an endothelial cell and surrounded by a muscle cell. A cell-free area occurs around the pump; it is crossed by a suspension apparatus of processes of the endothelial cell and microfilaments. Endothelial flaps at the in- and outlets of the pump control the fluid flow. *ac* Afferent capillary; *ct* connective tissue; *en* endothelial cell; *ep* process of endothelial cell; *ha* light area around the micropump; *mc* muscle cell; *mf* microfilament; *mp* micropump; *nu* nucleus; *vc* valve cell; *ve* vein. (**A, B** Schultze 1969; **C, D** Vogel & Mattheus 1998)

tion. (2) The connection to the eye arose from the posterior nostrils; choana are an evolutionary novelty of the Tetrapoda. The problem is still not resolved (JANVIER 1998).

Dipnoi – Tetrapoda

Dipnoi

The lungfish are inhabitants of fresh waters of the southern hemisphere. *Neoceratodus forsteri* (Australia) is the adelphotaxon of the Lepidosirenida with *Lepidosiren paradoxa* (South America) and four *Protopterus* species (Africa) (Fig. 82 A–D).

Neoceratodus forsteri. The existence of muscular fleshy fins from the ground pattern of the Sarcopterygii is primitive, but not the special construction of the fin skeleton (see below). Furthermore, the prevailing gill breathing is considered as a plesiomorphy. In contrast, the existence of an unpaired lung in the right body half is apomorphous.

Lepidosirenida. *Lepidosiren paradoxa, Protopterus aethiopicus, Protopterus annectens* and two further species. Paired lungs are plesiomorphous. However, in comparison with *Neoceratodus*, purely lung breathing is apomorphous. When the habitat waters dry up, members of the Lepidosirenida bury themselves in the ground and survive with maintenance of air breathing. An excellent synapomorphy of *Lepidosiren* and *Protopterus* is the transformation of the paired pectoral and pelvic fins to thread-like elements.

Autapomorphies (Fig. 78 → 4)

- Tooth plates.
  A pair of massive tooth plates each in the upper and lower jaw. Surface with dentine crests in a ray-like arrangement. Grinding system to break up hard objects (Fig. 82 E).
- Nose with choana.
  The anterior external openings of the nose lie on the border of the upper lip, the posterior openings = choana are in the roof of the mouth. The agreement with the choana of the Tetrapoda is interpreted as convergence (p. 215).
- Reduction of bone substance.
  The endoskeleton of the Dipnoi consists predominately of cartilage.
- Construction of the paired fins.
  In *Neoceratodus forsteri*, the pectoral and pelvic fins have a long central axis of many mesomers and numerous biserially arranged radialia (Figs.

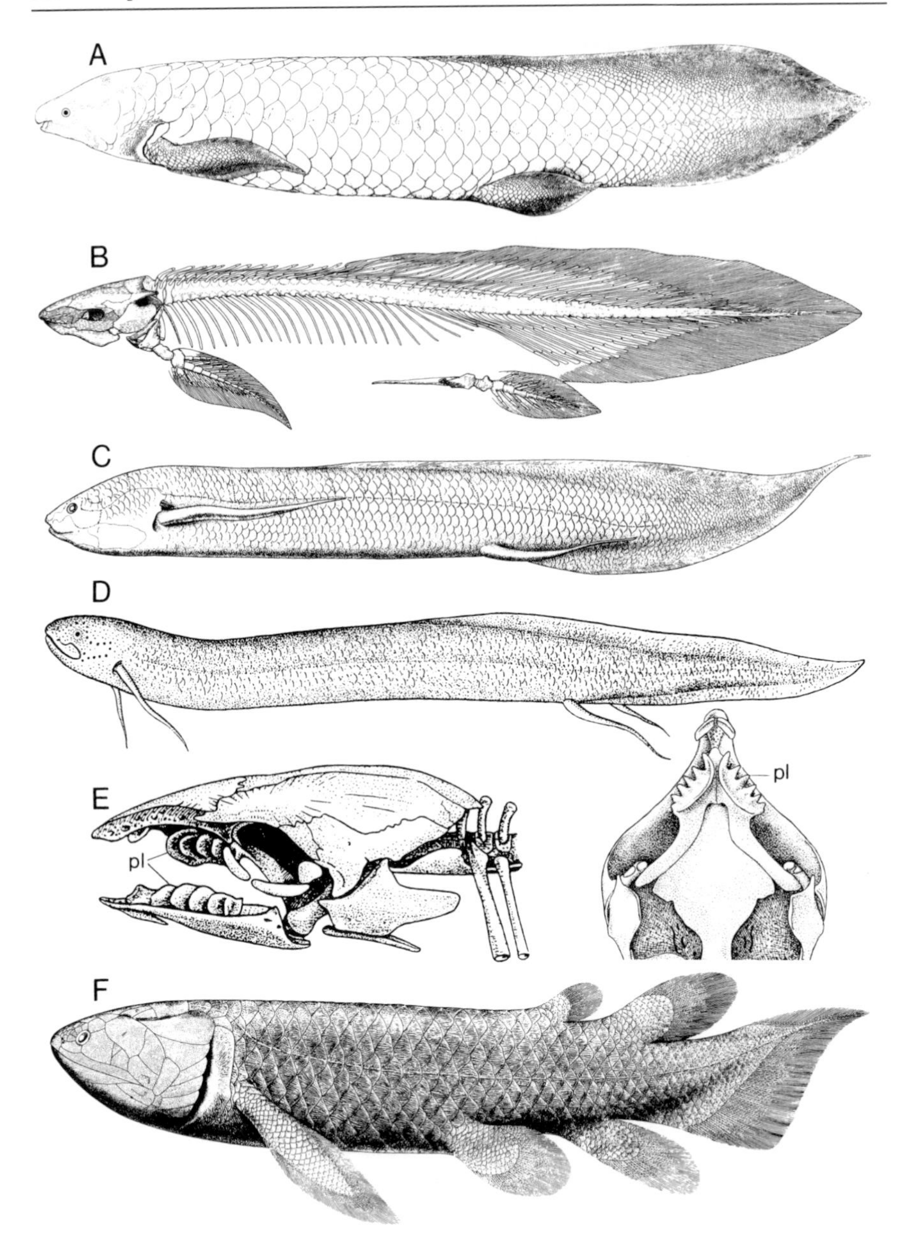

Fig. 82. Dipnoi (Sarcopterygii). **A** *Neoceratodus forsteri.* Australian lungfish. **B** Skeleton of *Neoceratodus.* **C** *Protopterus annectens.* African lungfish. **D** *Lepidosiren paradoxa.* South American lungfish. **E** *Neoceratodus.* Skull for demonstration of the tooth plates (*pl*) as autapomorphy of the Dipnoi. *Left* from the side, *right* from below. **F** † *Holoptychius* († Porolepiformes). Stem lineage member of the Dipnoi. (**A–C, F** Jarvik 1980; **D** Arambourg & Guibé 1958; **E** Starck 1979)

77 B, 82 B). This special construction presumably evolved first in the stem lineage of the Dipnoi. "The archipterygial type of fin skeleton is derived in every respect" (AHLBERG 1991, p. 276). The designation Archipterygium should no longer be used.
- In the stem lineage of the Dipnoi, a diphycercal tail fin with an unpaired border that runs continuously around the fin evolved. The † Porolepiformes as stem lineage members of the Dipnoi demonstrate the primitive development with heterocercal tail fin, with two separate dorsal fins and an anal fin (Fig. 82 F; ground pattern of the Gnathostomata).

Tetrapoda

The majority of characters of the quadrupeds are unequivocally linked with the elemental change of the habitat from water to dry land. Particularly conspicuous is the evolution of quadrupedal locomotion – the transformation of paired swimming fins to four legs for walking. In this process, five rays (digits) arose at the anterior and posterior extremities from five radials of the fins.

The hypothesis of the evolution of pentadactyl legs seems simple, however, fossil remains have given cause for irritation. There were apparently stem lineage members of the Tetrapoda with more than five digits; reconstructions of † *Acanthostega* have revealed eight fingers on the anterior extremities (COATES & CLACK 1990). This is perhaps spectacular – and has indeed led to the discussion of a large number of digits for the definition of the Tetrapoda (JANVIER 1998) and that, in turn, provokes fundamental considerations for the establishment of ground patterns. Nature may have tried out many things in the stem lineage of the Tetrapoda, but this is not relevant for our task of determining the character pattern of the last common stem species of the recent Tetrapoda before splitting into the stem lineages of the adelphotaxa Amphibia and Amniota. Here, there are not more than five digits, either in the Amphibia or in the Amniota. Thus, we can postulate with desirable certainty five digits on the anterior and posterior extremities of the last common stem species of the recent tetrapods and accordingly set this number in the ground pattern of the Tetrapoda.

Autapomorphies (Fig. 78 → 5; 84 → 1)

- Quadrupedy and pentadactyly.
  Evolution of four legs each with five rays from the paired pectoral and pelvic fins. The first mesomer of the extremities of the Sarcopterygii transforms anteriorly to the humerus, posteriorly to the femur. The first radial and the second mesomer convert to radius and ulna or, respectively, tibia and fibula. The phalanges originate from distal radialia of the fins.

- Stratum corneum.
  Deposition of keratin in the cells of the peripheral layer of the epidermis. One layer of cornified cells is realized in adults of the Amphibia.
- Autostylic skull.
  Firm linkage of the palaoquadrate arch with the neurocranium.
- Skull with unpaired condyle.
  A joint condyle composed of the basioccipital and the lateral exoccipitalia lies beneath the foramen magnum of the skull; it effects the connection to the vertebral column.
- Middle ear with stapes and tympanum.
  The hyomandibula from the hyoid arch is transformed into an ossiculum and incorporated in the space designated as middle ear (tympanic cavity). The stapes passes sound waves from the tympanic membrane through the tympanic cavity to the oval window at the inner ear. The tympanic membrane is interpreted as a transformation product of the reduced operculum, the middle ear as a derivative of the spiracular canal; the internal connection to the intestine remains intact as the Eustachian tube (auditory canal).
- Uniform vertebrae (?)
  All recent Tetrapoda possess uniform vertebrae with a center of ossification, while stem lineage members of the Tetrapoda are characterized by vertebrae with two centers, an anterior hypocentrum and a posterior pleurocentrum. A subject of dispute is whether the uniform vertebrae of the recent Tetrapoda evolved once from the pleurocentrum or possibly independently twice – in the Amphibia from the hypocentrum and only in the Amniota from the pleurocentrum. The solution is not important for the systematization of the Tetrapoda; however, in the latter case a uniform vertebra would have to be deleted from the list of Tetrapoda autapomorphies.
- Ribs with two connections to vertebral column.
  The ribs of the Tetrapoda have two proximal processes for articulation with the vertebrae, the dorsal tuberculum and the ventral capitulum. The sister group Dipnoi shows the plesiomorphous alternative of a single rib connection (capitulum).
- Dorsal extension of the pelvic girdle and connection with the vertebral column. The pelvis consist of ilium, ischium and pubis. The dorsal ilia make contact with a pair of modified sacral vertebrae.
- Nose with choana and lacrimal-nasal duct (p. 215).
- Lungs as the sole respiratory organ in the adult.
- Trachea and larynx.
  The trachea arose through lengthening of the unpaired connecting duct between the gut and the paired lungs. To separate the crossing pathways for air flow and food transport the larynx evolved at the beginning of the trachea, a closure apparatus with a cartilaginous laryngo-tracheal skeleton.

- Gills reduced; only present in larvae of the Amphibia. Gill slits closed, no operculum.
- Heart with interauricular septum.
  Complete division of atrium by a septum into a right and a left atrium.
- Eye glands and eyelids to moisten the eye surface and maintenance of the wetting.

On Fossil Records of the Choanata

The † Porolepiformes and the † Osteolepiformes are well known because together as the † "Rhipidistia" they form a subtaxon of † "Crossopterygii" in traditional classifications. † Porolepiformes and † Osteolepiformes, however, belong in two different stem lineages – the former in the stem lineage of the Dipnoi and the latter in the stem lineage of the Tetrapoda (Figs. 79, 83). This statement has been recently validated in detail (AHLBERG 1991; JANVIER 1998) and irrevocably removes the Rhipidistia and also the Crossopterygii as valid units from the phylogenetic system of the Choanata; the names are to be deleted because they refer to paraphyletic accumulations of the fossil Sarcopterygii on the basis of plesiomorphous features.

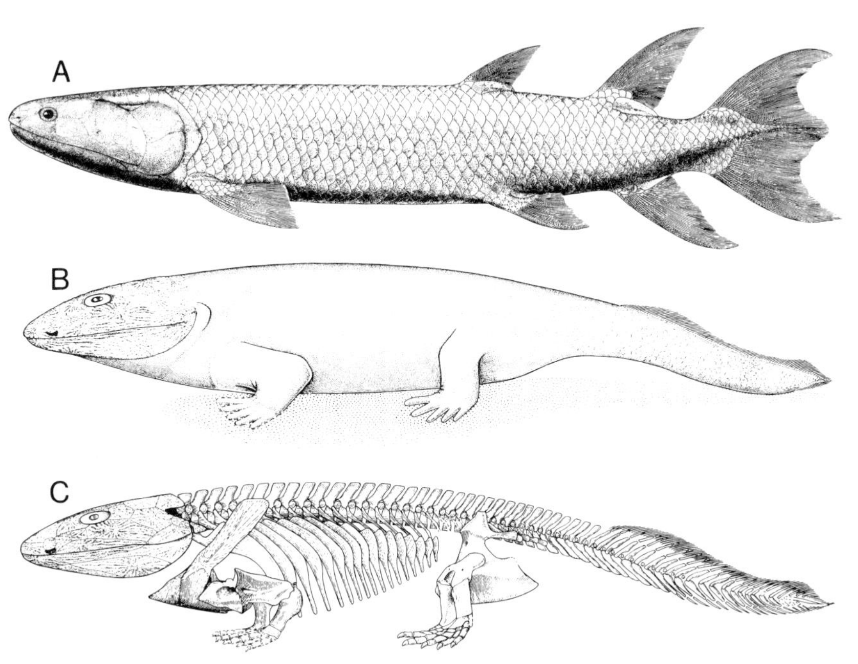

Fig. 83. Stem lineage members of the Tetrapoda (Sarcopterygii). **A** † *Eusthenopteron foordi* (Osteolepiformes). Habitus. **B, C** † *Ichthyostega* (Ichthyostegalia). Habitus and skeleton. (Jarvik 1980)

Apomorphous agreements between the † Porolepiformes and the Dipnoi are more than four mesomers in the pectoral fins as well as branched posterior radialia in the second dorsal fin. The Porolepiformes are characterized as a monophylum by "dendrodontic teeth" – a particular form of folded tooth; bone substance (osteodentine) is forced between the folds in the pulp cavity and can fill the latter completely (Fig. 81 B).

We come to the apomorphous agreements between the † Osteolepiformes and Tetrapoda. Both have only one pair of external nasal openings; this makes the existence of a Tetrapoda choana in the Osteolepiformes probable. In the fins of the Osteolepiformes, there are skeletal elements that can be homologized with humerus, radius and ulna or, respectively, femur, tibia and fibula of Tetrapoda legs. A characteristic apomorphous feature of the Osteolepiformes are enlarged scales at the base of the paired and unpaired fins.

In the stem lineage of the Tetrapoda, the † Osteolepiformes and † Panderichthyida are followed by a series of fossil taxa such as † *Acanthostega*, † *Ichthyostega* and † *Tulerpeton* in which the decisive transformation of the muscular swimming fins to walking legs with fingers and toes took place. These are late stem lineage representatives of the Tetrapoda with the ossiculum stapes (=hyomandibula) in the middle ear.

Systematization of the Vertebrata · Second Section

Tetrapoda
- **Amphibia**
  - **Batrachia**
    - **Urodela**
    - **Anura**
  - **Gymnophiona**
- **Amniota**
  - **Sauropsida**
    - **Chelonia**
    - **Diapsida**
      - **Lepidosauria**
        - **Rhynchocephalia**
        - **Squamata**
      - **Archosauria**
        - **Crocodylia**
        - **Aves**
  - **Mammalia**
    - **Monotremata**
    - **Theria**
      - **Marsupialia**
      - **Placentalia**

Amphibia – Amniota

The Amphibia and Amniota are discussed as highest-ranking adelphotaxa of the Tetrapoda. They are distinguished alternately by a combination of primitive and derived characters.

The elementary plesiomorphy of the Amphibia is the biphasic life cycle with change of habitat. The aquatic larva with gill breathing undergoes metamorphosis to a principally terrestrial organism with lung breathing. However, the latter remains closely linked with the aquatic milieu – probably because of a weak cornification of the epidermis (skin breathing) and a correspondingly low protection against desiccation.

On the basis of the primitive lifestyle with biphasic life cycle, the Amphibia have evolved a whole series of characteristic features. At this point, I mention in advance the reduction of the fifth finger, the division of the teeth into crown and pedicle, the existence of large palate windows or the connection of the skull with the vertebral column via two joint condyles.

The Amniota have achieved the complete emancipation from water as habitat. A decisive step was the evolution of the embryonic membranes amnion and serosa. Aquatic larvae with gill breathing disappeared. Eggs with shells could be laid on land. The direct development now took place in the amniotic cavity without metamorphosis – so to speak in its own tiny pond. The young animal hatched from the egg is equipped with a rigid horny skin, the multilayered stratum corneum of the epidermis for life on land.

On the other hand, there are various plesiomorphies in the ground pattern of the Amniota. In comparison to the mentioned characters of the Amphibia, these are the pentaradiate anterior extremities, the simple, uniform teeth, the closed palate and an unpaired occipital condyle.

Amphibia

It is necessary to start with a nomenclature problem of general importance for phylogenetic systematics. The recent Amphibia are often given the name Lissamphibia. The term was coined by HAECKEL (1866) for a union of the Urodela and Anura, later extended by GADOW (1896) to the Gymnophiona and adopted in the current literature (MILNER 1988; BOLT 1991; TRUEB & CLOUTIER 1991; SUMIDA & MARTIN 1997; JANVIER 1998).

The name Lissamphibia, however, is completely superfluous, may lead to confusion, and should be deleted. It is just as unnecessary as the name Neotetrapoda for the recent Tetrapoda[12] with the adelphotaxa Amphibia

[12] Or recent + fossil Tetrapoda with "five digits or less in both fore and hind limbs" (JANVIER 1998, p. 236).

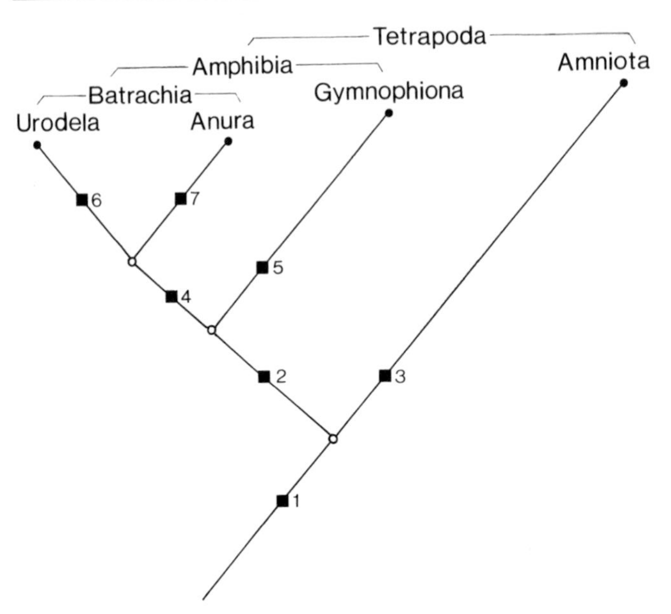

Fig. 84. Diagram of kinship relations of the Tetrapoda with the Amphibia and Amniota as sister groups and the highest-ranking adelphotaxa relationships in the Amphibia

and Amniota or the name Neornithes for the recent Aves with the sister groups Palaeognathae and Neognathae.

Lissamphibia is a synonym for Amphibia because there is no separate adelphotaxon among the recent vertebrates for an entity with this name. The Lissamphibia stand opposed to the entire fossil Amphibia – correctly without a proper name, because they do not form a monophylum. The problem is solved when the fossils are ordered in sequence in the stem lineage of an entity with the name Amphibia – namely, before splitting of the last common stem species in the lineages of the Batrachia (Urodela + Anura) and Gymnophonia. MILNER (1988) has presented a corresponding sequence of selected taxa of the paraphyletic † "Temnospondyli" in the stem lineage of the Amphibia.

Autapomorphies (Fig. 84 → 2)

A comprehensive analysis of the characters of the Amphibia and high-ranking subtaxa by PARSONS & WILLIAMS (1963) has been followed by some reviews (MILNER 1988; BOLT 1991; TRUEB & CLOUTIER 1991).

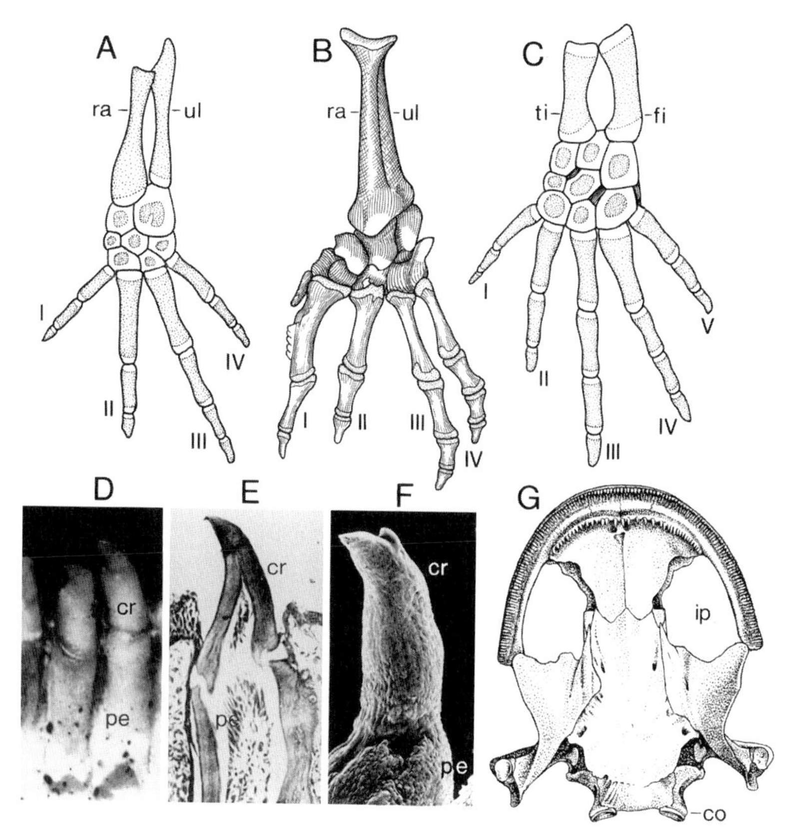

Fig. 85. Apomorphous characters in the ground pattern of the Amphibia. A, B Forelimb with four fingers. A *Triturus cristatus* (Urodela). B *Rana esculenta* (Anura). C Hind limb of *Triturus cristatus* with plesiomorphous number of five toes. D–F Two-part teeth (crown and pedicle). D *Amphiuma means* (Urodela). E *Schistometopum gregorii* (Gymnophiona). Longitudinal section with pulp cavity. F *Bombina orientalis* (Anura; SEM photograph). Tooth crown with two tips. G *Andrias* (*Megalobatrachus*) *japonicus*. Skull from ventral. Large interpterygoidal window. Two joint condyles. *co* Occipital condyle; *cr* crown of tooth; *fi* fibula; *ip* interpterygoidal window; *pe* pedicle of tooth; *ra* radius; *ti* tibia; *ul* ulna. (A–C, G Starck 1979; D, E Parsons & Williams 1962; F Tesche & Greven 1989)

- Anterior extremity with four fingers (Fig. 85 A–C).

  In agreement Urodela and Anura have only four fingers. From the ground pattern of the pentadactyl Tetrapoda extremities, the fifth finger has been reduced in the stem lineage of the Amphibia, although an embryonal precursor is retained.

  Since anterior extremities with four fingers were already present in Paleozoic stem lineage members († *Eryops*), this state can be unequivocally set as an apomorphy in the ground pattern of the Amphibia. The complete reduction of extremities in the stem lineage of the Gymnophiona must have proceeded from an organism with four digits in the forelimbs.

- "Divided" (pedicellate) teeth with sharp points (Fig. 85 D–F).
  In the Urodela, Anura and Gymnophyona, the simple, cone-shaped teeth are divided into two sections: a basal foot (pedicle) and a distal crown. The uncalcified or weakly calcified circular groove is a weak point at which the crown can break off (PARSONS & WILLIAMS 1962).
  In addition, two tips on the crown of the tooth (bicuspidate teeth) belong in the ground pattern of the Amphibia.
- Papilla amphibiorum.
  Spots of neuroepithelium in the sacculus that respond to sound waves (changes of pressure in the perilymph). Only found in the Amphibia.
  The papilla amphibiorum is interpreted as a part of the papilla neglecta that is displaced from the utriculus into the sacculus (FRITSCH & WAKE 1988).
- Large interpterygoidal window in the palate.
  A closed roof of the mouth belongs in the ground pattern of the Tetrapoda - as is realized in the Amniota. Extensive holes in the palate of a flattened skull have occurred in the stem lineage of the Amphibia with strong reduction of the pterygoids (Fig. 85 G).
- Skull with two condyles.
  From the unpaired condyle in the ground pattern of the Tetrapoda (p. 220), the basioccipital is reduced. The remaining exoccipitals move apart and form paired condyles (Fig. 85 G).
- Operculum as additional middle ear bone.
  Second sound conducting element besides the stapes. The operculum lies in the oval window behind the footplate of the stapes; it is linked to the shoulder girdle by the opercular muscle.

The three traditional taxa Urodela (Caudata), Anura (Salientia) and Gymnophiona (Apoda) are each well corroborated as monophyla. Furthermore, the question of the phylogenetic relationships between the three taxa is answered. Urodela and Anura are adelphotaxa; they are combined under the name Batrachia (Paratoidia)[13]. Moreover, the Batrachia then form the sister group of the Gymnophiona.

[13] The old name Batrachia is to be preferred (JANVIER 1998) over Paratoidia (GARDINER 1982, 1983).

Batrachia – Gymnophiona

Batrachia

Autapomorphies (Fig. 84 → 4)

- Auditory system for low frequencies.
  The ear of adult Urodela and Anura includes a low-frequency hearing system that is unique among the vertebrates (MILNER 1988). Sound waves from the ground take the following pathway: forelimb → shoulder girdle → opercular muscle – opercular ossification in oval window (see above) → by means of fluid movements to the papilla amphibiorum of the sacculus in the inner ear. (The Anura possess, in addition, a high-frequency hearing system to pick up sound waves from the air.)
- Lack of macula (papilla) neglecta in the ear labyrinth.
- Modification of the pronephros for the transport of sperm.
- ? Lack of bone scales.
  Skin scales occur in the Gymnophiona; they are possibly primitive. In this case, their common absence in the Urodela and Anura could form an autapomorphy of the Batrachia.

Urodela – Anura

Urodela

In habitus with long trunk and tail (Fig. 86B), over the relatively short extremities and with the highly water-dependent lifestyle, the salamanders and newts exhibit pronounced plesiomorphous traits of the Amphibia and Batrachia. These circumstances again give rise to the widespread difficulty to convincingly justify the "relatively primitive" subtaxa of extensive, supraordinated entities as monophyla.

For the Urodela, there is also the repeated evolution of neotenous, lifelong aquatic species. In these cases, it is not possible to check the expression of adult features.

Autapomorphies (Fig. 84 → 6)

- Vertebra with two-part rib supports (Fig. 86C).
  The vertebrae have two closely placed processes on each side for attachment to the two heads of the rib.
- Atlas with tuberculum interglenoideum (Fig. 86D).
  The first vertebra, designated as the atlas, represents the neck region of the vertebral column; it articulates with the condyles of the skull via two large, lateral processes.

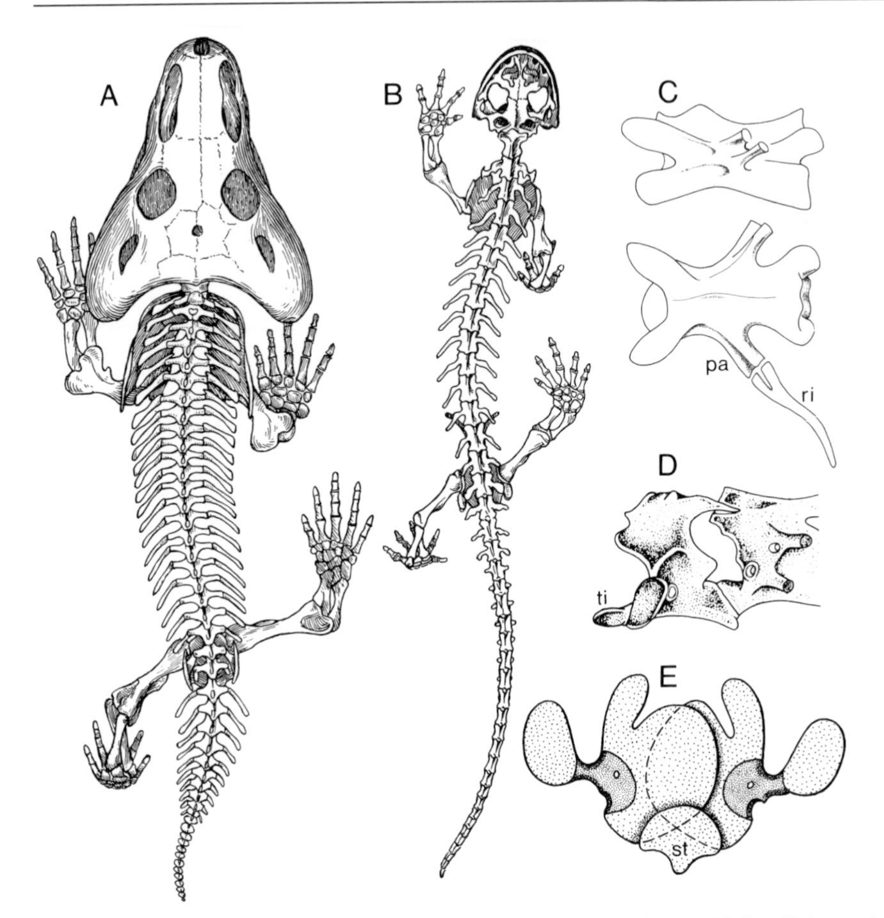

Fig. 86. Amphibia. **A** † *Trematops.* Stem lineage member of the Amphibia with five digits on the fore- and hindlimbs. The reduction of the fifth finger occurred in the stem lineage of the Amphibia. **B–E** Urodela. **B** Skeleton of *Salamandra.* Plesiomorphies are the extremely short limbs and the long tail. However, only four fingers on the forelimbs (autapomorphy of the Amphibia). **C** *Salamandra maculosa.* Trunk vertebra from the side and from above. On each side laterally two processes for articulation with the ribs. **D** *Eurycea.* First vertebra with rostrally oriented process set underneath (tuberculum interglenoideum). **E** *Hynobius peropus.* Reduced shoulder girdle of two cartilaginous plates and sternum. *pa* Parapophyses; *ri* rib; *st* sternum; *ti* tuberculum interglenoideum. (**A** Schaeffer 1941; **C, E** Starck 1979; **D** Wake 1979)

However, only now do we come to the specific character of the Urodela. The tuberculum interglenoideum is a process directed forwards with joint surfaces that reach into the foramen magnum of the skull (WAKE 1979).

- Reductions in the region of the shoulder girdle.
  The dermal shoulder girdle (cleithrum, clavicle, interclavicle) is completely absent. The girdle skeleton consists of two cartilage plates with processes directed ventrally (Fig. 86 E; STARCK 1979).
- Lack of tympanum and Eustachian tube in middle ear.

In comparison with the Anura, the Urodela with around 450 species forms a species-poor taxon. Geographically, the salamanders and newts are mainly limited to the northern hemisphere with a few settlements in the upper Amazon region being the only exceptions.

Andrias (Megalobatrachus) japonicus (Japanese giant salamander) reaches a length of 1.5 m. The adult remains in water after metamorphosis.

Ambystoma mexicanum (Axolotl) from Mexico. Classic example for neoteny. With experimental transformation in the adult state.

Salamandra with *S. salamandra* (fire salamander) and *S. atra* (alpine salamander, viviparous) in central Europe.

Triturus with four central European species: *T. cristatus* (crested newt), *T. vulgaris* (smooth newt), *T. alpestris* (alpine newt) and *T. helveticus* (palmate newt).

Proteus anguinus (European olm). Neotenous cave inhabitant of southern Europe.

Amphiuma means (two-toed amphiuma). From Virginia to Florida and Mississippi. Eel-like body with tiny limbs that each only carry two digits.

Siren lacertina (greater siren) in the east of North America. Hind limbs and pelvis completely reduced. Gills present life-long.

Anura

The enormous jumping capacity is certainly the most conspicuous feature in the behavior of the Salienta. Its evolution is connected with the acquisition of many novelties in the construction of the skeleton – especially the vertebral column and pelvis with hind limbs.

The jumping capacity probably occurred as a means of rapid flight into the protective milieu of water. However, there are various toads that cannot jump. This is apparently a secondary behavior within the Anura. On the North Sea island Amrum, I have often been fascinated by the distance-covering gait of the running toad; *Bufo calamita* is not able to jump.

The following autapomorphies are illustrated in Fig. 87.

Autapomorphies (Fig. 84 → 7)

- Shortening of the vertebral column.
  Seven or less vertebrae in front of the sacral vertebra.
- Urostyle (os coccygis).
  Fusion of postsacral vertebrae to a spine-like bone rod.
- Rod-like extended ilium of the pelvic girdle.
- Hind limbs longer than forelimbs.
- Hind limbs with following apomorphies:
- 1. Os cruris = fusion of tibia and fibula.

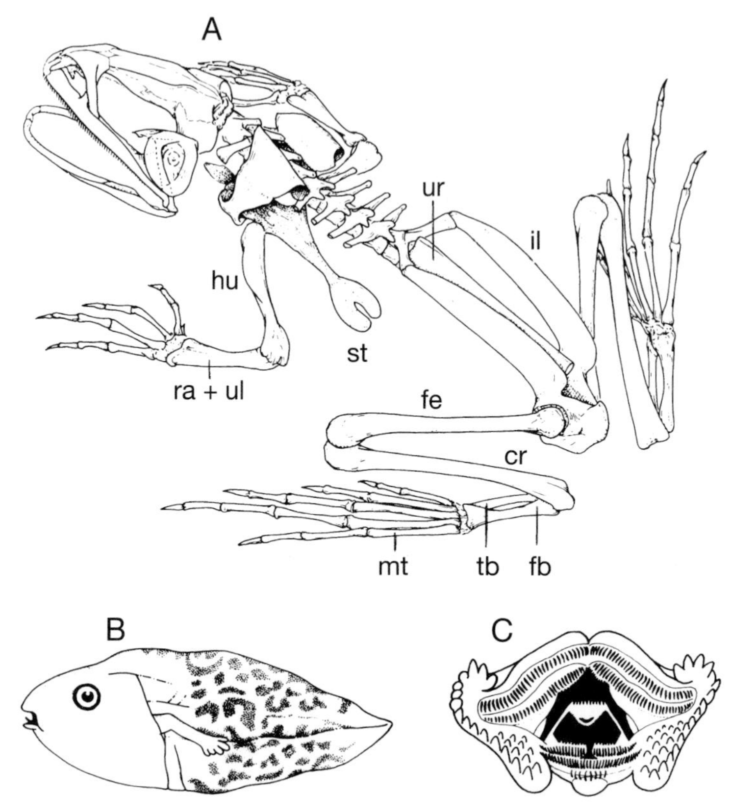

Fig. 87. Anura. **A** Skeleton of *Rana*. **B, C** Tadpole of *Hyla boulengeri*. **B** Side view. **C** View of mouth with horny jaw and horny teeth. *cr* Os cruris (tibia + fibula); *fe* femur; *fb* fibulare; *hu* humerus; *il* ilium; *mt* metatarsalia; *ra + ul* fusion product of radius and ulna; *st* sternum; *tb* tibiale; *ur* urostyle. (**A** Wake 1979; **B, C** Laurent 1986)

- 2. Lengthening of the proximal tarsalia (tibial and fibular).
- Forelimbs.
  Fusion of radius and ulna to a uniform bone rod.
- Larvae (tadpoles).
  Horny jaw and horn teeth. No genuine dentine teeth.

The Anura encompass around 4000 species; the monophylum is distributed worldwide. There is as yet no phylogenetic system for the Anura, and none for the Urodela. I will mention a few taxa, especially some from the central European fauna.

Ascaphus (North America) and *Leiopelma* (New Zealand) possess amphicoelous vertebrae; this is possibly a plesiomorphy.

Pipa pipa (Surinam toad, South America) with development of young animals in pouches on the back (♀) and *Xenopus laevis* (African clawed toad, laboratory animal) are secondarily tongueless frogs.

Alytes obstetricans (midwife toad), *Bombina bombina* (fire-bellied toad) and *Bombina variegata* (yellow-bellied toad) have a disk-shaped tongue.

Pelobates fuscus (garlic toad) possesses large webs on the hind limbs.

Rana with the widespread species *R. temporaria* (grass frog), *R. arvalis* (moor frog) and *R. esculenta* (edible frog).

Bufo with *B. bufo* (common toad), *B. viridis* (green toad) and *B. calamita* (natterjack). *Hyla arborea* (tree frog) has adhesive disks at the tips of the fingers and toes.

Gymnophiona

The terrestrial lifestyle in the soil forms a central apomorphy in the ground pattern of the Gymnophiona. The stem species of the recent Gymnophiona was undoubtedly a subterranean organism that burrowed through the soil or penetrated the substrate like an earthworm. Evolution has responded to the conquest of dark depths with reductive changes of the eyes. The old German name “Blindwühle” (blind burrower) well describes the two correlated features - even when the perception of different light intensities with the modified eye is probable.

Of course, numerous further autapomorphies are directly related to the subterranean existence - above all the lack of limbs.

Autapomorphies (Fig. 84 → 5)

- Worm-shaped, stretched trunk with ringing of the skin and an increased number of vertebrae.
  The number of presacral vertebrae varies between 95 and 285 (MILNER 1998). A precise value for the ground pattern is not possible.
- Shortened tail with position of the anus at the posterior end.
- Complete reduction of limbs, shoulder and pelvic girdles.
- Partial reduction of the eyes.
  The eyes are overgrown by thick skin. In the course of various transformations, the retina and optic nerve have been retained. Morphology of the eyes and behavior allow the assumption that at least a light/dark perception exists (WAKE 1985).
- Paired tentacles.
  Lateral sensory tentacles on the head; retractable into grooves. The modified musculus retractor bulbi from the transformed eye serves as the retractor (WAKE 1985).
- Phallodeum.
  An unpaired, extendable copulatory organ in the male (Fig. 88 C).
- Possession of more than 200 intersegmental lymphatic hearts under the skin.

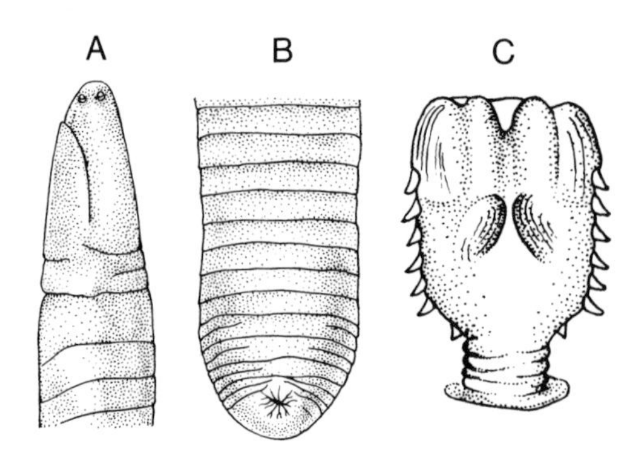

Fig. 88. Gymnophiona. **A, B** *Oscaecilia ochrocephala* with ringing of the body. **A** Head in side view. Tentacle groove between eye and mouth. **B** Posterior end with anal opening. Ventral view. **C** *Scolecomorphus uluguruensis*. Copulatory organ with spines. (Laurant 1986)

With circa 160 species, the Gymnophiona are characteristic inhabitants of the tropics (South America, Africa, Asia).

Ichthyophis, Rhinametra, Scolecomorphus, Caecilia, Siphonops, Typhlonectes (secondarily aquatic, flattening of posterior end to a rudder).

I refer to the attempt at a phylogenetic systematization of the entity by LESCURE et al. (1986).

Amniota

Autapomorphies (Fig. 84 → 3; 89 → 1)

- Monophasic life cycle with direct development.
  Eggs with shells laid on land. No larva, no metamorphosis.
- Meroblastic cleavage of yolk-rich eggs.
  The development of the eggs leads to the formation of a multilayered germ disk on the uncleaved yolk.
- Amnion and serosa.
  Ectodermal folds of the germ fuse over the core embryo to two embryonal membranes. The inner amnion encloses the embryo in the amniotic cavity. The external serosa encloses the entire germ (Fig. 90 A).
- Allantois.
  An extensive evagination of the ventral cloacal wall reaches far out from the core embryo and pushes itself between amnion and serosa. Embryonal urine organ (storage of uric acid), respiration and resorption organ.
- Yolk sac.
  The entoderm of the germ grows around the uncleaved yolk to form a closed yolk sac.
- Stratum corneum.
  Epidermis with multilayered callus of cells that have died due to deposition of keratin (Fig. 90 B).
- Convex occipital condyle, semicircular; well ossified.

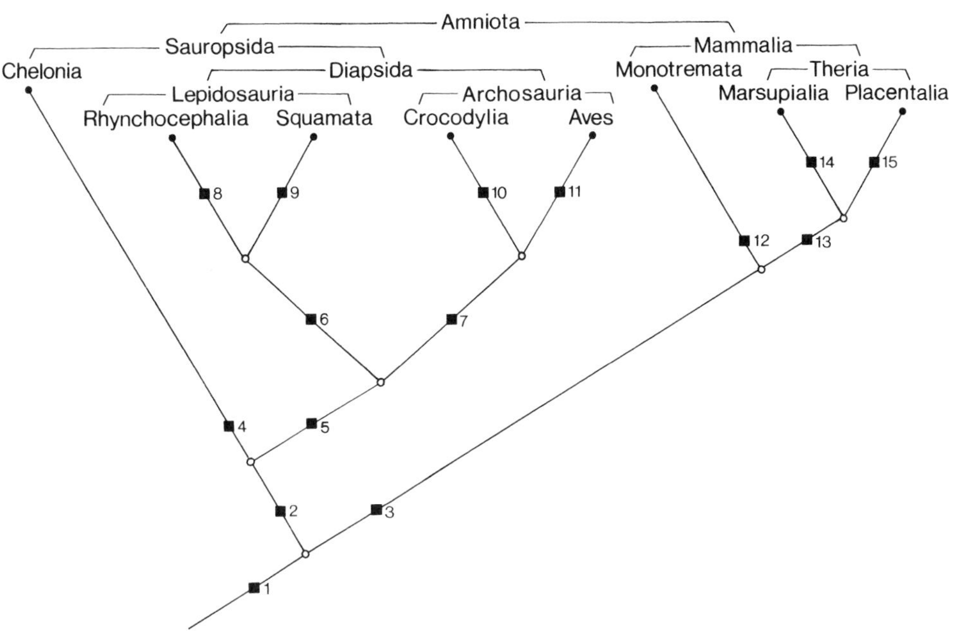

Fig. 89. Diagram of phylogenetic relationships of the Amniota with the sister groups Sauropsida and Mammalia and their further division in adelphotaxa

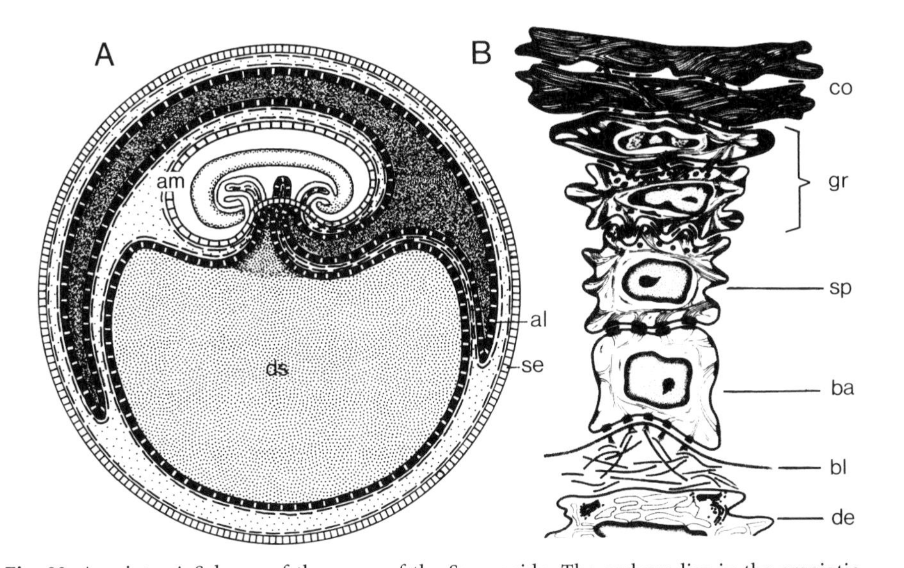

Fig. 90. Amniota. **A** Scheme of the germ of the Sauropsida. The embryo lies in the amniotic cavity; it is enclosed by the amnion. The serosa surrounds the entire germ. The allantois spreads out beneath the serosa. Yolk sac as a structure of the entoderm. **B** Epidermis with stratum corneum. Layers of the epidermis of the Mammalia in the example of a newly born mouse. *al* Allantois; *am* amnion; *ba* stratum basale; *bl* basal lamina; *co* stratum corneum; *de* dermis; *ds* yolk sac; *gr* stratum granulosum; *se* serosa; *sp* stratum spinosum. (**A** Fioroni 1987; **B** Fusenig 1986)

(A concave condyle is postulated for the ground pattern of the Tetrapoda.)
- Large exoccipitalia join over the basioccipital and under the occipital hole of the skull (foramen magnum).
- Evolution of an unpaired copulatory organ in the stem lineage of the Amniota (cf. p. 231).
  Convergence to the just discussed copulatory organ of the Gymnophiona. Otherwise the organ must have arisen in the stem lineage of the Tetrapoda and then have been given up in the Batrachia; this is however not very probable.

Sauropsida – Mammalia

Sauropsida

The monophyly of an entity Sauropsida (=Reptilia) with the adelphotaxa Chelonia and Diapsida still appears to be a subject of controversy. In any case, a monophylum with this name is not so convincing and universally justifiable as is the case for the Tetrapoda in their entirety, for the Aves as a subtaxon of the Sauropsida or for the sister group Mammalia.

From a broad-based analysis of the kinship relations of the Amniota (GAUTHIER et al. 1988a,b), I have chosen those characters that were recognized by BENTON (1990) and JANVIER (1998) as possible autapomorphies of an entity Sauropsida. In this case, I must also list very special features that can only be explained within the framework of comprehensive osteology.

In the modern phylogenetic literature the names Sauropsida and Reptilia are considered to be synonyms and are used arbitrarily. In general, however, the name Reptilia continues to represent a collection of turtles (tortoises), the tuatara, lizards and snakes as well as crocodiles – and that just with the exclusion of the birds. "Reptilia" in this encirclement is a paraphylum and thus has no place in the phylogenetic system of the Amniota.

The name Reptilia may still be used for the commonly known reptiles. The name Sauropsida specifically created for the combination of reptiles and birds has clear advantages.

Autapomorphies (Fig. 89 → 2)

- Tabular is small or absent.
- Supratemporal is small or absent.
- Supraoccipital has an anterior crest.
- Suborbital window (foramen) in the palatine.
- A single coronoid in the lower jaw.

- Fusion of the first two neck vertebrae atlas and axis (epistropheus). Pleurocentrum of the atlas fuses with the intercenter of axis during development. (Plesiomorphous alternative in Amphibia and Mammalia: pleurocentrum of the atlas separate from the intercentrum of axis.)
- Lack of central in the region of the tarsal.
- External nasal glands lie outside the nasal capsule. (Plesiomorphous alternative in the Amphibia and Mammalia: the glands are inside the nasal capsule.)
- Eye muscles. Iris and ciliary muscle with cross-striated musculature. (Plesiomorphous alternative in Amphibia and Mammalia: smooth musculature.)

Chelonia - Diapsida

In the systematization of the Sauropsida, the hypothesis of an adelphotaxa relationship between the Chelonia with anapsid skulls (Fig. 91 C) and an entity Diapsida (Lepidosauria + Archosauria) with two pairs of temporal fenestrae in the skull is widely held today (GAUTHIER et al. 1988 a, b, 1988; LAURIEN & REISZ 1995; LEE 1995, 1997). However, there are also modern concepts about a relationship of the Chelonia with the Lepidosauria (DE BRAGA & RIEPPEL 1997) and the Archosauria (MINDELL et al. 1999) within the Diapsida.

First of all, the Chelonia can be most certainly identified as a monophylum. I mention in advance only the bone armour or the horny sheaths on the upper and lower jaws. The decisive question involves the interpretation of the closed roof of the skull in the Chelonia. Is the anapsid skull realized here a plesiomorphy in comparison with the remaining Saurospida or have the temporal fenestra disappeared secondarily, which would be required in the case of a closer relationship of the Chelonia with the Lepidosauria (sister groups) or the Archosauria? I tend to favor the more simple hypothesis of the existence of a closed roof of the skull in the ground pattern of the Saurospida, the retention of this state in the Chelonia as well as the single evolution of two postorbital temporal fenestra in the stem lineage of the Diapsida, which would corroborate the monophyly of an entity with this name.

The data on the construction of the heart and blood vessels in subtaxa of the Amniota going back to GOODRICH (1916) have become questionable for the analysis of phylogenetic relationships (BISHOP & FRIDAY 1988).

In addition, the extravagant classifications with the combination of fossil and recent organisms coalesce to a large extent in our systematization - that only accepts adelphotaxa relationships between entities with recent members – when the fossil taxa are not lifted to the level of recent monophyla, but are rather inserted in sequence in their stem lineages.

Chelonia

"And so we reach God's noblest creature – the turtle" (GAFFNEY & MEYLAN 1988). The monophyly of the turtles/tortoises has never been questioned. We concentrate the justification on some outstanding characters.

Autapomorphies (Fig. 89 → 4)

- Body armour or protective shell of bone (Fig. 91 B).
  Dorsal carapace of eight median neural plates (fused with the neural arches of eight trunk vertebrae), two rows of eight left and right abutting costalia (grown together with directly underlying ribs) and two rows of lateral marginal plates.
  Ventral plastron from transformed clavicle, interclavicle as well as 3–5 pairs of bone.
  Carapace and plastron are firmly bound laterally. The bone armour includes the shoulder and pelvic girdles. During ontogenesis, the scapulae were incorporated in the armour and moved under the ribs.
- Loss of teeth.
  Maxilla, premaxilla and dentary without teeth.
- Horny sheaths.
  Horny sheaths occur in place of teeth with which plant food can be ground.
- Parietal (pineal) opening absent.
- Reduction in the number of phalanges.
  Maximum number in the Chelonia: 2-3-3-5-3. (In ground pattern of the Amniota front 2-3-4-5-3, back 2-3-4-5-4.)

The traditional division of the Chelonia (Testudines) in the Pleurodira and Cryptodira encompasses two groups of turtles that can be justified as monophyla and at the same time represent adelphotaxa. The differing modes for retracting the head and neck under the protecting shell were not yet realized in stem lineage members of the Pleurodira and Cryptodira; they evolved first in the stem lineages of the two entities. Besides, GAFFNEY & MEYLAN (1988) proposed differing mechanisms of the jaw musculature as well as differences in the basicranial architecture as alternating autapomorphies of the Pleurodira and Cryptodira.

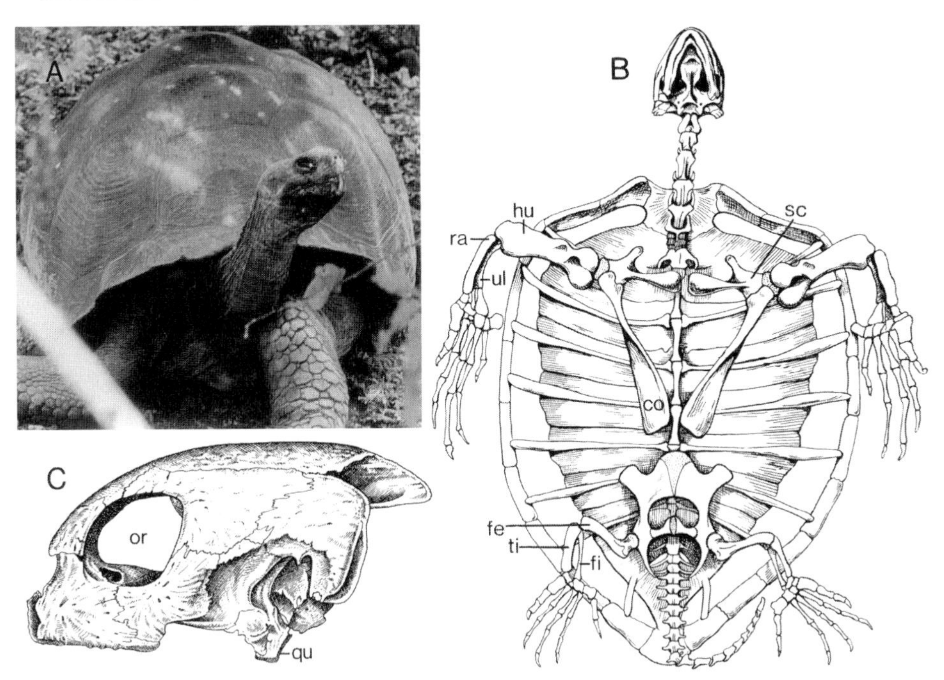

Fig. 91. Chelonia (Testudines). **A** *Geochelone elephantopus*. Giant tortoise of the Galapagos Islands (length of shell more than 1 m). **B** *Chelone*. Skeleton. Ventral part of shell removed. **C** *Chelonia mydas*. Anapsid skull in side view. *co* Coracoid; *fe* femur; *fi* fibula; *hu* humerus; *ma* marginal plate; *or* orbit; *qu* quadratum; *ra* radius; *sc* scapula; *ti* tibia; *ul* ulna. (**A** Original; **B, C** Young 1995)

Pleurodira

Autapomorphy. To take up a resting position the neck is bent to the side and placed between the back and belly shells.

Encompass barely one fifth of the around 220 recent turtles. Fresh water inhabitants of the southern continents.

Pelomedusa, Pelusios, Podocnemis, Chelus with *C. fimbratus* (mata-mata), *Chelodina, Hydromedusa.*

Cryptodira

Autapomorphy. The neck is drawn back in the sagittal plane and thereby takes on an S-shape.

Terricolous. *Testudo graeca* (Iberian tortoise, southern Europe), *Geochelone elephantopus* (giant tortoise of the Galapagos Islands).

Limnetic. *Chrysemis* (North America), *Emys orbicularis* (European pond turtle), *Trionyx* (representative of the soft-shelled turtles, predominantly northern hemisphere).

Marine. *Chelonia mydas* (green turtle), *Dermochelys coriacea* (leathery turtle), *Eretmochelys imbricata* (hawksbill turtle).

Diapsida

Autapomorphies (Fig. 89 → 5)

- Two temporal fenestrae.
  Here, it is not the holes in the skull as such, but rather the surrounding bone that constitutes the feature (GAFFNEY 1980). The upper window is surrounded by parietal, postorbital and squamose bones, the lower window of postorbital, squamose, jugal and quadratojugal bones. The temporal arch between the windows of postorbital and squamose bone is unique for the Diapsida (Fig. 92A, B).
- Suborbital window.
  Under the orbits on the underside of the skull are a pair of further windows. Surrounded by palatine, maxilla and ectopterygoid.
- The neck vertebrae are longer than the vertebrae in the middle of the back.

The Diapsida decompose into the adelphotaxa Lepidosauria (Rhynchocephalia + Squamata) and Archosauria (Crocodylia + Aves).

Lepidosauria – Archosauria

Lepidosauria

Autapomorphies (Fig. 89 → 6)

- Teeth superficially linked to the borders of the jaws.
  (Plesiomorphous alternative: teeth embedded in dental grooves are postulated for the ground pattern of the Amniota.)
- The condyles of the mandible are only formed from articular.
- Thyreoid window.
  In the pelvic girdle, there is a broad opening between pubis and ischium.
- Caudal autotomy.
  Preformed breaking point in the basal tail vertebrae.
- Astralagus and calcaneus of the tarsus fuse in young animals.
- Bending of metatarsal 5 in two planes.
- Cloacal slit placed transversely.
  (Primitive longitudinal slit in the Chelonia and Crocodylia.)

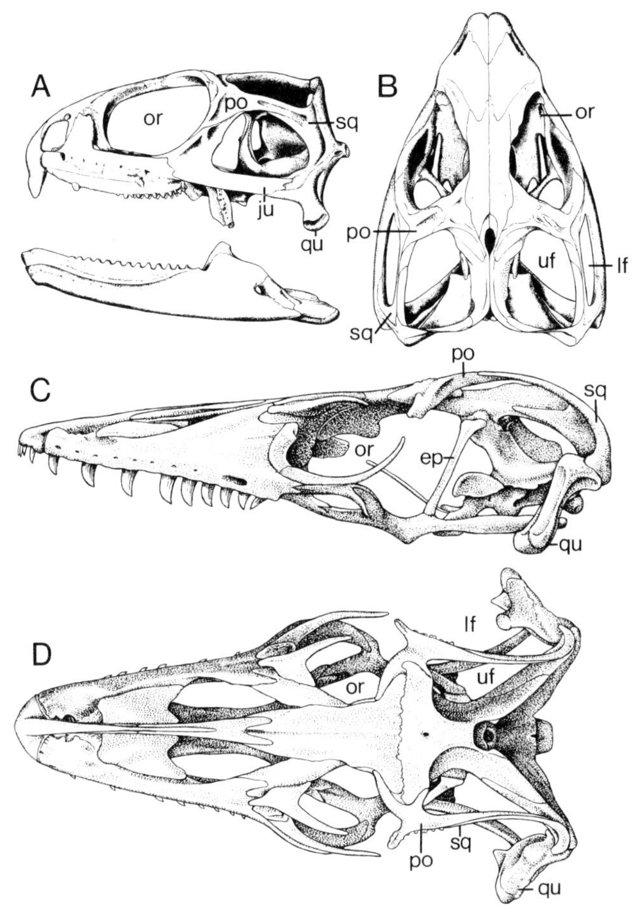

Fig. 92. Lepidosauria. Skull with construction of the temporal fenestra. **A, B** *Sphenodon punctatus* (Rhynchocephalia). Skull from the left and from dorsal. **C, D** *Varanus varanus* (Squamata). Skull from left and from dorsal. *ep* Epipterygoid; *ju* jugal; *lf* lower temporal window; *or* orbita; *po* postorbital; *qu* quadratum; *sq* squamosum; *uf* upper temporal window. (Starck 1979)

Rhynchocephalia – Squamata

Rhynchocephalia

The Rhynchocephalia are represented by two species in the recent fauna. The nocturnal tuatara *Sphenodon punctatus* and *Sphenodon guntheri* live on small islands off the coast of New Zealand (DAUGHERTY et al. 1990; POUGH et al. 1998).

The monophylum Rhynchocephalia has survived since the Trias; besides *Sphenodon* it includes a series of fossil taxa. The name tuatara is made up from the Maori tua=back and tara=spine. The animal exhibits a slim bone rod between the temporal fenestrae, which gave rise to the German name "Brückenechse". The existence of these two windows and the firm connection between the quadratum and the skull are pronounced plesiomorphous characters in comparison with the adelphotaxon Squamata.

Autapomorphies (Fig. 89 → 8)

- Fusion of teeth of the premaxilla that are longer than the teeth of the maxilla. This results in a beak-like extension of the upper jaw.
- Lack of lacrimal.
- Lack of external auditory canal and tympanic membrane.

Squamata

In colloquial language, lizards and snakes can together be unequivocally justified as a monophylum. Even so, the Lacertilia (Sauria) and Ophidia (Serpentes) are not sister groups. The "Lacertilia" (lizards) are an artificial collection of Squamata without any apomorphous characters. "The 'Lacertilia' is a paraphyletic group, and we recommend that use of the taxonomic term be avoided" (ESTES et al. 1988, p. 189). "The taxon 'Lacertilia' should be abandoned and taxa defined and named solely on the basis of monophyly" (SCHWENK 1988, p. 591). The latter applies to the Ophidia which, however, represent nothing more than a lower-ranking subtaxon of the Squamata.

Autapomorphies (Fig. 89 → 9)

- Skin with horny scales.
  Scales with differentiated micropatterns on the surface (Fig. 93 A).
- Tongue split distally into two parts.
- Lower temporal window opens ventrally (Fig. 92 D).
  The opening is a result of the complete reduction of the quadratojugal and a strong regression of the squamosum.
- Mobile quadratum.
  The loss of the lower temporal window is coupled with dissolution of the quadratum from the union of skull bones. The quadratum is now loosely attached to the skull.
- Hemipenis and hemiclitoris.
  The tip of the male copulatory organ is split in two (Fig. 93 B, C). This is unique among the Sauropsida (BÖHME 1988). The wide distribution of paired extendable and erectile structures in the root of the tail of females has been recently discovered. Named hemiclitoris, they are a smaller, mirror image of the male hemipenis (BÖHME 1995).

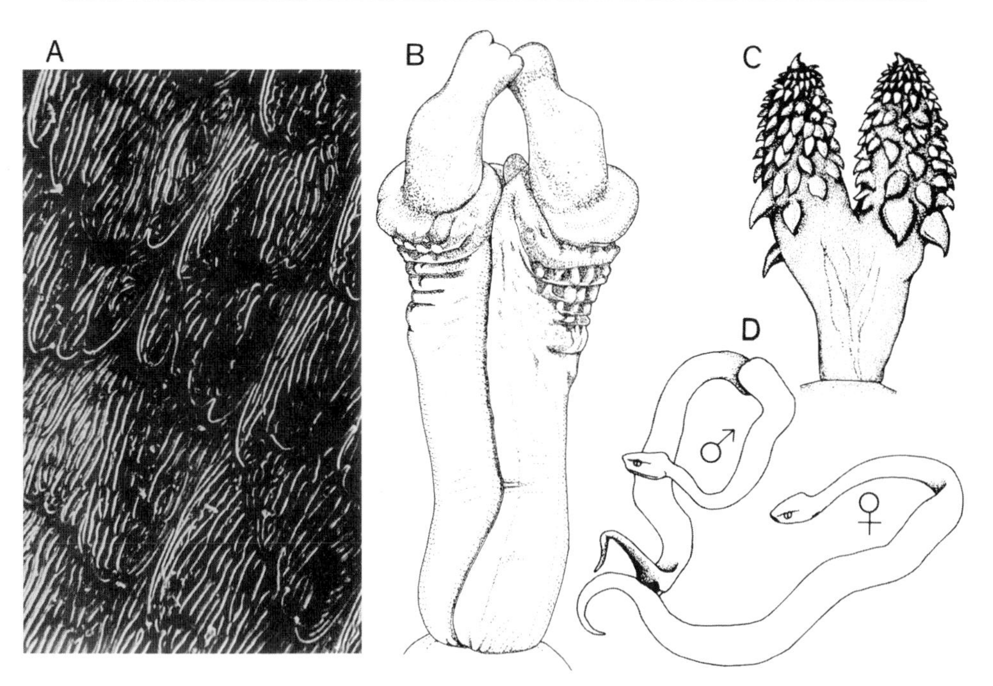

Fig. 93. Squamata. **A** *Natrix natrix* (grass snake). Section of a dorsal scale of the epidermis with microscopic ornamentation (REM photograph). **B** Hemipenis of the Komodo dragon *Varanus komodoensis*. With a median sulcus as transport groove for sperm. **C** *Vipera berus* (common viper). Hemipenis with dense coverage by spines. **D** *Vipera lebetina*. Coupling with the sexual partner by means of the hemipenis. (**A** Landmann 1986; **B–D** Böhme 1988)

For interpretation of the paired copulatory organs in the Squamata, the following considerations from the ambient of the Amniota may be put forward. An unpaired copulatory organ is present in the Chelonia and Crocodylia, in the ground pattern of the Aves (exists in the Ratitae) and in the Mammalia. This circumstance is explained most simply by the hypothesis of a single evolution of the organ in the stem lineage of the Amniota. Within the Lepidosauria a copulatory organ is absent in *Sphenodon punctatus* (Rhynchocephalia).

We have two possibilities for interpretation: (1) the unpaired organ was reduced in the stem lineage of the Lepidosauria, the hemipenis of the Squamata is a new formation; (2) the unpaired organ first regressed in the stem lineage of the Rhynchocephalia; the hemipenis arose from an evolutionary change of the unpaired copulatory organ. The latter seems to be the simpler explanation. Either way hemipenis and hemiclitoris represent an autapomorphy of the Squamata.

Systematization

In the first consequent phylogenetic systematization of the Squamata, the Iguanina and the Scleroglossa represent the highest ranking adelphotaxa (ESTES et al. 1988). I will place them in the following survey with a simple listing of subordinated taxa. Widely distributed central European species are given in parentheses.

Iguanina

"Iguanidae", "Agamidae", Chamaeleontidae.

Scleroglossa

Gekkota
  Gekkonidae, Pygopodidae.
Scincomorpha
  Scincidae, Cordylidae, Xantusiidae, Lacertidae (wall lizard *Podarcis muralis*, sand lizard *Lacerta agilis*, green lizard *L. viridis*, common lizard *L. vivipara*), Teiidae, Gymnophthalmidae.
Anguimorpha
  Xenosauridae, Anguidae (blindworm *Anguis fragilis*), Helodermatidae, Varanidae, *Lanthonotus borneensis*.

ESTES et al. (1988) handled the taxa Dibamidae, Amphisbaenia and Serpentes (Ophidia) (grass snake *Natrix natrix*, smooth snake *Coronella austriaca*, common viper *Vipera berus*) as Scleroglossa incerta sedis.

The Iguanina with fused frontalia and the Scleroglossa with a horny tongue are recognized in the newest contributions to the phylogenesis of the Squamata as valid, high-ranking system entities – the kinship relationships of the just mentioned Dibamidae, Amphisbaenia and Serpentes, however, are still a subject of controversy (HALLERMANN 1998; LEE 1998; CALDWELL 1999; RIEPPEL 1994, 2000). I cannot discuss this within the limits of a textbook nor repeat the validations for the monophyly of the diverse taxa mentioned because they mostly concern highly specific osteological characters. However, as most zoologists will expect at this point at least a justification of the snakes as a monophylum, I will list a few purported **autapomorphies of the Ophidia (Serpentes)** (RIEPPEL 1988, 1994; ESTES et al. 1988):

Compact skull capsule through complete closure of the lateral skull wall.

Exoccipitalia meet dorsally of the foramen magnum (exclusion of the supraoccipital from the edge of the occipital hole).

Vomer and septomaxilla form a closed capsule for Jacobson's organ.

Coverage of the footplate of the stapes by a circumfenestral crista.

Dentalia on the symphysis only loosely linked (enormous extensibility during feeding).

More than 120 vertebrae in front of the cloaca.

Other well-known "snake characters" such as the lack of extremities or the reduction of one lung cannot be evaluated as autapomorphies since they occur widely among highly lengthened Squamata and possibly form synapomorphies with certain other taxa of the Squamata.

Archosauria

Autapomorphies (Fig. 89 → 7)

- Thecodont dentition in the jaws.
  The existence of deep tooth grooves in the upper and lower jaws is a derived ground pattern character that evolved among the † "Thecodontia" in the stem lineage of the Archosauria.
  Thecodonty in the Mammalia must have occurred convergently.
- Flattened teeth.
  Teeth not round, but rather laterally compressed.
- Preorbital window (antorbital window).
  Opening in the side of the skull between nasal vestibule and the eye cavity.
- Mandibular window.
  Opening in lower jaw between dental, angular and supraangular (Figs. 94 C, 95 A).
- Fourth trochanter.
  Additional button-like muscle insertion on the femur.
- Pubis and ischium.
  Not plate-like, but rather rod-like lengthened and oriented posteriorly (Fig. 95 D, E). Together with the ilium a triradiate structure is formed.
- Fifth toe reduced.
  In the Crocodylia small (Fig. 94 E), completely absent in the Aves.
- Loss of urinary bladder.
- Transparent nictitating membrane in the eye.
- Eustachian tubes fused medially.
- Pulmonary diaphragm.
  Uniform pleuroperitoneal cavity (ground pattern of the Amniota, Lepidosauria, Chelonia) is divided by a septum into an unpaired peritoneal cavity and the two pleural cavities.
- Ventricles of the heart completely separated.

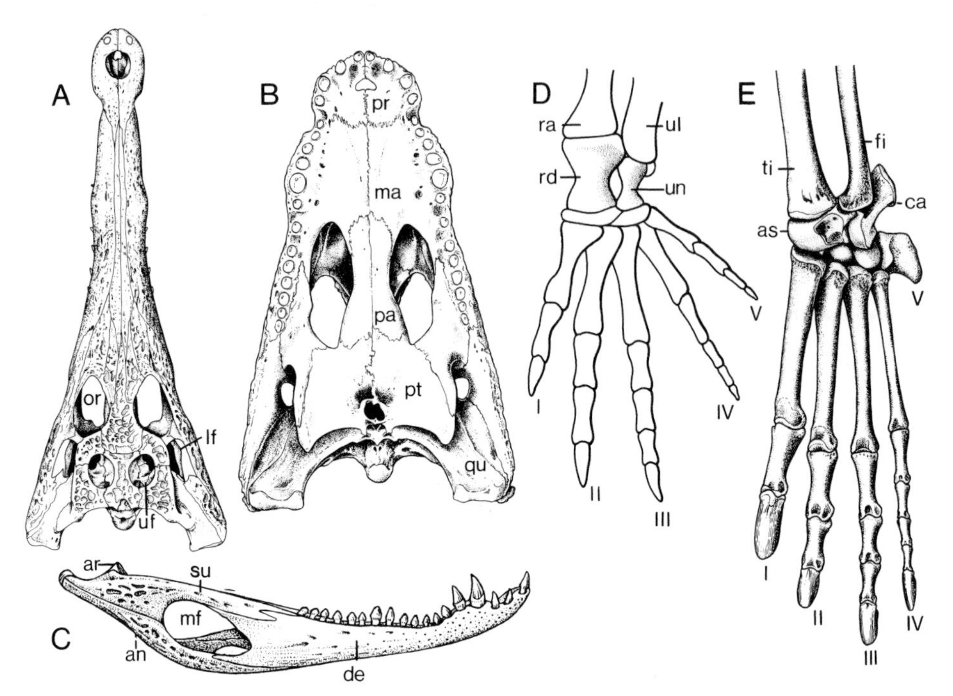

Fig. 94. Archosauria. Crocodylia. **A** *Crocodylus cataphractus* (long-snouted crocodile). Skull from dorsal. Two pairs of postorbital temporal windows as autapomorphy of the Diapsida. **B** *Osteolaemus tetraspis* (dwarf crocodile). Skull from ventral. Secondary palate from premaxilla, maxilla, palatinum and pterygoid as autapomorphy of the Crocodylia. **C** *Caiman latirostris* (broad-nosed caiman). Right lower jaw from lateral. Mandibular window as autapomorphy of the Archosauria. **D** *Crocodylus*. Hand skeleton. Lengthening of radiale and ulnare form an autapomorphy of the Crocodylia. **E** *Crocodylus porosus* (saltwater crocodile). Foot skeleton. Ball-and-socket joint between tarsal bones astragalus and calcaneus. Fifth toe reduced to a small piece of bone. *an* Angular; *ar* articular; *as* astragalus; *ca* calcaneus; *de* dental; *fi* fibula; *lf* lower temporal window; *ma* maxilla; *mf* mandibular window; *or* orbita; *pa* palatinum; *pr* premaxilla; *pt* pterygoid; *qu* quadratum; *ra* radius; *rd* radiale; *su* supraangular; *ti* tibia; *uf* upper temporal window; *ul* ulna; *un* ulnare. (Starck 1979)

Crocodylia - Aves

Crocodylia

The Crocodylia are secondarily hemiaquatic organisms. Various autapomorphies are linked with this lifestyle (HENNIG 1983).

Autapomorphies (Fig. 89 → 10)

- Unpaired nasal opening.
  Union of the two nasal openings on the tip of the snout. Elevated position, closable.

- Frontalia and parietalia fused.
- Loss of preorbital window in skull.
- Secondary palate.
  Premaxilla, maxilla, palatinum and pterygoid form a roof with which the choana are displaced posteriorly (Fig. 94 B).
- Long tail compressed at the sides.
- Lack of clavicle.
- Lengthening of the metacarpal bones radial and ulnar. Fourth and fifth fingers partially reduced, they have no claws.
- Joint between astragalus and calcaneus.
  In the tarsus astragalus and calcaneus articulate with double ball-and-socket joints. The foot together with the calcaneus can be rotated against the astragalus (STARCK 1979).
- Pubis without connection to acetabulum.
  Only ileum and ischium form the joint groove for the femur.
- Anus developed as longitudinal slit (convergence with the Chelonia).

Crocodylus, Alligator, Caiman, Gavialis, Tomistoma.

Aves

Let us start once again with an elementary conclusion that is valid for every monophyletic entity – perhaps it cannot be repeated often enough. The ground pattern autapomorphies of a monophylum belong in the character catalogue of the last common stem species of its recent members. In other words, the following autapomorphies characterize the feature pattern of the last common stem species of the Aves before splitting into the stem lineages of the Paleognathae and Neognathae. We will discuss the stem lineage of the Aves themselves and their stem lineage members later (p. 252).

Upon surveying the autapomorphies, we place the flying ability, the homoiothermy and the existence of feathers at the beginning (HENNIG 1983); as a complex of mutually linked features they determine numerous other characters in the organization of the birds (Fig. 95).

Autapomorphies (Fig. 89 → 11)

- Flying ability.
  The individuals of the stem species of the recent Aves were certainly organisms with the ability to fly. All flightless birds living today are secondarily unable to fly.
- Homoiothermy.
  Flight is associated with a fast metabolism that was achieved through the evolution of warm-bloodedness. In this connection with the flying

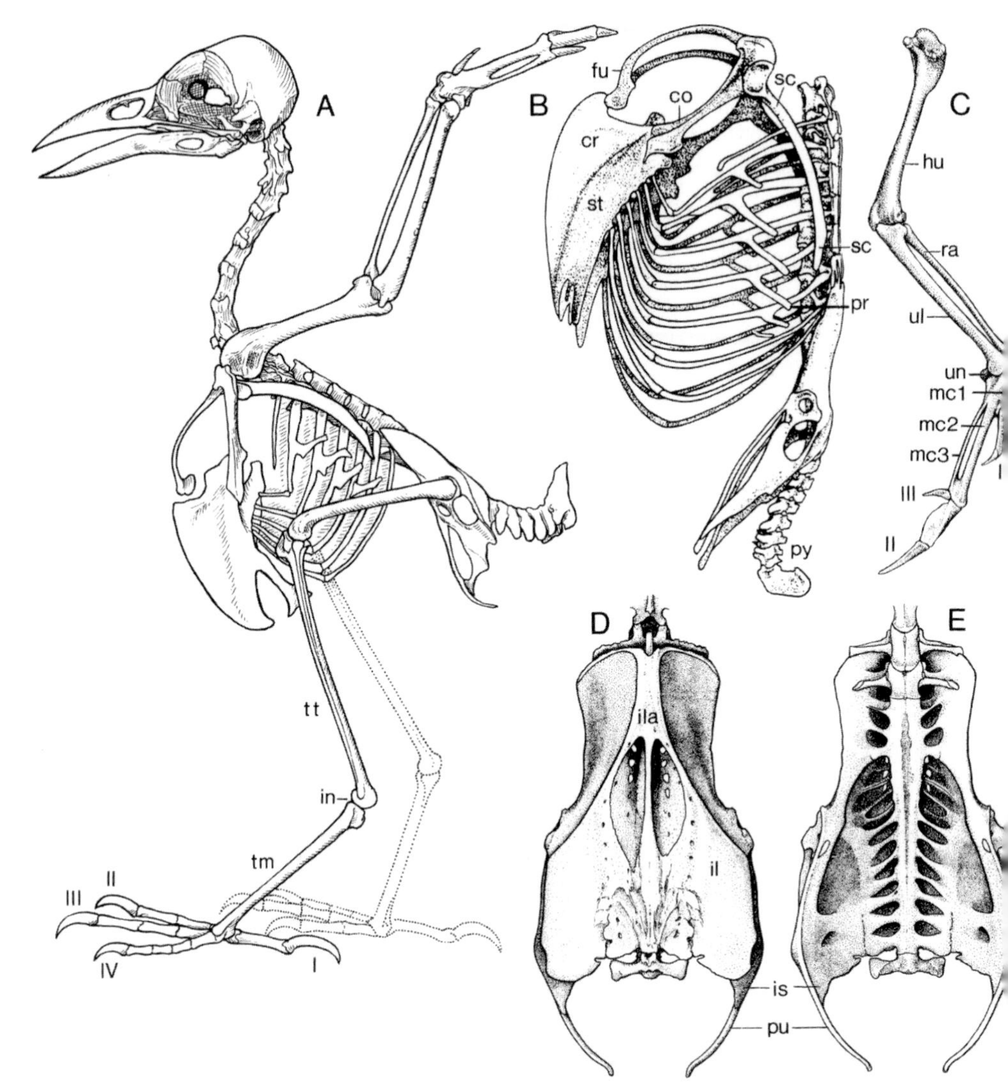

Fig. 95. Aves. **A** Skeleton of the magpie *Pica pica*. **B** Trunk skeleton of the herring gull *Larus argentatus*. **C** Wing skeleton of the purple gallinule *Porphyrio porphyrio*. Right wing from above. **D, E** Pelvis of the capercaillie *Tetrao urogallus*. **D** Dorsal view. **E** Ventral view. *co* Coracoid; *cr* crista sterni; *fu* furcula (fused clavicles); *hu* humerus; *il* ilium; *ila* fusion product of the ilia; *in* intertarsal joint; *is* ischium; *mc* metacarpal; *pr* uncinate process; *pu* pubis; *py* pygostyle; *ra* radius; *sc* scapula; *st* sternum; *tm* tarsometatarsus; *tt* tibiotarsus; *ul* ulna; *un* ulnare. (Starck 1979)

ability, homoiothermy of the birds must be interpreted as the result of a convergent evolution to the homoiothermy of the mammals.

- Feathers.

  Evolution of different feathers in the stem lineage of the Aves with double importance: (1) contour feathers (penna) for the flying apparatus,

(2) down (pluma) for protection against loss of warmth by the homoiothermic organism.

- Bill.
  Formation through elongation of the premaxilla.
- Lack of teeth - possession of horny sheaths - gizzard.
  The reduction of the teeth was followed by the evolution of horny sheaths on the bill and a gizzard inside the body.
- Fusion of temporal windows with each other and with the orbit.
  As result of reduction of the postorbital, the temporal arch between the two temporal windows that was adopted from the ground pattern of the Diapsida by the Archosauria has disappeared. In addition, since the original connection between the postorbital and jugal is lacking, the separation of the joined window from the eye cavity is also eliminated.
- Neognathy.
  In the neognathic palate, the pterygoid is divided into two, mutually articulated sections (p. 249). The anterior part is firmly joined to the palatinum, the posterior part articulates with the quadratum.
- Quadratum loosely joined with the neurocranium.
- Lack of ectopterygoid (transversum) in the roof of the palate.
- Prokinetic skull.
  The upper beak can be moved against the neurocranium. The flexion point between upper jaw skull and neurocranium lies in front of the eye cavities (preorbital).
- Saddle joint on the neck vertebrae.
  Saddle-like joint surfaces - convex in the sagittal plane, concave in the frontal plane - allow extensive bending of the neck.
- Rigid trunk skeleton with various apomorphous characters.
  Small number of free thoracic vertebrae (6–10) in front of the synsacrum. They are firmly linked with each other.
  Synsacrum. Several pre- and postsacral vertebrae fuse with the two primary sacral vertebrae.
  Pygostyle. Fusion of the last four to six tail vertebrae to a uniform bone piece.
  Long, thin scapula.
  Combination of the clavicles to a furcula. Sternum with crista sterni.
  Os innominatum. Fusion of the highly lengthened pelvic elements ilium, pubis and ischium to a bony unit.
  Thoracic vertebrae flattened, firmly linked to the sternum. Processus uncinati each lie on the following rib and are linked to this by connective tissue. This is a further contribution to the stability of the thorax.
  Lack of abdominal ribs (gastralia).

- Forelimbs as carriers of the remiges.
  The shoulder feathers emerge at the humerus, the secondary feathers at the strong ulna. Metacarpals II and III as well as the fingers following them are the carriers of the powerful primary feathers. The alula sit on finger I (thumb).
  The ulna as supporter of the secondary feathers is much more developed than the radius. In the ontogenesis fusion and transformation of the separately arranged carpalia occur in the carpus.
  Reduction of the rays IV and V.
  Metacarpals I, II and III fuse together.
  Finger I with one or two phalanges. Finger II with two (or three) phalanges. Finger III with one element.
- Construction of the posterior extremities for bipedal locomotion.
  Fusion of tibia and fibula. The fibula is only free proximally and participates here in the formation of the knee joint.
  Intertarsal joint between the bone parts tibiotarsus and tarsometatarsus.
  Tibiotarsus: fusion of the proximal tarsalia with each other and with the tibia.
  Tarsometatarsus: product of fusion of the distal tarsalia with the uniform metatarsus from metatarsalia II–IV. Metatarsal I remains free as a small skeletal part.
  Complete lack of the fifth toe. Partial reduction already present in the ground pattern of the Archosauria. Opposable thumbs. Rotation of the hallux to the back with opposition to the remaining three toes.
- Large eyes as central organs in the orientation system.
- Brain with enlargement of the integration sites tectum opticum, cerebellum and corpus striatum (basal ganglion) in the telencephalon.
- Syrinx as unique sound producing organ of the birds at the branching of the trachea into the two main bronchi.
- Division of the respiratory organ in the lungs as site of respiration and the air sacs in the body (ventilation apparatus) that even reach into cavities of the bones.
- Regression of the left efferent (from the right atrium) aortic arch.
- Oil glands (uropygial glands) in paired arrangement above the last tail vertebrae. The oily secretion keeps the feather keratin supple (STARCK 1982).

Systematization

Palaeognathae – Neognathae

The Palaeognathae with a rigid palate and the Neognathae with an intrapterygoidal joint in the palate are today hypothesized as the highest ranking sister groups of the recent Aves[14]. The accepted names, however, are not well chosen; the palaeognathic skull apparently represents an apomorphy within the birds, whereas the neognathic skull was adopted from the stem lineage of the Aves into the ground pattern of the birds and further taken over in the stem lineage of the Neognathae.

Palaeognathae

The flightless running birds of the southern continents (Ratitae) and the tinamou of South America (Tinamidae) form adelphotaxa of the Palaeognathae[15]. Their skulls are fundamentally similar and at the same time different from the skulls of other birds (CRACRAFT 1974).

Autapomorphies

- Palaeognathic skull.
  Suppression of the formation of an intrapterygoidal palate joint that belongs in the ground pattern of the Aves (neognathy, see above).
- Rhynchokinetics.
  Anterior displacement of the flexion position between neurocranium and upper jaw skull to the rostral third of the upper beak.
- Divided horny bill.
  The horny sheath of the bill in all species is divided into longitudinally oriented sections (Fig. 96) – in contrast to the uniform horn beak of most of the other birds (PARKES & CLARK 1966).
- Ilio-ischiadic window.
  In the pelvic girdle, the ileum and ischium do not fuse with each other; accordingly, there is a broad window between the bones.
  In the ground pattern of the Aves, the two bones are fused with retention of only a small window (see above).

The monophyly of the **Ratitae** is supported by 20 characters (LEE et al. 1997). Outstanding correlations with the secondary flightlessness include the loss of the crista sterni (insertion of flying musculature), the fusion of

[14] Support of the hypotheses by DNA hybridization (SILBLEY & AHLQUIST 1990).
[15] Support of the hypotheses by DNA hybridization (SILBLEY & AHLQUIST 1990).

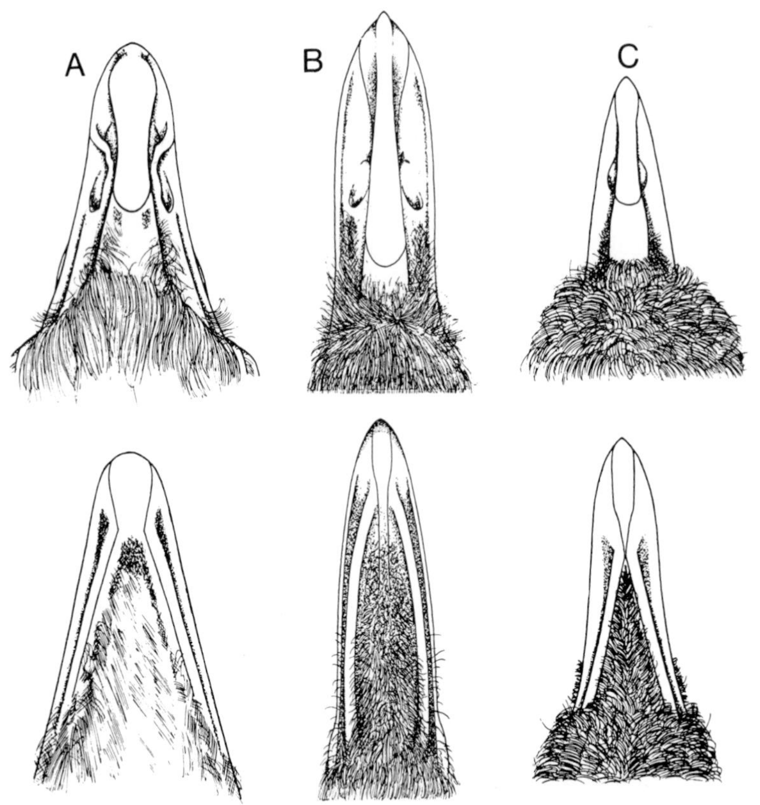

Fig. 96. Aves. Palaeognathae. Bill structure. **A** *Struthio camelus* (Ratitae). **B** *Rhea americana* (Ratitae). **C** *Crypturellus soui* (Tinamidae). *Above* Dorsal view of the bills. *Below* Ventral views. On the upper side, there is a flat crest with a swelling at the base, the anterior edge of which has a U or V shape. Ventrally, the bill is divided into three with a central, wedge-shaped piece in the tip. (Parkes & Clark 1966)

scapula and coracoid (scapulocoracoid) at an obtuse angle or the degeneration of wing elements (CRACRAFT 1974).

Living Ratitae are the ostrich (*Struthio camelus*) in Africa, the nandu (*Rhea americana, Pterocnemia pennata*) in South America, the cassowary (*Casuarius*, two species) in Australia and New Guinea, the emu (*Dromaius novaehollandiae*, second species extinct) in Australia and the kiwis (*Apteryx*, three species) in New Zealand.

The largest Ratitae disappeared from the Earth only after colonization of their habitats by man – the elephant birds of Madagascar with *Aepyornis maximus* and the moas of New Zealand with *Dinornis giganteus*.

Autapomorphies of the **Tinamidae** (Crypturi) probably constitute the lack of uncinate processes on the ribs, the very long lateral process at the posterior border of the sternum as well as the dorsal bone (notarium) as fusion of the last five thoracic vertebrae to a rigid bone element.

Otherwise, the Tinamidae are characterized by a series of plesiomorphies. In comparison with the listed autapomorphies of the Ratitae, these are the existence of wings, a crest on the sternum as well as scapula and coracoid as unfused bones. However, the Tinamidae are only poor flyers; they remain mostly on the ground.

With just about 50 species, the Tinamidae are limited to Central and South America. *Tinamus, Nothocerus, Nothoprocta.*

A closing remark about the Palaeognathae. Arguments against the evaluation of the longitudinal division of the horny bill as an autapomorphy of the Palaeognathae or the fusion of scapula and coracoid in correlation with flightlessness as an autapomorphy of the Ratitae have been put forward, namely that comparable phenomena occur here and there within subtaxa of the Neognathae (FEDUCCIA 1980). However, this is not relevant when we want to explain the existence of a divided bill as a common character of **all** Ratitae + Tinamidae and the rod-like fusion of scapula and coracoid as a common character of **all** Ratitae. The principle of parsimony demands here the hypothesis of a single evolution of the first mentioned feature in the stem lineage of the Palaeognathae as well as of the second feature in the stem lineage of the Ratitae.

Neognathae

The majority of the recent birds are generally united under the name Neognathae and placed as sister group alongside the above-mentioned Palaeognathae. Even so, the corroboration as a monophylum appears difficult – a problem that is usually ignored in textbooks. Of the characters discussed as autapomorphies (CRACRAFT 1988; HENNIG 1983), I will emphasize the construction of the quadratum.

Autapomorphy

- Quadratum with double condyles.
  Interpretation as apomorphy compared to the simple condyle that joins the quadratum to the squamosum in the Ratitae and Tinamidae.
  However, this character appears disputed; also for the Ratitae two condyles on the quadratum have been alleged (WHETSTONE 1983; THULBORN 1984).

On the Stem Lineage of the Aves

The Archosauria form a monophylum with the Crocodylia and Aves as highest-ranking subordinated sister groups. We have mentioned good reasons for this hypothesis above. In this situation, there are in principle three possibilities for the position of fossil Archosauria (provided that they are not to be classified as young fossils in subtaxa of the Crocodylia or Aves; Fig. 97).

A. Stem lineage of the Archosauria

The fossils belong in the stem lineage of the Archosauria when they exhibit even one autapomorphy of this entity, for example, the thecodont dentition or the preorbital window in the skull. The justification is more reliable when more, up to all autapomorphies that had evolved at the end of

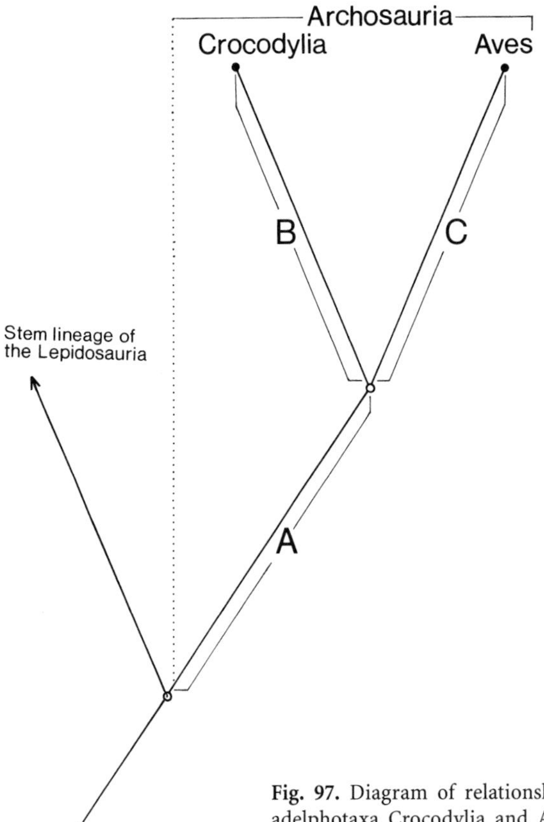

Fig. 97. Diagram of relationships of the Archosauria with the adelphotaxa Crocodylia and Aves. In the stem lineage concept **A** represents the stem lineage of the Archosauria, **B** the stem lineage of the Crocodylia and **C** the stem lineage of the Aves

the stem lineage of the Archosauria, are detectable. However, a representative of this stem lineage may not exhibit an autapomorphy of the Crocodylia or a derived character of the Aves.

B. Stem lineage of the Crocodylia

Members of the Archosauria known from fossil remains are identifiable as representatives of the stem lineage of the Crocodylia when they exhibit at least one autapomorphy from the ground pattern of this monophylum. This may be the unpaired nasal opening or the secondary palate, but just no apomorphy from the feature pattern of subordinated taxa of the crocodiles.

C. Stem lineage of the Aves

A corresponding argumentation holds for the birds as the sister group. Fossil Archosauria are then representatives of the stem lineage of the Aves when they possess one unique character, two or more, up to all, autapomorphies from the feature pattern of the last common stem species of the recent birds, but not derived characters of the subordinated units Palaeognathae or Neognathae.

We will look more closely at the **stem lineage of the Aves** and emphasize a central statement of the stem lineage concept (AX 1984–1989). For every monophylum with recent and fossil members, there is only one objective delineation that do justice to the kinship relationships of Nature. All fossils that are more closely related to recent members of the unit than to the sister group belong in the monophylum. In our example, all fossil taxa that are more closely related to the recent birds than to the Crocodylia are to be combined in a monophylum with the name Aves.

It is not relevant for this statement whether the fossil stem lineage members are similar or not to the modern representatives. One may find it difficult to accept gigantic dinosaurs such as † *Triceratops* or † *Tyrannosaurus* (Fig. 98) subsumed under the name Aves (GEE 2000) – in fact, they are representatives of the stem lineage of the birds and, as such, fossil members of the Aves (Fig. 99); moreover, *Tyrannosaurus* is purported to be even more closely related to the recent birds than † *Archaeopteryx* (THULBORN 1984).

However, let us put aside the provocation with the “Dinosauria” for a moment and start with the ubiquitously known ancestral bird *Archaeopteryx*.

In 1984, a scientific conference was held on a single fossil species – *Archaeopteryx lithographica* “the geologically oldest known bird” (HECHT et al. 1985, p. 7). It appears undisputed that this is a bird. One accepts without hesitation the extension of the name Aves to fossil remains and back into the Mesozoic era because *Archaeopteryx* possessed feathers and wings.

Fig. 98. † *Tyrannosaurus* ("Dinosauria"). Predator of 12 m length. North America. Member of the stem lineage of the Aves. (Charig 1983)

One also accepts tacitly that further outstanding bird features were "not yet" present – *Archaeopteryx* did not have a horny beak, had no saddle joints on the neck vertebrae and did not have a synsacrum in the pelvic region. So far so good – how should it be different? When the last common stem species of the recent Aves possessed 60 or more various autapomorphies that successively evolved in the stem lineage of the Aves, then it is to be expected that a certain species from the Mesozoic era would first exhibit a limited number of derived bird features. The argumentation becomes unacceptable, however, when one wants to define the extent and thus the limits of a unit Aves on the basis of the existence of feathers and accordingly to refer to *Archaeopteryx lithographica* as the ancestral bird or "oldest known bird". This is a subjective decision, a completely arbitrary intrusion in the sequence of the evolution of bird characters in the phylogenesis of the Aves – an artificial incision in a continuum of fossil taxa with growing relationships to the recent birds. The "Dinosauria" belong without doubt to these fossil stem lineage members.

"The Dinosauria, according to all recent cladistic analyses, form a monophyletic group" (BENTON & CLARK 1988, p. 315). "In other words, the 'Dinosauria' have not yet been shown to form a natural group; and it may well be that they never will – for the simple reason that the natural group in question may not have existed!" (CHARIG 1983, p. 15).

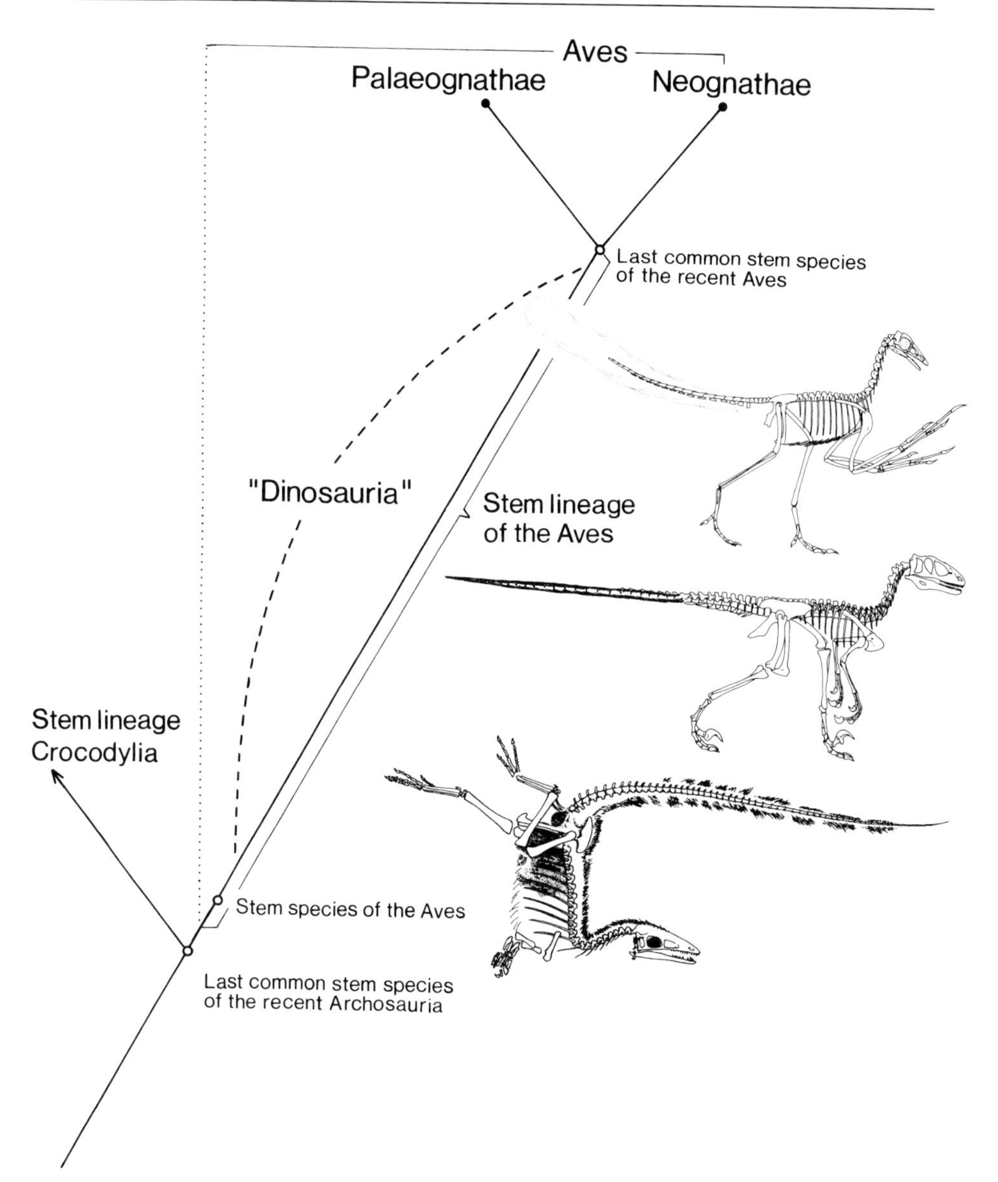

Fig. 99. Stem lineage of the birds. Beginning with the split of the last common stem species of the recent Archosauria into the stem species of the Crocodylia and the Aves. Finishing with the split of the last common stem species of the Palaeognathae and Neognathae. "Dinosauria" as a paraphyletic collection of fossil taxa are to be ordered sequentially in the stem lineage. Examples of representatives of the stem lineage of the Aves on the *right* side. *Below* † *Sinosauropteryx prima* (Compsognathidae; from Chen et al. 1998). *Middle* † *Deinonychius* (Dromaeosauridae). *Above Archaeopteryx lithographica* (Archaeopterygidae; from Carrol 1993). (Diagram original Ax)

Opinions cannot be further apart, however, the alternatives shrink to a formality of nomenclature when one can agree on the name of a monophylum in which certain fossils as Dinosauria Owen, 1842 stand beside the recent Aves Linnaeus, 1758.

"Birds are dinosaurs" (GAUTHIER 1986). In this apodictic formulation, the name Dinosauria stands for a monophylum in which the Aves constitute a subordinated taxon of recent organisms. If, in reverse, we extend the name Aves beyond *Archaeopteryx* to all fossil relatives of the currently living birds, then the "Dinosauria" are a paraphylum without a stem species common only to them. Their arrangement as stem lineage members of the Aves is "surely more tolerable than the alternative, which would be to have the Aves flying around as living dinosaurs" (AX 1989, p. 40). Things are so simple. Fuzzy formulations about the relation between "dinosaurs" and "birds", however, continue even in the most recent reviews.

"There is no longer reasonable scientific doubt that birds evolved from small theropod (carnivorous) dinosaurs sometime during or shortly before the Middle to Late Jurassic, over 150 million years ago" (PADIAN & CHIAPPE 1998, p. 2). The opposite result arises from this statement when one interprets the "theropod dinosaurs" as representatives of the stem lineage of the Aves. Biped, terrestrial, agile runners apparently mark a decisive stage in the phylogenesis of the Aves in the ambient of the evolution of feathers and wings (GEE 2000). In "dromaeosaurid theropods" without feathers († *Velicoraptor*), characteristic bird features such as the furcula (fused clavicles), air sacs and pneumatic bones have been found; i.e., they were evolved prior to the ability to fly. Possible first stages of feathers are filamentous structures in the epidermis of † *Sinosauropteryx* and † *Sinornithosaurus* (CHEN et al. 1998; UNWIN 1998; XU et al. 1999). Unequivocally feathered relatives of † *Archaeopteryx* are then † *Protarchaeopteryx* with a fan of feathers at the tip of the tail as well as † *Caudipteryx* with additional feathers (remiges) on the second finger of the anterior extremity (QIANG et al. 1998).

Quite another question concerns the character pattern of the stem species of the Aves at the beginning of the stem lineage. It seems legitimate to suspect that organisms with features of the "dromaeosaurids" formed the stem species or at least were in the vicinity of them. The problem can, however, hardly be solved without a binding answer to the next question. Besides the "Dinosauria" do any other fossil Archosauria belong in the stem lineage of the Aves. If so, in which section? The † Pterosauria have long been considered as candidates (BENNET 1996), even when they have evolved a completely different flying apparatus.

Mammalia

We have now reached the last large unit in the phylogenetic system of the Vertebrata – the monophylum Mammalia. The hypothesis of an adelphotaxa relation to the Sauropsida ("Reptilia" + Aves) could be better corroborated, but is generally accepted with good reasons today. The alternative hypothesis of a sister group relationship between the homoiothermic Aves and the homoiothermic Mammalia (GARDINER 1982, 1983) was just as generally rejected in the course of intensive discussion – finally on the basis of analyses of whole mitochondrial genomes (MINDELL et al. 1999).

For "mammal-like reptiles" from the Mesozoic era (KEMP 1982), high-ranking system units with names like Synapsida, Pelycosauria and Therapsida are in use. We can delete them when the paraphyletic collections of fossil mammals described by these names are arranged sequentially in the stem lineage of the Mammalia (p. 273).

Monotremata (Prototheria), Marsupialia (Metatheria) and Placentalia (Eutheria) form the three known, high-ranking subtaxa of the Mammalia. The latter two are now unambiguously united in the monophylum Theria. The alternative Marsupionta hypothesis with the Monotremata and Marsupialia as adelphotaxa has been irrevocably rejected (p. 258).

In place of the stereotypic, uninformative designations Proto-, Meta- and Eutheria, I use the easily remembered and also older names Monotremata, Marsupialia and Placentalia, simply because they make feature-oriented statements about the respective organisms. As for thousands of other names, objections can easily be made against them. The Monotremata within the Tetrapoda share the one hole for the urogenital system and anus as a plesiomorphy with the Sauropsida and the Amphibia. The marsupium possibly does not belong in the ground pattern of the Marsupialia, but rather could have first evolved within the unit. Moreover, besides the chorioallantoic placenta of the Placentialia with a profound connection between uterus and embryo, there is the simple yolk-sac placenta of the Marsupialia.

However, this does not alter the fact that the one hole of the Monotremata is a sensation in comparison with the other mammals, that the brood pouch is considered to be the predominant character of the Marsupialia and that the formation of a chorioallantoic placenta forms a unique feature of the Placentialia.

For the work of phylogenetic systematics, it would certainly be desirable when the names of all monophyla were unequivocally based on autapomorphies. This wish stands against the proven rules for the careful handling of already introduced names.

Autapomorphies (Fig. 89 → 3)

- Synapsid skull.
  With a lateral temporal window behind the eye cavity, surrounded by the bones postorbital, jugal and squamosum (Fig. 100 A).
  The temporal window is a character that evolved early, in the Carboniferous era, in the stem lineage of the Mammalia (Vol. I, Fig. 11). If the evaluation of the anapsid skull of the Chelonia as a plesiomorphy of the Sauropsida is correct, then there cannot be an evolutionary connection to the temporal windows of the Diapsida (p. 238). The lateral temporal window of the Mammalia must then have developed independently from the primarily windowless skull of the Amniota.
  Later, in the stem lineage of the Mammalia, the separating postorbital bone arch was reduced. The recent mammals have a wide open connection between eye cavities and temporal windows (Fig. 100 B).
- Secondary jaw joint.
  Joint between the squamosum of the skull and the dental as sole lower jaw bone. Late evolved autapomorphy (Jurassic); present in all recent Mammalia. Early stem lineage members have the primary jaw joint of the Tetrapoda consisting of quadratum and articulare (Fig. 100).
- Middle ear with three auditory ossicles and tympanicum.
  Hammer (malleus), anvil (incus) and stirrup (stapes) in middle ear cavity. The tympanicum is a dorsally open ring in which the tympanic membrane is stretched.
  In contrast, in the sister group Sauropsida and in the Amphibia, there is only one auditory ossicle in the middle ear – the columella auris (stapes); in addition, the tympanicum is absent.
  The purported homologies are shown in Fig. 100 C. The quadratum of the skull becomes the incus, the articular of the lower jaw the malleus (in which also the prearticular is taken up) and the angular the tympanic ring. The sole auditory ossicle from the ground pattern of the Tetrapoda is taken over in the Mammalia as the stapes in the middle ear.
- Heterodonty.
  Differentiation of the primarily homodont dentition of early members of the stem lineage (Fig. 100 A) into incisors, canines, premolars and molars (Fig. 101 A). The buccal teeth (premolar, molar) have more than one root.
- Diphyodonty [16].
  In connection with the evolution of heterodont dentition, a limitation of the dental regrowth occurred. The teeth-bearing vertebrates grow nu-

[16] The Marsupionta hypothesis (GREGORY), with the Monotremata and Marsupialia as adelphotaxa, was reactivated by KÜHNE (1973, 1977). One tooth of the embryonal dentition of *Ornithorhynchus anatinus* is purported to develop twice – just like the last premolars of the Marsupialia. This statement is wrong; there is no development of replacement teeth in *Ornithorhynchus* (LUCKETT & ZELLER 1989).

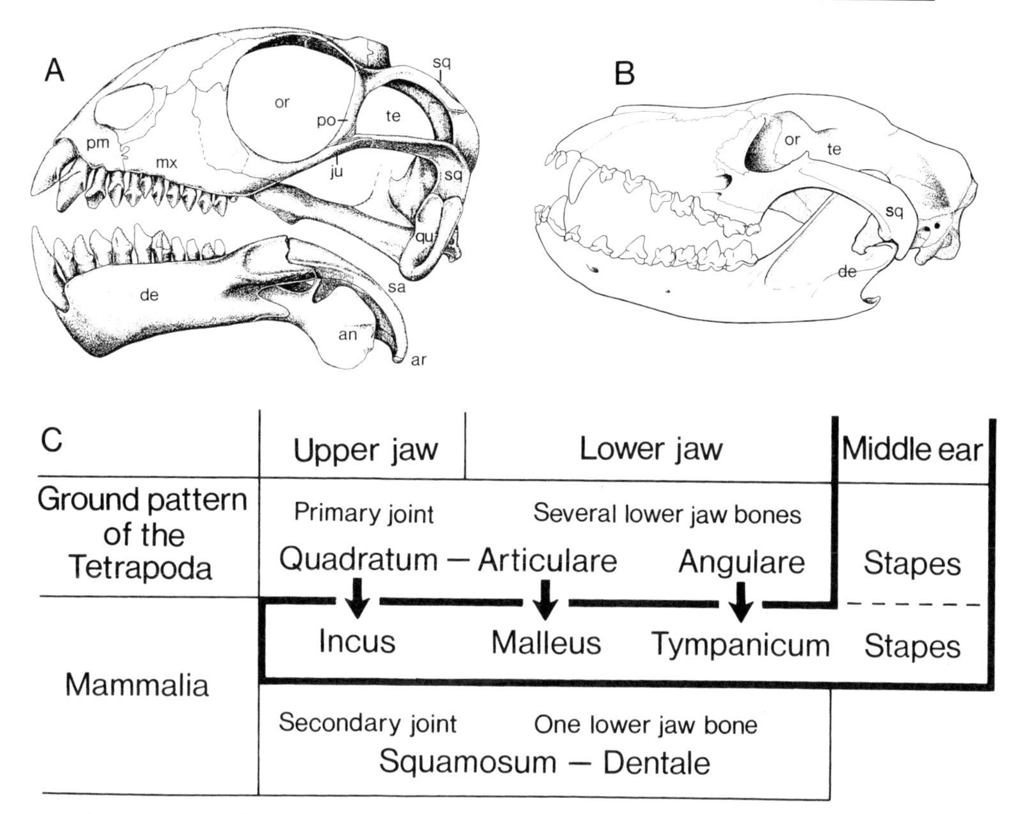

Fig. 100. Mammalia. Synapsid skull and alternative in the jaw articulation. **A** Stem lineage member † *Suminia getmanovi* ("Anomodontia") from the late Permian period. With bone strut between orbit and temporal window as well as primary jaw joint (quadratum + articular). **B** Tasmanian wolf *Thylacinus cynocephalus* (Marsupialia). Wide connection between eye cavity and temporal window (reduction of postorbital) as well as secondary jaw joint (squamosum + dental). **C** Scheme of the change from primary jaw joint of the Tetrapoda to the secondary joint of the Mammalia with migration of jaw joint bones into the middle ear. *an* Angular; *ar* articular; *de* dental; *ju* jugal; *mx* maxilla; *or* orbita; *pm* premaxilla; *po* postorbital; *pt* pterygoid; *qu* quadratum; *sa* suprangular; *sq* squamosum; *te* temporal window. (**A** Rybczynski 2000; **B** Starck 1995; **C** Ax 1988)

merous generations of teeth in the ground pattern; they are polyphyodont. In contrast, the Mammalia produce maximal two generations of incisors, canines and premolars – the generation of the milk teeth and the generation of permanent teeth. The posterior buccal teeth (molars) occur in principle only once (Fig. 101 C).

The classical change of dentition that we all experienced in our youth is, however, only realized in the Placentalia. In the dentition of the Marsupialia, only the last premolars are changed; in other words, here there are only four teeth with development in the form of milk and perma-

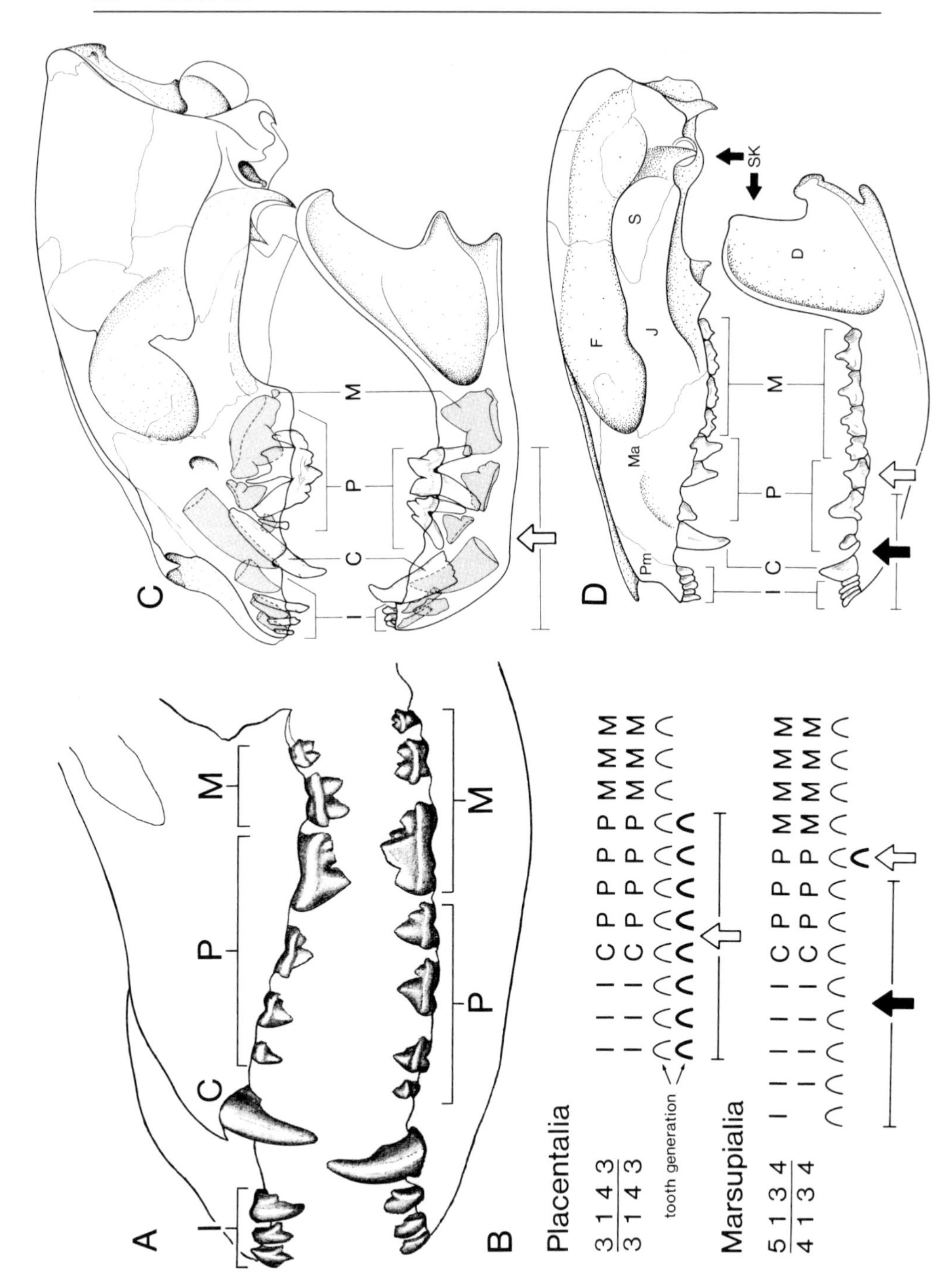
A
B
C
D
I
C
P
M
Placentalia
3 1 4 3
3 1 4 3
tooth generation
Marsupialia
5 1 3 4
4 1 3 4
F
J
S
Ma
Pm
D
SK

◀ **Fig. 101.** Mammalia. Heterodonty and change of teeth. **A** *Canis* (Placentalia). Differentiation of the dentition to incisors (*I*), canines (*C*), premolars (*P*) and molars (*M*). **B** The alternative diphyodonty – monophyodonty. Plesiomorphy (Placentalia): two generations of teeth with replacement of incisors, canines and premolars. Apomorphy (Marsupialia): no change of teeth with the exception of the last premolars. The tooth numbers each refer to the ground pattern of the Placentalia and Marsupialia. **C** *Panthera leo* (Placentalia). Young animal with milk teeth in function. Permanent teeth (*dark*) lie in the upper and lower jaws. **D** *Didelphis marsupialis* (Marsupialia). Adult. Solely the change of the last premolars (*open arrow*) is plesiomorphous here, apomorphous is the single development of incisors, canines and the remaining premolars (*black arrow*). *C* Canine; *D* dental; *F* frontal; *I* incisor; *J* jugal; *M* molar; *Ma* maxilla; *P* premolar; *Pm* premaxilla; *S* squamosum; *SK* secondary jaw joint. (Ax 1988)

nent teeth. All other teeth in the marsupials are formed only once (Fig. 101 D).
The state in the Marsupialia can be interpreted with good reasons as an apomorphy (AX 1988). The change of incisors, canines and premolars is to be set in the ground pattern of the Mammalia and treated as an autapomorphy under the name diphyodonty.
Independent of this, the teeth of adult Monotremata are completely reduced

- Homoiothermy.
  Evolution of a body temperature that is independent of the surroundings in correlation with the evolution of a hair covering against loss of warmth. The much discussed agreement between the mammals and birds must be interpreted as the result of convergent evolution when we proceed from the well-founded hypothesis of an adelphotaxa relationship between poikilothermic Crocodylia and the homoiothermic Aves (p. 243).
- Hair.
  Pure epidermal product of horn with differentiation into cuticle, cortex and medulla.
- Dermal glands.
  Of the many epidermal glands of the Mammalia, the polyptych sebaceous glands (hair follicle glands) as well as the monoptych sweat glands and mammary glands are emphasized.
- Anuclear erythrocytes.
  Expulsion of the cell nucleus from the nucleated erythroblasts. Improvement of the respiratory performance through gaining more space for haemoglobin molecules (STARCK 1982).
- Unpaired external nasal opening in the skull.
  Apomorphy in comparison with the paired nasal openings in the ground pattern of the Tetrapoda and in the adelphotaxon Sauropsida.

- Secondary palate.
  The bony part is formed by palatine processes of premaxilla, maxilla and palatinum. In contrast to the identically named secondary palate of the Crocodylia, the pterygoid is not involved primarily.
- Dicondylic skull.
  Possession of two condyles in contrast to the primary monocondyly in the ground pattern of the Tetrapoda and the sister group Sauropsida. The existence of two condyles in the Amphibia (p. 226) is with certainty the result of a convergent evolution.
- Number of phalanges 23333.
  Early reduction in the stem lineage of the Mammalia from the ground pattern of the Amniota (23454).
- Obturator foramen.
  Hole in the pubic bone for the passage of nerves and vessels.
- Rotation of the limbs under the body.
  Change from creeping to walking locomotion in the stem lineage of the Mammalia. Rotation of the limbs with orientation of the elbow joints backwards and the knee joints forwards (Fig. 107, upper right).
- Evolution of "marsupial bones" (p. 265).

Monotremata – Theria

The mammals once again illustrate the widespread, almost "judicial" relation in the composition of the highest ranking sister groups of extensive monophyla. The egg-laying Monotremata with only three species in the region of Australia and with numerous primitive characters stands opposed to the viviparous Theria with several thousand species distributed worldwide and a whole series of autapomorphies.

Monotremata

The best justification for the monophyly of a unit Monotremata (Prototheria) is provided by the development of a femoral gland with an outlet spur on the hind limb of the male. I place it at the beginning of the following characters.

Autapomorphies (Fig. 89 → 12)

- Crural gland and spur.
  In the males of the three species of Monotremata there is a special gland in the thigh and hip joint region (Vol. I, Fig. 9). A long duct links the

poison gland with a hollow spur on the foot next to the toes. Repulsion of rivals and demarcation of territory are the presumed functions.
- Skull without jugal bone.
- Secondary skull side wall in the orbitotemporal region.
  Independent evolution in comparison with the Theria (KUHN 1971; ZELLER 1989).
- Musculus detrahens mandibulae.
  The muscle runs from behind to the dental and serves as depressor and retractor of the jaw (STARCK 1995). Only present in the Monotremata.

Plesiomorphies

At this point, I will emphasize three states of affairs from the primitive features.
- Cloaca. Common opening of the urogenital system and the hindgut - taken over from the ground pattern of the Amniota (Fig. 103 A).
- Oviparous. Laying large, yolk-rich eggs. Meroblastic cleavage.
- Lack of teats. Exit of milk through the skin from which the secretion is licked or sucked up by the young animals (Fig. 102 C, D).

The three recent species (Vol. I, Fig. 9) from Australia, Tasmania and New Guinea are distributed between two supraspecific taxa - Tachyglossidae and Ornithorhynchidae. Their members are characterized alternately by apomorphies and plesiomorphies, whereby the prerequisites for an interpretation as adelphotaxa are fulfilled.

Tachyglossidae (spiny anteater or spiny echidna)

Anterior end with a tube-like snout that is covered with horn. Small terminal mouth opening from which a long, sticky tongue can be extended to catch prey.

Short-billed brevirostrate *Tachyglossus aculeatus*. Short snout and short legs. Insect eater.

Long-billed echidna *Zaglossus bruijni*. Long, curved "elephant trunk" and correspondingly longer limbs. Food sources: terrestrial Oligochaeta, beetle larvae.

Autapomorphies: Spiny covering. Fingers and toes with strong claws for digging in the ground. Long cleaning claws on toes 2 and 3 to clean the spiny coat.

Plesiomorphy: Terrestrial lifestyle.

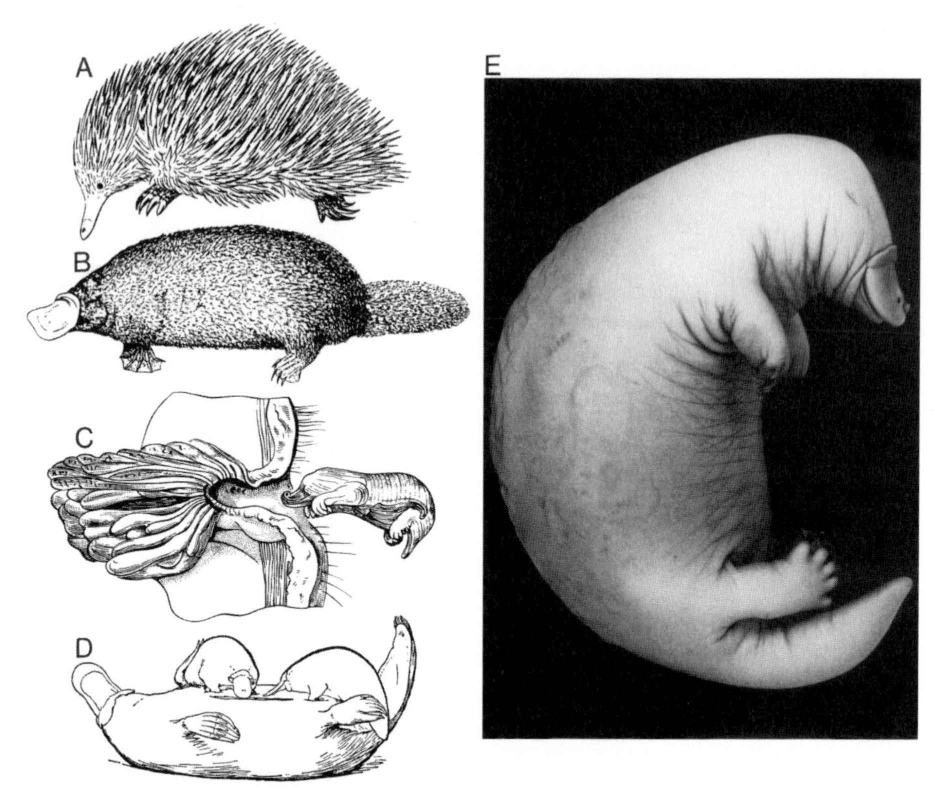

Fig. 102. Monotremata. **A** *Tachyglossus aculeatus.* **B** *Ornithorhynchus anatinus.* **C** Opening of the mammary glands of *Tachyglossus* into a groove of the skin. In front a young animal in the suckling position. **D** Young animals of *Ornithorhynchus* licking milk from the ventral side of a female. **E** *Ornithorhynchus anatinus.* Animal of nesting age of 1.8 cm length. (**A, B** Starck 1978; **C, D** Grassé 1955; **E** Zeller 1989)

Ornithorhynchidae (duck-billed platypus)

One recent species – the semiaquatic *Ornithorhynchus anatinus.* Duck-like bill with which food is taken up from water (Crustacea, Mollusca).

The discovery of fossil duckbills († *Obdurodon*, † *Steropon*) led to the establishment of a supraspecific taxon. The extinct species are members of the stem lineage.

Autapomorphies: Torpedo-shaped body, steering tail and limbs with webs between the digits in correlation with the habitat water.

Plesiomorphy: Soft hair covering without spines.

In conclusion, three characteristics have to be evaluated: the lack of teeth in adult Monotremata, the existence of a pouch (incubatorium) in the Tachyglossidae and the formation of marsupial bones (ossa epipubica) in both sister groups.

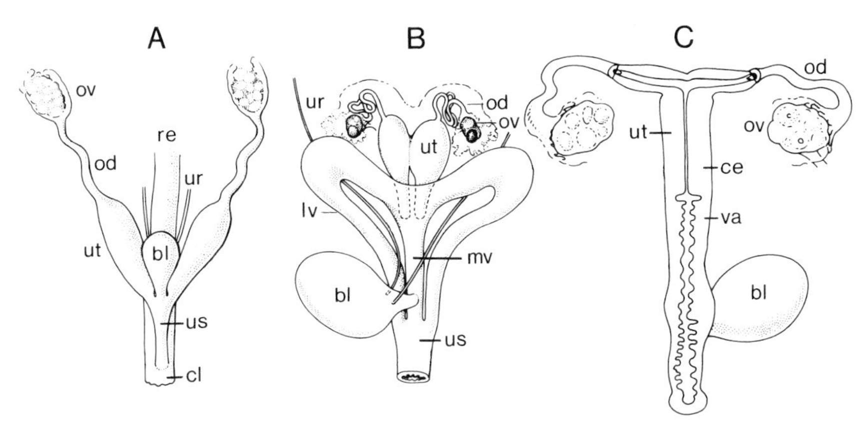

Fig. 103. Mammalia. Urogenital system. **A** Spiny anteater *Tachyglossus aculeatus*, ♀ (Monotremata). Plesiomorphy: fusion of urogenital sinus and hindgut in the cloaca. **B** Kangaroo *Macropus eugenii*, ♀ (Marsupialia). Ventral view. Secondary median vagina as outgrowth of the vaginal sinus. Autapomorphy of the marsupials. **C** Shrew *Crocidura russula*, ♀ (Placentalia). Apomorphous uterus simplex and unpaired vagina. *bl* Urinary bladder; *ce* cervix; *cl* cloaca; *lv* lateral vagina; *mv* median vagina; *od* oviduct; *ov* ovary; *re* rectum; *ur* ureter; *us* urogenital sinus; *ut* uterus; *va* vagina. (Renfree 1993)

Teeth are completely reduced in the Tachyglossidae; they are absent throughout the entire life cycle. In contrast, in the ontogenesis of *Ornithorhynchus anatinus*, there are precursors of several teeth in the upper and lower jaws that disappear in the course of development. The notion to postulate the start of the reduction of teeth already in the stem lineage of the Monotremata seems reasonable. However, it must be rejected because a permanent dentition was found in the fossil † *Obdurodon dicksoni* (Ornithorhynchidae; ARCHER et al. 1993). This state of affairs results in the hypothesis of a twofold independent reduction of teeth in the stem lineages of the Tachyglossidae and the Ornithorhynchidae.

The female of *Ornithorhynchus anatinus* lays her eggs in an egg chamber of the lair on the banks of the water (Fig. 102E). In contrast, *Tachyglossus aculeatus* and *Zaglossus bruijni* develop a breeding pouch during the reproduction period to take up the eggs. There are two reasons opposing the assumption of a homology with the marsupium of the marsupials. The pouch develops as an unpaired structure in the Monotremata, from a paired precursor in the Marsupialia. In addition, the pouch only occurs in a subtaxon of the Monotremata. The pouch of the Tachyglossidae is distinguished terminologically by the name incubatorium as compared to the name marsupium for that of the marsupials.

Marsupial bones occur in both sexes. Moreover, marsupial bones are also found in *Ornithorhynchus anatinus* without a pouch. In other words, they functionally have obviously nothing to do with the breeding pouch, but must rather be considered as supporting bones of the muscular abdominal wall (STARCK 1979, 1995).

Marsupial bones insert at the anterior end of the pubic bone; the name ossa epipubica is better than ossa marsupii. Their identical existence in the Monotremata and Marsupialia supports a homology. However, the lack of marsupial bones in the recent Placentalia inevitably leads to the question of whether this is a symplesiomorphy or a synapomorphy. Fortunately, ossa epipubica have been found in representatives of the stem lineage of the Placentalia (MARSHALL 1979). Thus, the answer is: marsupial bones evolved in the stem lineage of the Mammalia were passed on to the stem lineages of the Monotremata and Theria, and were later reduced in the stem lineage of the Placentalia. The common possession of marsupial bones in the Monotremata and Marsupialia is a symplesiomorphy (AX 1988).

Theria

Autapomorphies (Fig. 89 → 13)

- Abandonment of the cloaca.
  Separation of hindgut exit and urogenital sinus.
- Viviparity.
- Mammary papilla.
  Opening of the mammary glands in teats as local differentiations of the skin.
- Spirally twisted cochlea in inner ear.
  Apomorphy compared with the cochlear duct of the Monotremata that exhibits hardly more than one loop.
- Simple connection between sternum and shoulder blade.
  Only clavicle. Coracoids reduced from the ground pattern of the Mammalia, in contrast retained in the Monotremata.
- Scapula with fossa supraspinata.
  Division of the shoulder blade into two bone lamella – the fossa infraspinata and the fossa supraspinata. The latter is interpreted as a new formation of the Theria.
- Coracoid process.
  The coracoid process sits on the shoulder blade as remnant of the regressed metacoracoid.
- Facial hairs.
  As they are only present in the Marsupialia and Placentalia, the evolution of tactile hairs on the head in the stem lineage of the Theria is hypothesized.

We will now take a glance at the heterodont dentition of the Theria, consisting of incisors (I), canines (C), premolars (PM) and molars (M) (Fig. 101 B). From a comparison between the Marsupialia and the Placentalia, the following minimal tooth formula can be given for their last common Theria stem species.

5I 1C 4PM 4M
4I 1C 4PM 4M

Marsupialia $\frac{5134}{4134}$ Placentalia $\frac{3143}{3143}$

The Marsupialia retain the large number of incisors in the ground pattern as a plesiomorphy. In contrast, the reduction of one premolar each in the upper and lower jaw is derived.

The Placentalia remain plesiomorphous possessing four premolars. The maximal three incisors and three molars in the upper jaw and lower jaw are apomorphies.

Marsupialia – Placentalia

Marsupialia

In their characteristic disjunctive distribution, the Marsupialia (Metatheria) currently live as about 80 relatively uniform, mouse- to cat-sized species in South America (late migration of the opossum *Didelphis virginiana* to North America) and 180 species with a fascinating diversity of forms in the region of Australia.

In habitats primarily not colonized by Placentalia, excellent copies of placental mammals evolved twice, i.e., marsupialian tiger († *Thylacosmilus*) in South America or the native cats (*Dasyurus*) and the Tasmanian wolf (*Thylacinus cynocephalus*) in Australia.

Autapomorphies (Fig. 89 → 14)

- Monophyodonty.
  Only the last premolar is exchanged (p. 259). In other words, milk and permanent teeth occur in succession in only four positions (two each in upper and lower jaw). All other teeth are persistent milk teeth which can be concluded from the precursors of a second set of teeth on the lingual side of the functioning teeth (STARCK 1995).
- Vaginal sinus.
  The Marsupialia possess paired vaginas. In comparison with the Placentalia, this is evaluated as a plesiomorphy. In contrast, the cranial fusion

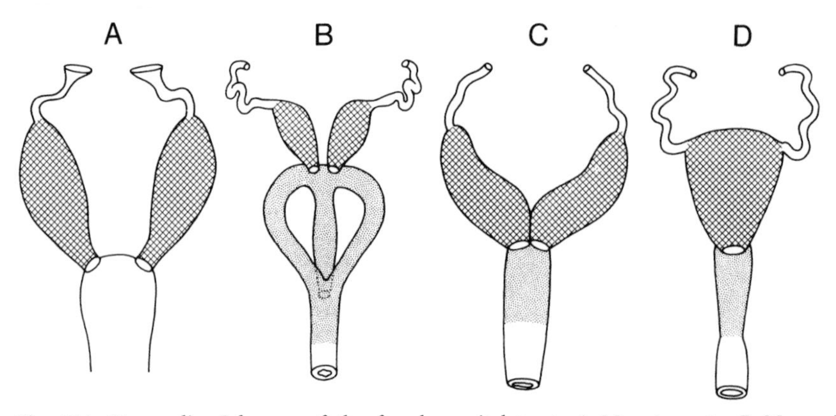

Fig. 104. Mammalia. Schemes of the female genital tract. **A** Monotremata. **B** Marsupialia (*Macropus*). **C** Placentalia with uterus bipartitus (e.g., ant bear *Orycteropus afer*). **D** Placentalia with uterus simplex (e.g., Simiae; *Crocidura* see above). *White* (*above*) uterine tubes. *Shaded* Uterus; *dotted* vagina. (Starck 1995)

of the vaginas to an unpaired vaginal sinus into which the uteri open is apomorphous. This grows to an unpaired diverticulum between the vaginas through which birth occurs (Figs. 103 B, 104 B).
- Reduction of a premolar in upper and lower jaw (see above).

We have already discussed the marsupium in comparison with the incubatorium of the Tachyglossidae (Monotremata). The marsupial pouch gives the unit its name and is usually considered the major element in the organization of the marsupials. However, it is disputed whether the marsupium can be taken as an autapomorphy in the ground pattern of the Marsupialia. The brooding pouch is absent in some Didelpidae, in the Caenolestidae and in some Dasyuridae. If one wants to identify this state of affairs as a primary deficit (MARSHALL 1979, 1984), the consequence is the extravagant assumption of a repeated, independent evolution of the pouch within the Marsupialia.

Otherwise, the Marsupialia exhibit a series of primitive characters in the reproductive behavior.

Yolk-sac placenta or choriovitelline placenta belong in their ground pattern. Here, there is only a simple apposition of the germ wall of chorion and yolk sac entoderm on the uterus wall. Thus, this restricted possibility of nourishing the embryo results in a short duration of the gravidity.

The young animals are accordingly very immature at birth. What was "neglected" in the womb has to be made up for in a long postnatal period on the teat. This includes unique features that could have arisen within the framework of a principally plesiomorphous mode of development first in the stem lineage of the Marsupialia as autapomorphies. The newly born animal has powerful claws on the forelimbs; they serve for movement from the birth opening to the teats and disappear later. After gripping the teat, the mouth turns into a roundish suckling organ through tissue outgrowths while the teat swells in the oral cavity. Together, this results in a firm anchoring of the young animal to the mother (Fig. 105 A, B)

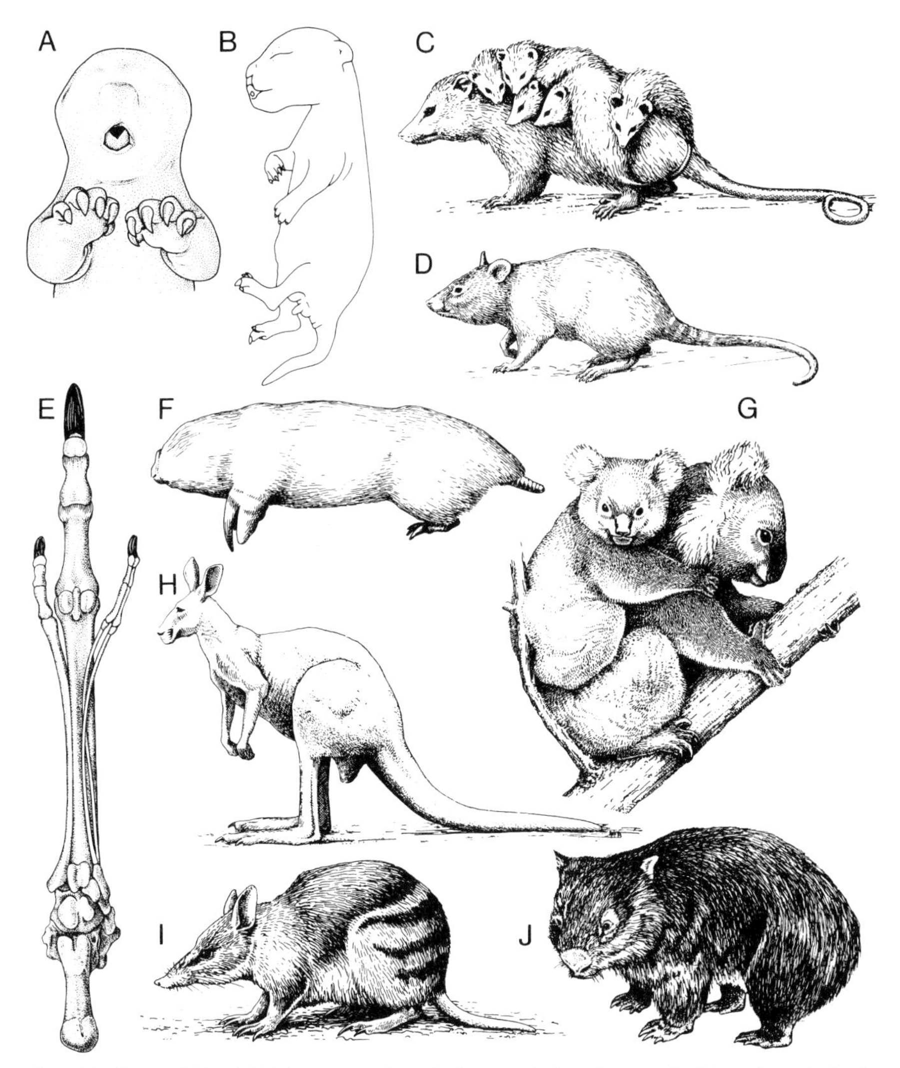

Fig. 105. Marsupialia. **A** *Trichosurus vulpecula* (possum). Pouch animal of 2 cm length. Suckling mouth with tongue nipple. Forelimbs with claws. **B** *Wallabia bicolor* (swamp wallaby). Nest young animal of 4.3 cm length. **C, D** Members of the monophylum Didelphida. **C** *Didelphis virginiana*. Opossum (Didelphidae). **D** *Caenolestes obscurus* (Caenolestidae). **E** *Macropus giganteus* (kangaroo). Skeleton of the right foot from the sole. Syndactyly: fusion of toes II and III. **F–J** Members of the monophylum Syndactyla. **F** *Notoryctes typhlops* (marsupial mole). **G** *Phascolarctos cinereus* (koala). **H** *Macropus rufus* (red kangaroo). **I** *Perameles gunni* (bandicoot). **J** *Vombatus ursinus* (wombat). (**A, B, E** Starck 1995; **C, D, F, G–J** Bourlière 1955)

Systematization

SZALAY (1993, 1994) presents a classification of the Marsupialia in which fossil taxa stand in identical rank with their purported recent relatives and have assigned correspondingly coordinated categories. This measure led to a plethora of new taxa and categories of the "infraclass" Metatheria (Marsupialia); there are the categories Cohort, Order, Semiorder, Suborder, Semisuborder, Superfamily, Family, Subfamily and Tribe. In addition, more than two taxa with identical categories appear on single hierarchical levels. The Metatheria were arranged into three equal-ranking taxa – Cohort Holarctidelphia, Cohort Ameridelphia and Cohort Australidelphia. In the Order Didelphida we find four suborders – Suborder Archimetatheria, Suborder Sudameridelphia, Suborder Glirimetatheria, Suborder Didelphimorpha. Corresponding arrangements exist at the levels of Family, Subfamily and Tribe – the examples must suffice.

For rearrangements in a phylogenetic system of the Marsupialia, we must remember the following methodological principles.

1. In the case of units with recent members, only taxa that also have recent species can be considered as sister groups.
2. Only two taxa can appear as sister groups in identical rank of a single hierarchical level.
3. Fossil taxa are to be included as stem lineage members in units with recent species.
4. All categorical designations are to be eliminated.

Under these premises, I take the highest-ranking taxa with recent representatives of the uppermost two levels of the system hierarchy from the classification of SZALAY (loc. cit.). Subordinated units in the conventional rank of families are assigned to these taxa.

Metatheria (Marsupialia)
  Didelphida
    Caenolestidae, Didelphidae
  Australidelphia
   Gondwanadelphia
    Microbiotheriidae
    Dasyuridae, Myrmecobiidae, Thylacinidae
   Syndactyla
    Notoryctidae, Peramelidae, Phalangeridae, Petauridae, Tarsipedidae, Macropodidae, Phascolarctidae, Vombatidae

I can present justifications of monophyly for the beginning and end of this list.

In the South American taxa Didelphidae and Caenolestidae (Fig. 105C, D), the agreement in pair formation of sperm in the epididymis is interpreted as a synapomorphy (L.G. MARSHALL 1984).

On the other hand, a large unit of Australian marsupials is characterized by the development of a cleaning foot; the second and third toes are fused with the exception of the claws (Fig. 105E). The assumption of a single evolution of syndactyly led to the creation of the taxon Syndactyla (Fig. 105F–J).

The seemingly obvious hypothesis of a monophyly of the Australian Marsupialia is not compatible with the combination of the South American Microbiotheriidae (one recent species *Dromiciops australis*) with the Australian dasyurids (Dasyuromorphia) in a unit Gondwanadelphia.

Placentalia

As the result of a fantastic phylogenetic development, the Placentalia (Eutheria) constitute the majority of the mammals. They have conquered all large biotopes of the Earth and made use of all possible sources of food. The key to the "success" of the Placentalia was apparently the intensive brood care through the evolution of the intimate contact between mother and embryo in the allantochorioic placenta.

Autapomorphies (Fig. 89 → 15)

- Trophoblast and chorioallantoic placenta.
  The trophoblast is an envelope and nourishment tissue that arises from the extraembryonal ectoderm of blastocysts in the ontogenesis. The embryoblast is displaced to the inside of the germ.
  The trophoblast combines with the allantois of the germ to form an allantoid placenta. The chorion arises from the trophoblast and the adjacent somatopleura. Bulges of the chorion penetrate into the wall of the uterus; branches of the allantois migrate into these villi.
- Corpus callosum.
  Extensive fiber connection between the dorsal regions of the neopallium in the brain – "a major brain structure found only in placentals" (L.G. MARSHALL 1979, p. 374).
- Monodelphy.
  In the Monotremata and Marsupialia, oviduct and uterus have separate openings; in the Marsupialia paired vaginas follow (p. 267).
  In contrast, in the Eutheria, the oviducts open via the uterus jointly in an unpaired vagina – irrespective of whether this is a primitive uterus duplex or the apomorphous uterus simplex (Figs. 103C, 104C, D).

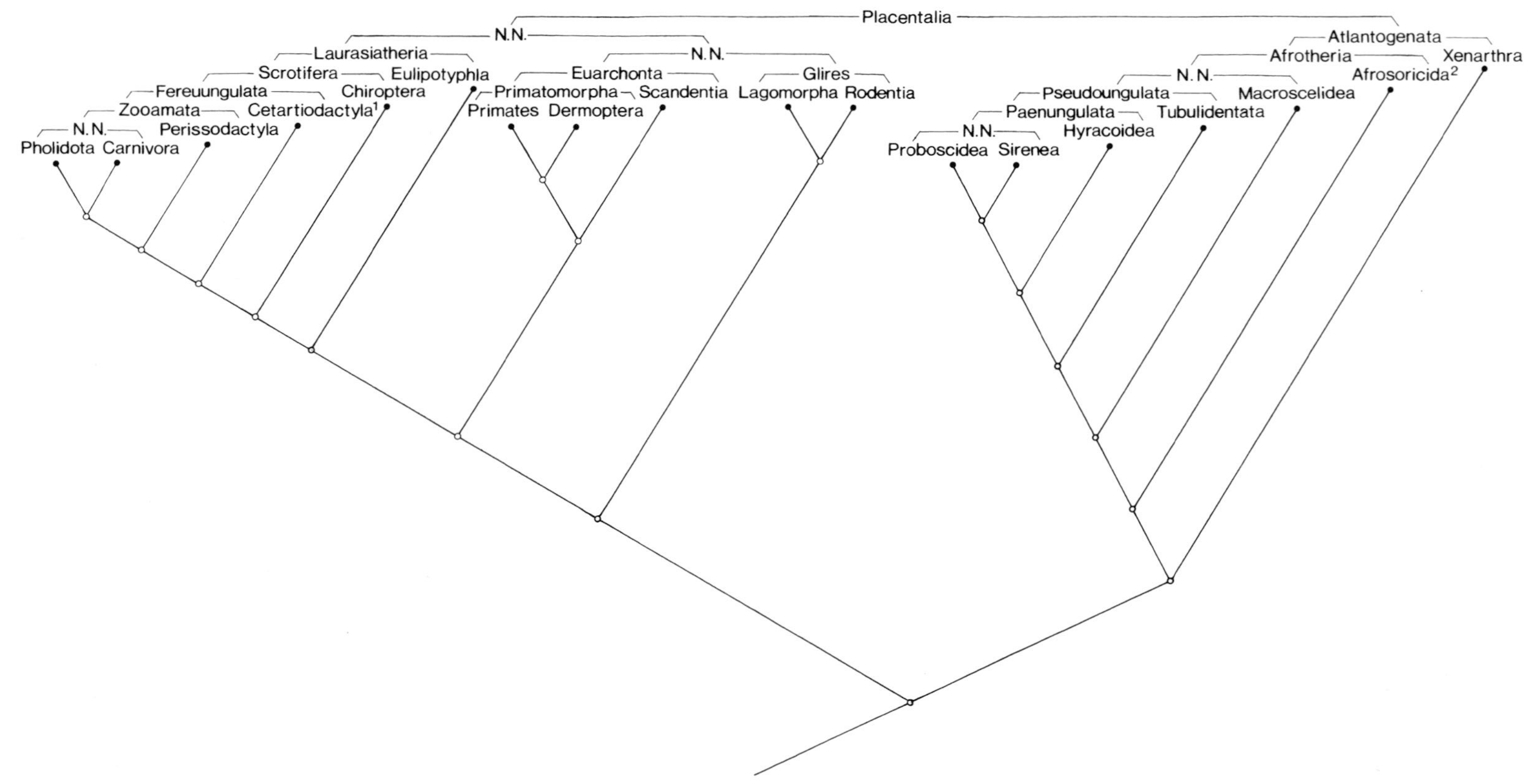

Fig. 106. Diagram of phylogenetic relationships of the Placentalia (Eutheria). [1]Cetacea + Artiodactyla; [2]Tenrecidae + Chrysochloridae. (Redrawn from Waddell et al. 1999)

- Reduction of marsupial bones in the stem lineage (p. 266).
- Reduction of the number of incisors and molars (p. 267).

Systematization

One reaches 30 "orders" of the Placentalia when a series of fossil units are added with the category order to the recent taxa (STORCH & WELSCH 1997). Most of the units with recent members can be securely justified as monophyla, but with equal certainty they are not equivalent units on one and the same level in the hierarchy of the system of the Placentalia as could be suggested by the homonymous label order.

"Particularly, the higher eutherian mammal radiation still presents many problems" (NOVACEK 1993, p. 3). Even so, there is now a first draft of a consequent phylogenetic system of the Placentalia; however, it is still inconsistent in the continued use of the category order (WADDELL et al. 1999). I have reformulated the genealogical tree of "interordinal relationships of placental mammals" as a diagram of phylogenetic relationships with consistent arrangement of sister groups on identical levels in the system hierarchy (Fig. 106). Now, the widely differing levels of diverse taxa in the ranking of customary orders are apparent at a glance. Thus, the Pholidota + Carnivora in the unit Laurasiatheria are placed two levels below the Primates + Dermoptera. In the Atlantogenata, the Xenarthra form an unusual, high-ranking taxon; the Proboscidea + Sirenea follow as sister groups only after five successive levels of subordination.

In my opinion, the authors still owe us a detailed validation of their "present best estimate of placental relationships" on the basis of a combination of morphological and molecular data. On the way to a comprehensive phylogenetic system of the Placentalia, the possibly stimulating effect of the dendrogram that "may well have more correct clades than any published to date" (WADDELL et al. 1999, p. 4) could support its inclusion in a textbook without conclusive justification of all taxa as monophyla. Since we have reached the end, something like this will not happen again.

On the Stem Lineage of the Mammalia

With more than 60 autapomorphies, the Mammalia form an excellently corroborated monophylum. The constitutive characters have developed successively in the history of the Earth during the course of 150 million years from the Upper Carboniferous to the boundary Jurassic/Cretaceous eras (Vol. I, Fig. 11).

The above two statements are undisputed. However, it is still a subject of much controversy where to place the segment in the phylogenesis of the Amniota in which the autapomorphies evolved that were present in the last

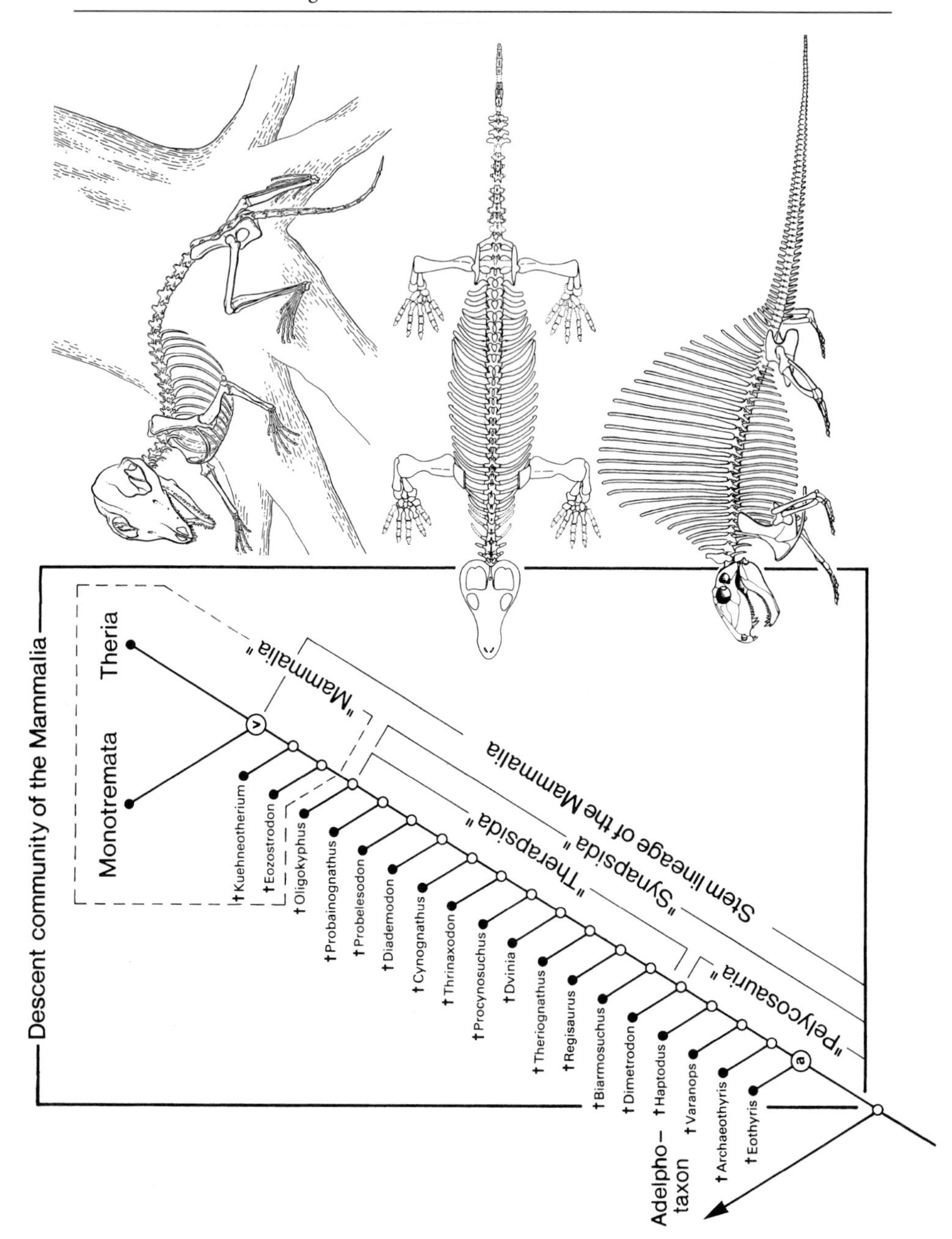
Descent community of the Mammalia
Monotremata
Theria
"Mammalia"
Stem lineage of the Mammalia
"Synapsida"
"Therapsida"
"Pelycosauria"
†Kuehneotherium
†Eozostrodon
†Oligokyphus
†Probainognathus
†Probelesodon
†Diademodon
†Cynognathus
†Thrinaxodon
†Procynosuchus
†Dvinia
†Theriognathus
†Regisaurus
†Biarmosuchus
†Dimetrodon
†Haptodus
†Varanops
†Archaeothyris
†Eothyris
Adelpho-
taxon
v
a

◀ **Fig. 107.** Stem lineage of the Mammalia. Starting with the split of the last common stem species of the recent Amniota into the stem species *a* of the Mammalia and the stem lineage of the adelphotaxon Sauropsida. Ending with the split of the last common species *v* of the recent Mammalia into the lineages of the Monotremata and Theria. Sequential incorporation of a selection of 18 fossil species as stem lineage representatives according to the increasingly closer relationship with the recent Mammalia (after Kemp 1982). The inclusion of new taxa in the stem lineage as well as the rearrangement of fossil units according to new knowledge are possible without problems and without any changes in our hypothesis about a sister group relationship between the Monotremata and Theria. "What is a mammal?" Explanation of the three positions described in the text. (1) When starting with the stem species *a*, namely, all fossil and recent taxa are together in a unit. However, when this is given the name Synapsida all recent Mammalia become a subtaxon of the Synapsida. (2) When starting with the stem species *v*, on the contrary, all previous fossil relatives are excluded from the taxon. (3) The definition of a boundary in between the fossil remains (here between † *Oligokyphus* and † *Eozostrodon*) is an arbitrary measure on the basis of subjectively chosen characters. The stem lineage concept avoids the inadequacies of these three positions. Here, all fossil taxa between the stem species *a* and *v* are members of the stem lineage of the Mammalia and, thus, members of the descent community Mammalia. Names such as "Pelycosauria," "Therapsida" and "Synapsida" cover certain sections of the stem lineage; they designate paraphyletic collections of stem lineage members. Examples of representatives of the stem lineage on the right side. *Below* † *Dimetrodon* (Permian) with postorbital temporal window as an autapomorphy of the Mammalia. *Middle* † *Procynosuchus* (Permian). The reptile-like spreading of the limbs with horizontal orientation of humerus and femur is primitive. *Above* † *Megazostrodon* (Triassic). Apomorphous characters are the rotation of the limb under the body and the open connection between eye cavity and temporal window. (Ax 1988)

common stem species of the recent Mammalia. For the controversially discussed question "what is a mammal?" (DESUI 1991), there are three principally different positions that I will describe exemplarily.

Firstly, HOPSON (1991) accepted for the scope of the recent Amniota the sister group relationship between the Sauropsida and Mammalia with just these names. Upon integration of the pertinent fossil remains, however, the latter receive the name Synapsida that was created for fossil Amniota. The recent Mammalia become a subunit of the Synapsida, just as the recent Aves become a subunit of the Dinosauria in the formulation "birds are dinosaurs" (p. 256).

The extreme opposite is the position that wants to limit the name Mammalia to the Monotremata + Theria and their last common stem species (ROWE 1993). All that which belongs in the lineage of the Mammalia after the splitting of the Amniota into the Sauropsida and Mammalia is lifted, staggered in time, taxon for taxon as sister group to the level of the recent mammals and named with them as monophylum. The taxon † *Morganucodon* then forms with the Mammalia the unit of the Mammaliaformes, the † *Tritylodontia* and all successors with the Mammalia the unit Mammaliamorpha. The attempt to assign a name to every putative adelphotaxa relationship between recent and fossil units rapidly leads to an indeterminable plethora of new names – and just for this reason ROWE (loc. cit.)

probably left most of the branches on the way from the † Cynodontia to the recent Mammalia unnamed.

Hence, we come to the third possibility – the definition of a boundary between Mammalia and non-Mammalia in the middle of the fossil remains. HOPSON (1991) pointed out eight characters for such a measure – above all, the secondary jaw joint with an ovoid condyle on the dental and a concave joint cavity on the squamosum as well as the differentiation of buccal teeth into premolars (with a single change of teeth) and molars (without any change).

One may choose twice as many or even completely different characters from the syndrome of 60 derived features that have all arisen as evolutionary novelties in the stem lineage of the Mammalia. In every case, the selection remains a completely arbitrary, subjective decision – the result of an arbitrary intervention in a continuum of fossil taxa with increasingly closer relationships with the recent Mammalia – exactly as we have pointed out for the stem lineage of the Aves (p. 253).

With the stem lineage concept, we overcome the weaknesses of all three positions. At the risk of lapsing into the "weakness of repetition"[17], I will formulate for the last time the following statement for the example of the Mammalia.

All species or species groups known from fossil remains that evolved in the lineage leading to the recent Mammalia after splitting of the Amniota into the Sauropsida and Mammalia are more closely related to the Mammalia than to any other Amniota. This is an undisputed truism from which we derive the legitimation to include the respective fossils as stem lineage members in the monophylum Mammalia. Of course, we only recognize the relationship when the fossil taxon in the concrete case exhibits at least one of the numerous autapomorphies of the mammals capable of fossilization; this may be the early evolved synapsid skull, but also any other character. At the other end of the stem lineage, it is irrelevant when the certainly late evolution of feeding the young animal with milk occurred. The often repeated objection that one cannot classify an animal as a mammal when it was not (yet) able to suckle is invalid. Names designate taxa; they do not provide a definition or characteristics of the respective units.

We then arrange the fossil Mammalia in the stem lineage in order of the increasing number of autapomorphies – and that without naming the correspondingly increasing relations between certain fossils and the recent Mammalia.

Union of fossil stem lineage members to taxa with names like Synapsida, Pelycosauria or Therapsida automatically leads to paraphyla (Fig. 107) –

[17] Frida von Uslar-Gleichen to Ernst Haeckel (1903). Das ungelöste Welträtsel. N. ELSNER (Ed.) (2000) Wallstein-Verlag. Göttingen.

and these have no place in a phylogenetic system of organisms. Finally, the stem lineage members of the mammals are not "mammal-like reptiles" as stated in the title of KEMP's volume (1982), but rather Mammalia. If one really needs an informal name then "reptile-like mammals" would be more appropriate.

Chaetognatha

A Monophylum of the Bilateria with Unclarified Adelphotaxa Relation

Even with all the successes of phylogenetic systematics there are, unfortunately, still cases where our studied object stubbornly resists the formulation of plausible hypotheses of phylogenetic relationship. The Chaetognatha are just such a case (GHIRARDELLI 1995) - and it is my destiny to end this textbook with an unsolved problem.

Habitat

About 100 species of glass-clear, transparent arrow worms represent marine plankter with a worldwide distribution. Vertically, the Chaetognatha reach into the bathypelagic zone and they extend horizontally to brackish seas; for example, species of the taxon *Sagitta* have been observed in the Baltic Sea through to the Gulf of Danzig (GERLACH 2000). *Sagitta* and diverse taxa with similar names (*Aidosagitta* to *Zonosagitta*) as well as *Eukrohnia* and *Heterokrohnia* make up the majority of planktonic species (BIERI 1991). Inhabitants of the pelagic zone can float motionless in water; they slowly sink thereby and suddenly dash away by means of dorsoventral undulations.

Around 20 species inhabit the sea bottom on diverse hard substrates. Adhesive papilla on the ventral epidermis of the trunk in *Spadella* and lateral adhesive organs on the tail of *Paraspadella* serve for anchoring to the bottom (Fig. 108 B). Head and trunk are lifted obliquely to take up food.

The lifestyle as plankter in the pelagic zone most certainly belongs in the ground pattern of the Chaetognatha. Otherwise, the evolution of fins as stabilizers for movement in free water would hardly be explainable. On the evolutionary change to life on the bottom the fins are retained in the epibenthic organisms.

Body division

The Chaetognatha reach lengths between 3 mm and 5 cm (SHINN 1997). The discussion of the morphological construction of the body in two or

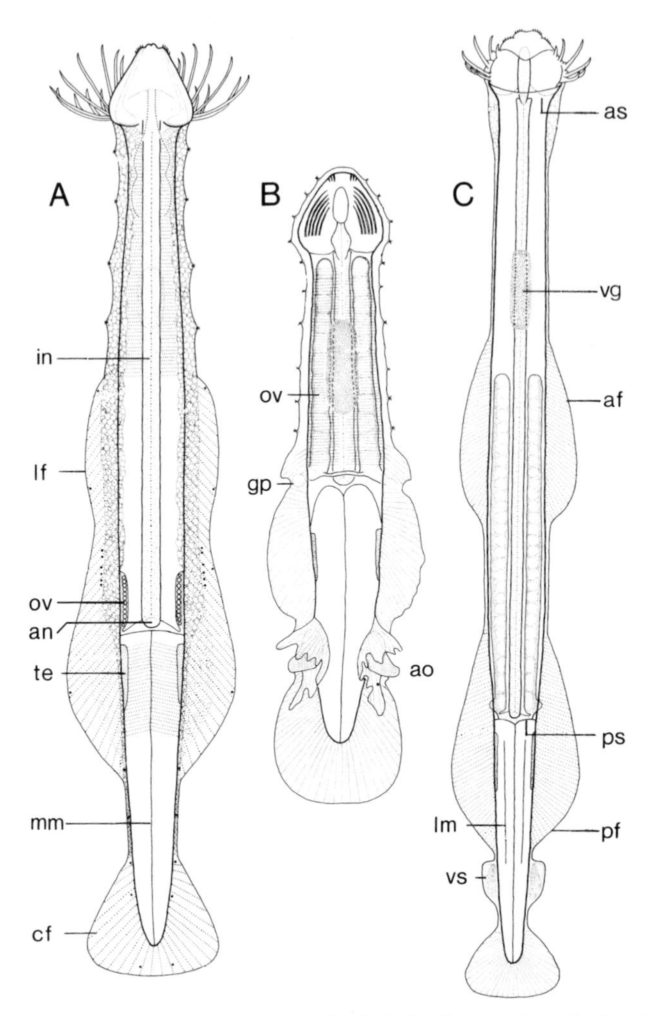

Fig. 108. Chaetognatha. **A** *Heterokrohnia involucrum*. Dorsal view. **B** *Paraspadella gotoi*. Ventral view. Grasping spines hidden under a skin fold (hood). Adhesive organs lateral on tail. **C** *Adhesisagitta hispida*. Dorsal view. *af* Anterior lateral fin; *an* anus; *ao* adhesive organ; *as* anterior septum; *cf* caudal fin; *gp* female genital pore; *in* intestine; *lf* lateral fin; *lm* lateral mesentery; *mm* median mesentery; *ov* ovary; *pf* posterior lateral fin; *ps* posterior septum; *te* testis; *vg* ventral ganglion; *vs* seminal vesicle. (Shinn, in F.W. Harrison, E.E. Ruppert (eds.) (1997) Microscopic Anatomy of Invertebrates, Vol. 15: Hemichordata, Chaetognatha, and the Invertebrate Chordates, p. 104. Reprinted by permission of John Wiley & Sons Inc.)

three sections is not very fruitful for the clarification of phylogenetic relationships. We follow the formal descriptive division into head, trunk and tail, but avoid the term segment because the body sections of the Chaetognatha have nothing to do with the segments (metamers) of the Articulata. A survey is presented as an introduction to the following discussion.

First section Head	Second section Trunk	Third section Tail
Septum		Septum
Head coelom Grasping spine Teeth Hood Eyes Corona ciliata Mouth opening Oesophagus Six ganglia	Trunk coelom One or two pairs of lateral fins; can extend on to the tail. Gut with ventral anus in front of the posterior septum. Ventral ganglion as large nerve center Two ovaries	Tail coelom One horizontal tail fin that encompasses the posterior end. Two testes Two seminal vesicles

Coelom

We start with the coelom of the Chaetognatha because the organ is usually considered to be important as a result of the historical burden and appreciation attributed to the term coelom.

SHINN & ROBERTS (1994, p. 60) redefined the word coelom "to include any body cavity completely lined by a mesodermal derived epithelium." This corresponds in principle to our even broader definition of the secondary body cavity (coelom) as a cellularly lined cavity between ectoderm and endoderm (Vol. I, p. 116). The Chaetognatha accordingly and undisputedly possess a coelom. It must be particularly emphasized, however, that the presented definitions are not linked to statements about a possible homology of coelom compartments in various units of the Bilateria. We can, therefore, consider the coelom of the Chaetognatha without prejudice (SHINN 1997; SHINN & ROBERTS 1994).

The Chaetognatha have three sets of coelom sacs with the following construction in the adult.

The head is mostly filled by the cranial ganglion, the oesophagus and, above all, musculature to move the anterior end, the grasping spines and the teeth. Accordingly, only a very small coelom compartment exists here. Trunk and tail contain spacious coelom compartments lined with epithelial muscle cells. In the strong parietal epithelium four large muscle bands of cells with cross-striated myofibrils pass through the body. The inner, visceral epithelium, in contrast, consists of flat cells with smooth fibrils; they form a delicate circular musculature around the gut.

In addition, in the adult, there are flat, nonmuscular "peritoneocytes" which lie partially on the above-described epithelia in the trunk and tail. They are not present in hatching young animals; the ontogenetic development is unknown.

Short dorsal and ventral mesenteries fix the gut to the body wall in the trunk and divide the trunk coelom into a left and a right half. There is also a median mesentery in the tail. Incomplete lateral mesenteries are added that further divide the tail coelom.

The picture is supplemented in two points by studies on freshly hatched young animals (hatchlings; SHINN & ROBERTS 1994).

1. Besides the anterior septum, the posterior septum between trunk and tail is already present. This opposes the widely held assumption of a dimeric ground organization with head and rest of the body in which the posterior septum develops later.
2. There is continuity between the coelomatic cell aggregates and coelomatic spaces of the hatchling and the adult. Older observations about an early fusion of embryonal coelom cavities and their later reorganization are obsolete.

   The described coelom forms an autapomorphy of the Chaetognatha; this will be validated below. First, we present a selection of the many impressive, unique characters.

Autapomorphies

- Coelom.
  In the arrangement with a small head coelom as well as paired trunk and tail compartments.
- Multilayered or striated epidermis.
  Present in practically the entire body. Only on the underside of the head and the inner side of the hood is there a simple, monolayered epithelium. The multilayered epidermis of the Chaetognatha is an exception among the "invertebrate" animals; of course, it has nothing to do with the epidermis of the Craniota. There is no stratum germinativum and no cornification in the arrow worms. Mitoses or epidermal stem cells are not known (SHINN 1997). The outer cells form a thin layer of a secretion that sticks to the body surface.
- Grasping spines and teeth (Fig. 109).
  The monolayered epidermis produces a flattened cuticle with thickenings to lateral and ventral head plates. Specialized cells of this epidermis form the curved grasping spines and short teeth that flank the mouth laterally and frontally. In contrast to the development of annelid chaeta from a single chaetoblast, each grasping spine and each tooth is formed by several cells. Long epidermal pulp cells penetrate into a pulp cavity; they produce the tip and shaft of the spine. The base is secreted by a second population of so-called anchor cells. The hard substance of the grasping spines and teeth contains α-chitin.

Grasping spines and teeth serve to trap prey. They insert a paralytic poison in the prey and transport it subsequently to the oesophagus.

- Hood (preputium).
  Grasping spines and teeth are hidden under a skin fold that arises dorsolaterally from the head. Protractors draw them together under the mouth leaving only a small diaphragm. Retractor muscles release the graspers to catch prey.
- Lateral fins and tail fin.
  Folds from the upper and lower, multilayered epidermis that enclose a thick layer of extracellular matrix. Reinforcement by fin rays; these are extended epidermis cells with filaments.
- Eyes.
- Two eyes on the dorsal side of the head, embedded in the ECM of the body wall.
  The eye consists of a large, central pigment cell with indentations and 70 to over 600 sensory cells. Their receptoral processes reach in the indentations; they are differentiated cilia with lamellar membranes. Proximally adjacent conical bodies possibly have a dioptric function. They are unique to the eyes of the Chaetognatha.
- Corona ciliata (Fig. 109 A, E).
- Ring of ciliary and glandular cells in the dorsal epidermis that extends from the head to the trunk. The corona has the only motile cilia on the body surface.
- Ciliary fence receptors.
  Short, straight rows of 50–300 sensory cells with stiff cilia that are distributed in large numbers transverse or parallel to the longitudinal axis of the body.
  Recognition of prey by perception of vibrations over short distances is the presumed function.
- Nervous system with six ganglia in the head and a large ventral ganglion in trunk (Fig. 109 E).
  A brain-like center lies as an unpaired, subepidermal cerebral ganglion in the dorsal part of the head. A retrocerebral organ of unknown function is embedded in the posterior end of the ganglion.
  A total of four pairs of longitudinal nerves (connectives) run forwards to the vestibular ganglia, lateral to trunk ganglion as well as back to the eyes and the corona ciliata. The ventral vestibular ganglia are connected with smaller, more inwards lying oesophageal ganglia. A commissure beneath the oesophagus closes the vestibular ganglia together and so leads to a nerve ring in the head. A small suboesophageal ganglion may be linked with this commissure.
  The rectangular ventral ganglion of the trunk is the largest nerve center in the body. About 12 pairs of radial nerves exit at the sides. A pair of

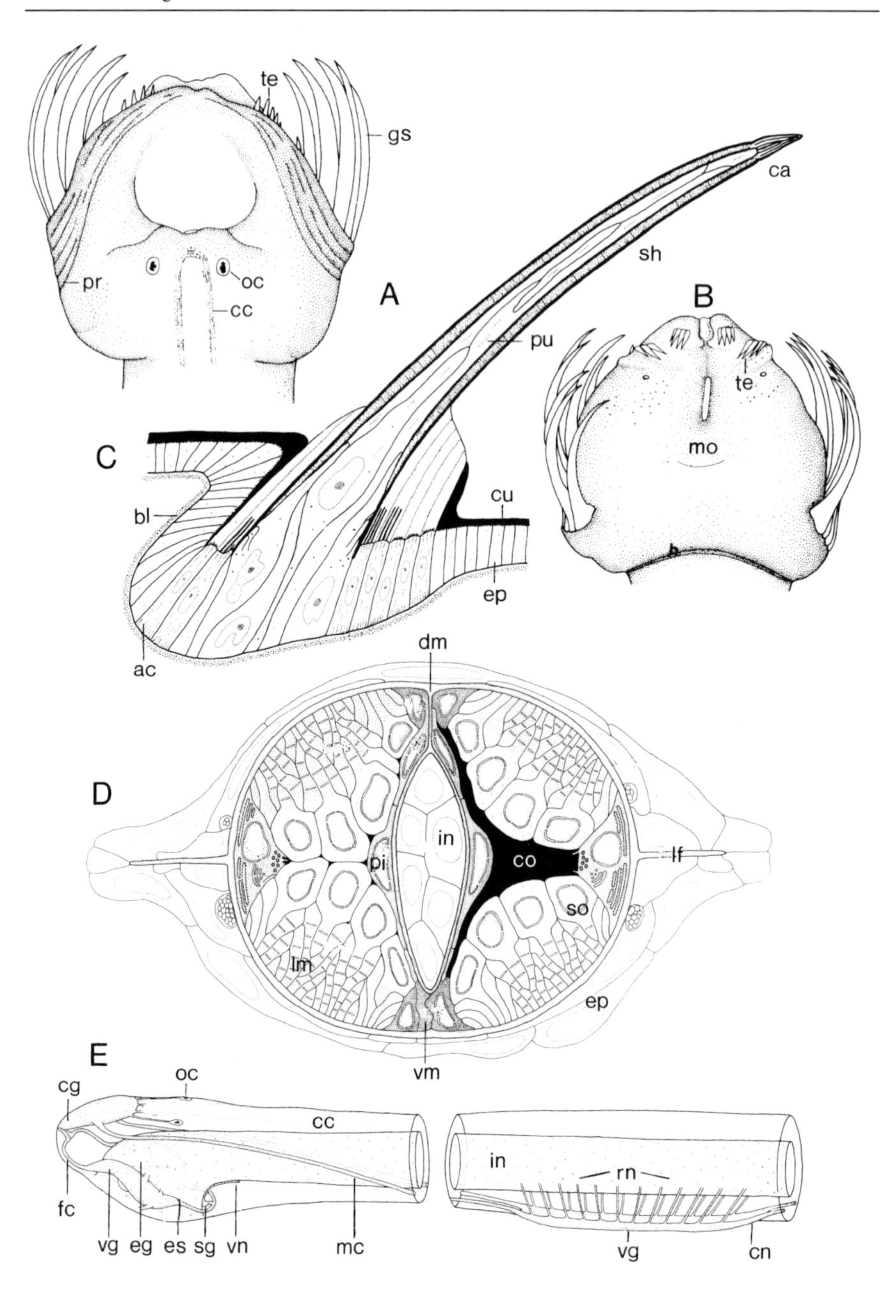
te
gs
ca
sh
pr
oc
cc
A
B
pu
te
mo
C
cu
bl
ep
ac
dm
D
in
pi
co
lf
so
lm
ep
vm
E
oc
cg
cc
in
rn
fc
vg
eg
es
sg
vn
mc
vg
cn

◀ **Fig. 109.** Chaetognatha. **A** Head in dorsal view. Hood (preputium) drawn back. **B** Head in ventral view. Teeth at anterior end. **C** *Parasagitta elegans*. Grasping spines in longitudinal section. **D** *Adhesisagitta hispida*. Cross section scheme through the body of a freshly hatched young animal. Four groups of longitudinal muscles in parietal coelothelium. The *left* side shows coelom cavities in the form of small slits, as they are realized in hatchlings in the largest part of the trunk and in the anterior third of the tail. *Right* side with extensive coelom cavities in the end of the trunk and in the posterior two thirds of the tail. **E** Scheme of the arrangement of the ganglia and main nerves. *ac* Anchor cell; *bl* basal lamina; *ca* cap of grasping spine; *cc* corona ciliata; *cg* cerebral ganglion; *cn* caudal nerve; *co* coelom cavity; *cu* cuticle; *dm* dorsal median cells; *ec* extracellular matrix; *eg* oesophageal ganglion; *ep* epidermis; *es* oesophageal commissure; *fc* frontal connective; *gs* grasping spine; *in* intestine; *lf* lateral fin; *lm* longitudinal muscle; *mc* main connective; *mo* mouth opening; *oc* ocellus; *pi* perintestinal mesodermal cells (visceral coelom epithelium); *pr* preputium (hood); *pu* pulp of the grasping spine with epidermis cells; *rn* radial nerves; *sg* suboesophageal ganglion; *sh* shaft of grasping spine; *so* somatic coelom epithelium with longitudinal musculature; *te* teeth; *vg* vestibular ganglion; *vm* ventral median cells; *vn* ventral oesophageal nerve. (**A, B** Kapp 1991; **C, E** Shinn in: F.W. Harrison, E.E. Ruppert (eds.) (1997) Microscopic Anatomy of Invertebrates, Vol. 15: Hemichordata, Chaetognatha, and the Invertebrate Chordates, pp. 115, 151. Reprinted by permission of John Wiley & Sons, Inc.; **D** Shinn & Roberts 1994)

caudal nerves runs from the posterior corners of the ganglion to the tail. In comparison with the Radialia (Epineuralia), SALVINI-PLAWEN (1988) emphasized the basiepithelial position of the nerve system in the Chaetognatha. This is, however, a useless plesiomorphous feature. There are no apomorphous agreements with other Bilateria concerning the special construction of the nervous system.

The Chaetognatha do not have nephridial organs. They are hermaphrodites with the differentiation of ovaries in the posterior trunk and testes in the tail. Finally, the Chaetognatha are characterized by a direct development; there is no stage with larval features that could be considered as larva in their life cycle.

Lack of nephridia, hermaphroditism and direct development are widespread among the Bilateria – and this with alternatives. Before knowledge of a sister group is available, it cannot be decided to what degree these are plesiomorphies or apomorphies in the Chaetognatha.

"Molecular" Relationships

Detailed morphological analyses through to the ultrastructure level have not revealed agreements with any other taxon of the Bilateria that could be interpreted as synapomorphies.

Under these circumstances, I am astonished at the nonchalance in the presentation of relationship hypotheses on the basis of molecular data. I present three examples with the formulation of three different adelphotaxa relationships of the Chaetognatha.

In HALANYCH (1966a), the Nematoda are the sister group of the Chaetognatha. The author attempts to present a scenario in the evolution of the Nematoda to the Chaetognatha – albeit with some very questionable comparisons. Thus, the grasping spines of the Chaetognatha are placed in relation to the head setae of the Nematoda or the membranous copulatory bursa with ribs on the tail of certain male Nematoda are viewed as precursors of the Chaetognatha fins.

In LITTLEWOOD et al. (1998b), the Gnathostomulida are the adelphotaxon of the Chaetognatha. In this case, a comparison of the hard structures at the anterior end of Chaetognatha with the jaw and basal plate in the oral cavity of the Gnathostomulida was emphasized – even if with the noteworthy comment "the features alone do not persuade an acceptance of the new hypothesis and, like all molecular systematic studies, the strength of interpretation relies on comparative morphological data to make biological sense" (LITTLEWOOD et al. 1998b, p. 77).

A corresponding lack is no problem for GIRIBET et al. (2000). In their "favored tree", the Nemertodermatida (Plathelminthes) and the Chaetognatha appear as adelphotaxa, although this grouping is difficult "to justify on the basis of morphological/anatomical characters" (GIRIBET et al. 2000, p. 556). The DNA sequence of *Nemertinoides elongatus* deposited in the Gen Bank is an artifact (addendum), but this apparently does not influence the relationship hypothesis.

One must await a process of clarification among the supporters of "molecular" phylogenies with a separation of the chaff from the wheat before seriously considering contradictory publications of this nature.

Comparison with the Radialia

Last, but not least, the agreements with oligomeric Radialia in early ontogenesis and in the coelom arrangement must be discussed.

In ontogenesis (Fig. 110), there are at first homologous agreements in the radial cleavage pattern, in the adjacent blastula and in the formation of a gastrula by invagination. However, these are common plesiomorphies of the Chaetognatha and the Radialia that we postulated earlier for the ground pattern of the Bilateria. The symplesiomorphies do not say anything at all about a possible relationship with the Radialia, however, they would self-evidently not hinder an ordering of the Chaetognatha in the Radialia.

The agreement in the deuterostomy, i.e., the new formation of a definitive mouth opposite to the blastopore in the Chaetognatha and Deuterostomia, seems to be more serious because this is without doubt an apomorphous agreement. If we propose it as a synapomorphy, two possible positions for the Chaetognatha result: (1) The Chaetognatha form the sister

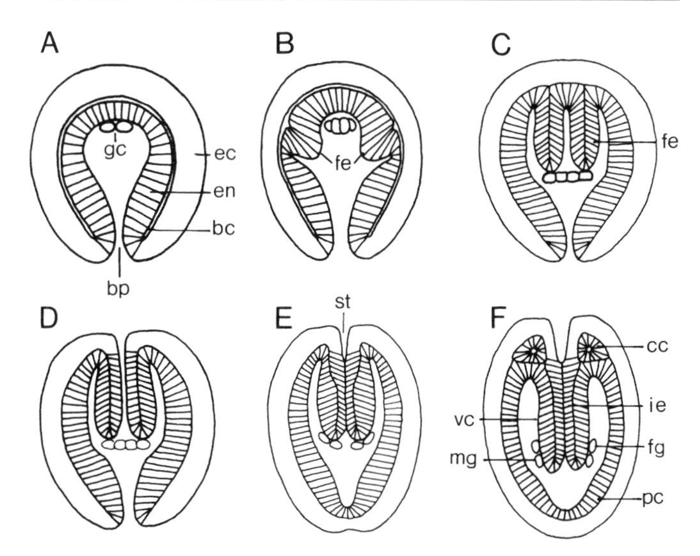

Fig. 110. Chaetognatha. Scheme of some stages of the early ontogenesis. **A** Gastrula with a small blastocoel develops through a total radial cleavage. Primordial germ cells are formed from the entoderm. **B** Infoldings of the entoderm opposite the blastopore initiate the genesis of the coelom. **C** The infolded epithelium grows further in the direction of the blastopore. **D** Breakthrough of the stomodaeum opposite the blastopore. **E** Closure of the blastopore. **F** Rostral lacing of the head coelom beside the stomodaeum. Intestine epithelium arises from the inside wall of the entoderm fold. The outside wall together with the peripheral entoderm forms the coelom epithelium. *bc* Blastocoel; *bp* blastopore; *cc* head coelom; *ec* ectoderm; *en* entoderm; *fg* female germ cell; *fe* fold in entoderm; *gc* germ cell; *ie* intestine (inside epithelium of the entoderm fold); *mg* male germ cell; *pc* parietal coelom wall (of peripheral entoderm of the gastrula); *st* stomodaeum; *vc* visceral coelom wall (of outside epithelium of the entoderm fold). (Kapp 2001)

group of the Deuterostomia in their entirety. (2) The Chaetognatha are the sister group of a specific subtaxon within the Deuterostomia, i.e., the sister group of the Echinodermata or the Stomochordata.

These two hypotheses are not compatible with the elaboration of the ground pattern of the Deuterostomia (p. 93). The stem species of the Deuterostomia was a sedentary organism with a tentacle apparatus as filter system as well as a division of the body into three sections with three sets of coelom sacs. The formal agreement in the number of coelom cavities is opposed by fundamental differences in their structure and function in the comparison of real organisms. In *Rhabdopleura*, a "basal" representative of the Deuterostomia or at least of the Stomochordata, the protocoel forms a resistance for the action of the shield-like creeping disk on the head, carries the mesocoel the tentacular filter apparatus of the collar, and serves the metacoel for movement of the trunk within the dwelling tubes. There are no arguments at all for the derivation of the predatory Chaetognatha of the pelagial zone with trunk and tail coeloms as resistances for locomotion in free water from benthal, microphagic Deuterostomia with a dipleura larva and with an organization pattern of *Rhabdopleura, Cephalodiscus* or even *Balanoglossus.*

In addition, there is the unique, incomparable formation of the coelom in the ontogenesis of the Chaetognatha (KAPP 1997, 2001). At the stage of the gastrula, infoldings of the entoderm next to the prospective mouth initiate the genesis of the coelom. The infolded epithelium pushes itself further in the direction of the blastopore; the inside wall forms the gut epithelium, the outside wall together with the peripheral entoderm of the gastrula form the wall of the trunk and tail coeloms. This process and the rostral lacing of the head coelom have nothing to do with the enterocoely of the Deuterostomia.

The presented facts and considerations lead inevitably to the assumption of a nonhomology of the body cavities designated as coelom in Chaetognatha as compared with the Deuterostomia; they must have evolved independently of each other. Moreover, this leads a posteriori to the interpretation of deuterostomy as a convergence.

The new mouth opening opposite the blastopore evolved independently in the stem lineages of the Chaetognatha and Deuterostomia.

As a result, the requirements of phylogenetic systematics for a validated identification of the sister group of the Chaetognatha still remain unfulfilled. The question of the phylogenetic relation of the arrow worms has not yet been answered.

References

ADRIANOV, A. V. & V. V. MALAKHOV (1994). Kinorhyncha: Structure, development, phylogeny and taxonomy. Nauka Publishing, Moscow, 1-260.

ADRIANOV, A. V. & V. V. MALAKHOV (1996). Priapilida (Priapulida): Structure, development, phylogeny, and classification. KMK Scientific Press Ltd. Moscow, 1-266.

AHLBERG, P. E. (1991). A re-examination of sarcopterygian interrelationships, with special reference to the Porolepiformes. Zool. J. Linn. Soc. **103**, 241-287.

AHLRICHS, W. H. (1993a). Ultrastructure of the protonephridia of Seison annulatus (Rotifera). Zoomorphology **113**, 245-251.

AHLRICHS, W. H. (1993b). On the protonephridial system of the brackish-water rotifer Proales reinhardti (Rotifera, Monogononta). Microfauna Marina **8**, 39-53.

AHLRICHS, W.H. (1995). Ultrastruktur und Phylogenie von Seison nebaliae (Grube 1859) und Seison annulatus (Claus 1876). Hypothesen zu phylogenetischen Verwandtschaftsverhältnissen innerhalb der Bilateria. Cuvillier Verlag. Göttingen.

AHLRICHS, W. H. (1997). Epidermal ultrastructure of Seison nebaliae and Seison annulatus, and a comparison of epidermal structures within the Gnathifera. Zoomorphology **117**, 41-48.

ALBERTI, G. & V. STORCH (1988). Internal fertilisation in a meiobenthic priapulid worm: Tubiluchus phillipinensis (Tubiluchidae, Priapulida). Protoplasma **143**, 193-196.

ALBRECHT, H., EHLERS, U. & H. TARASCHEWSKI (1997). Syncytial organisation of acanthors of Polymorphus minutus (Palaeacanthocephala), Neoechinorhynchus rutili (Eoacanthocephala), and Moniliformis moniliformis (Archiacanthocephala) (Acanthocephala). Parasitol. Res. **83**, 326-338.

ALLDREDGE, A. L. (1976). Appendicularians. Scientific American **235**, 94-102.

AMSELLEM, J. & C. RICCI (1982). Fine structure of the female genital apparatus of Philodina (Rotifera, Bdelloida). Zoomorphology **100**, 89-105.

ARAMBOURG, C. & L. BERTIN (1958). Super-ordres des Holostéens et des Halécostomes (Holostei et Halecostomi). Traité de Zoologie. **XIII**, 2173-2203. Paris.

ARAMBOURG, C. & J. GUIBÉ (1958). Sous-classe des Dipneustes (Dipneusti). Traité de Zoologie **XIII**, 2522-2540. Paris.

ARCHER, M., MURRAY, P., HAND, S. & H. GODTHELP (1993). Reconsideration of monotreme relationships based on the skull and dentition of the Miocene Obdurodon dicksoni. In F.S. SZALAY, M. J. NOVACEK & M. C. McKENNA (Eds.). Mammal phylogeny. Mesozoic differentiation, multituberculates, monotremes, early therians, and marsupials. 75-107. Springer-Verlag. New York, Berlin, Heidelberg.

AX, P. (1956). Monographie der Otoplanidae (Turbellaria). Morphologie und Systematik. Akad. d. Wiss. u. d. Lit. Mainz. Abhand. d. Math.-Naturw. Kl. Jg. 1955, **13**, 1-298.

AX, P. (1984). Das Phylogenetische System. Systematisierung der lebenden Natur aufgrund ihrer Phylogenese. G. Fischer. Stuttgart, New York.

AX, P. (1985). Stem species and the stem lineage concept. Cladistics **1**, 279-287.

AX, P. (1987). The phylogenetic system. The systematization of organisms on the basis of their phylogenesis. J. Wiley & Sons. Chichester, New York, Brisbane, Toronto, Singapore.

AX, P. (1988). Systematik in der Biologie. Darstellung der stammesgeschichtlichen Ordnung in der lebenden Natur. G. Fischer. Stuttgart, New York.

AX, P. (1989). The integration of fossils in the phylogenetic system of organisms. Abh. naturwiss. Ver. Hamburg (NF) **28**, 27–43.

AX, P. (1993). Turbanella lutheri (Gastrotricha, Macrodasyoida) im Brackwasser der Färöer. Microfauna Marina **8**, 139–144,

AX, P. (1996) Multicellular Animals. A new Approach to the Phylogenetic Order in Nature. Vol. I. Springer, Berlin Heidelberg New York.

AX, P. (2000) Multicellular Animals. The Phylogenetic System of the Metazoa. Vol. II. Springer, Berlin Heidelberg New York.

BAKER, A. N., ROWE, F. W. E. & H. E. S. CLARK (1986). A new class of Echinodermata from New Zealand. Nature **321**, 862–864.

BALSER, E. J. & E. E. RUPPERT (1990). Structure, ultrastructure, and function of the preoral heart-kidney in Saccoglossus kowalevskii (Hemichordata, Enteropneusta) including new data on the stomochord. Acta Zool. **71**, 235–249.

BARRINGTON, E. J. W. (1965). The biology of Hemichordata and Protochordata. Oliver & Boyd. Edinburgh, London.

BARTOLOMAEUS, T. (1993). Die Leibeshöhlenverhältnisse und Nephridialorgane der Bilateria – Ultrastruktur, Entwicklung und Evolution. Habilitationsschrift. Univ. Göttingen.

BARTOLOMAEUS, T. (2001): Ultrastructure and formation of the body cavity lining in Phoronis muelleri (Phoronida, Lophophorata). Zoomorphology **120**, 135–148.

BARTOLOMAEUS, T. & P. AX (1992). Protonephridia and metanephridia – their relation within the Bilateria. Z. zool. Syst. Evolut.-forsch. **30**, 21–45.

BARTOLOMAEUS, T. & P. GROBE (2000): Towards a phylogenetic system of the Bryozoa. Zoology **103** (Suppl. III): 99.

BEAUCHAMP, P. DE (1960). Classe des Brachiopodes. Traité de Zoologie **V**, 1380–1499. Paris.

BEAUCHAMP, P. DE (1965). Classe des Rotifères. Traité de Zoologie **IV**, 3, 1225–1379. Paris.

BELYAEV, G. (1974). A new family of abyssal starfishes. Zoologicheskii Zhurnal **53**, 1502–1508.

BENITO, J. & F. PARDOS (1997). Hemichordata. In F. W. HARRISON & E. E. RUPPERT (Eds.). Microscopic Anatomy of Invertebrates **15**, 15–101. Wiley-Liss, Inc. New York.

BENNETT, S. C. (1996). The phylogenetic position of the Pterosauria within the Archosauromorpha. Zool. J. Linn. Soc. **118**, 261–308.

BENTON, M. J. (1990). Phylogeny of the major tetrapod groups: morphological data and divergence data. J. Mol. Evol. **30**, 409–424.

BENTON, M. J. & J. M. CLARK (1988). Archosaur phylogeny and the relationships of the Crocodylia. In M. J. BENTON (Ed.). The phylogeny and classification of the tetrapods. Vol. I. Syst. Ass. Spec. Vol. **35 A**, 295–338. Clarendon Press. Oxford.

BEREITER-HAHN, J. (1984). Cephalochordata. In J. BEREITER-HAHN, A. G. MATOLTSY & K. SYLVIA RICHARDS. Biology of the Integument. 1 Invertebrates. Springer. Berlin, Heidelberg, New York, Tokyo.

BERRILL, N. J. (1950). The Tunicata. With an account of the British species. Ray Society, London.

BERRILL, N. J. (1955). The origin of the vertebrates. Oxford University Press, London.

BIERI, R. (1991). Systematics of the Chaetognatha. In Q. BONE, H. KAPP & A. C. PIERROT-BULTS (Eds.). The biology of chaetognaths. 122–136. Oxford University Press. Oxford, New York, Tokyo.

BIRD, A. F. & J. BIRD (1991). The structure of nematodes. Academic Press, Inc. San Diego.

BISHOP, M. J. & A. E. FRIDAY (1988). Estimating the interrelationships of tetrapod groups on the basis of molecular sequence data. In M. J. BENTON (Ed.). The phylogeny and classification of the tetrapods. Vol. I. Syst. Ass. Spec. Vol. **35A**, 33–58. Clarendon Press, Oxford.

BLAKE, D. B. (1987). A classification and phylogeny of post-Palaeozoic sea stars (Asteroidea: Echinodermata). J. Nat. Hist. **21**, 481–528.

BLAXTER, M. L., DE LEY, P., GAREY, J. R., LIU, L. X., SCHELDEMAN, P., VIERSTRAETE, A., VANFLETEREN, J. R., MACKEY, L. Y., DORRIS, M., FRISSE, L. M., VIDA, J. T. & W. K. THOMAS (1998). A molecular evolutionary frame work for the phylum Nematoda. Nature **392**, 71–75.

BLOME, D. & F. RIEMANN (1994). Sandy beach meiofauna of Eastern Australia (Southern Queensland and New South Wales). III. Revision of the nematode genus Onyx Cobb, 1891, with a description of three new species (Nematoda) (Desmodoridae). Invertebr. Taxon. **8**, 1483–1492.

BOADEN, P. J. S. (1968). Water movement – a dominant factor in interstitial ecology. Sarsia **34**, 125–136.

BOARDMAN, R. S., CHEETHAM, A. H. & P. L. COOK (1983). Introduction to the Bryozoa. In R. S. BOARDMAN & A. H. CHEETHAM & D. B. BLAKE & O. L. JOHN UTGAARD KARKLINS & P. L. COOK & P. A. SANDBERG & G. LUTAUD & T. S. WOOD (Eds.). Part G. Bryozoa. Revised. Vol. I: Introduction, Order Cystoporata, Order Cryptostomata, 3–48. Treatise on Invertebrate Palaeontology. The Geological Society of America, Inc. & The University of Kansas. Boulder Colorado, and Lawrence, Kansas.

BOBIN, G. (1977). Interzoecial communications and the funicular system. In R.M. WOOLLACOTT & R. L. ZIMMER (Eds.). Biology of bryozoans, 307–333. Academic Press. New York, San Francisco, London.

BOCK, W. J. (1963). The cranial evidence for ratite affinities. Proc. 13th Intern. Ornithol. Congr. 39–54.

BÖHME, W. (1988). Zur Genitalmorphologie der Sauria: Funktionelle und stammesgeschichtliche Aspekte. Bonn. zool. Monogr. **27**, 1–176.

BÖHME, W. (1995). Hemiclitoris discovered: a fully differentiated erectile structure in female monitor lizards (Varanus ssp.) (Reptilia; Varanidae). J. Zool. Syst. Evol. Research **33**, 129–132.

BOLT, J. R. (1991). Lissamphibians origins. In H.-P. SCHULTZE & L. TRUEB. Origins of the higher groups of tetrapods. Controversy and consensus. 194–222. Cornell University Press. Ithaca.

BONE, Q. (1958). The asymmetry of the larval Amphioxus. Proc. Zool. Soc. London **130**, 289–293.

BONE, Q. (1998a). The biology of pelagic tunicates. Oxford University Press. Oxford, New York, Tokyo.

BONE, Q. (1998b). Locomotion, locomotory muscles, and buoyancy. In Q. BONE (Ed.). The biology of pelagic tunicates. Oxford University Press. Oxford, New York, Tokyo.

BOURLIÈRE, F. (1955). Ordre des Marsupiaux. Systématique. Traite de Zoologie **XVII**, 143–171. Paris.

BRAGA, M. de & O. RIEPPEL (1997). Reptile phylogeny and the interrelationships of turtles. Zool. J. Linn. Soc. **120**, 281–354.

BRANDENBURG, J. & G. KÜMMEL (1961). Die Feinstruktur der Solenocyten. J. Ultrastr. Res. **5**, 437–452.

BREIMER, A. (1978). General Morphology. Recent Crinoids. In C. MOORE & C. TEICHERT (Eds.). Treatise on Invertebrate Palaeontology. Part T, Echinodermata 2, Vol. **1**, 9–58. Lawrence. University of Kansas Press.

BRIEN, P. (1948). Embranchement des Tuniciers. Morphologie et reproduction. Traité de Zoologie **XI**, 553–894. Paris.

BRIEN, P. (1960). Classe des Bryozoires. Traité de Zoologie **V**, 1053–1379. Paris.

BRUSCA, R. C. & G. J. BRUSCA (1990). Invertebrates. Sinauer Associates, Inc. Sunderland, Massachusetts.

BUNKE, D. & P. SCHMIDT (1976). Philodina citrina (Rotatoria). Organisation und Fortpflanzung. Inst. f. d. wiss. Film, 1–10. Göttingen.

BURDON-JONES, C. (1952). Development and biology of the larva of Saccoglossus horsti (Enteropneusta). Phil. Trans. Royal Soc. London, Ser. B, **236**, 553–589.

BURDON-JONES, C. (1956). Observations on the enteropneust, Protoglossus koehleri (Caullery & Mesnil). Proc. Zool. Soc. London **127**, 35–58.

BURIGHEL, P. & R. A. CLONEY (1997). Urochordata: Ascidiacea. In F. W. HARRISON & E. E. RUPPERT (Eds.). Microscopic Anatomy of Invertebrates **15**, 221–347. Wiley-Liss, Inc. New York.

CALDWELL, M. W. (1999). Squamate phylogeny and the relationships of snakes and mosasauroids. Zool. J. Linn. Soc. **125**, 115–147.

CAMERON, C. B., GAREY, J. R. & B. J. SWALLA (2000). Evolution of the chordate body plan: new insights from phylogenetic analysis of deuterostome phyla. Proc. Natl. Acad. Sci. **97**, 4469–4474.

CAMPBELL, A.C. (1983). Form and function of pedicellariae. In M. JANGOUX & J. M. LAWRENCE (Eds.). Echinoderm studies **1**, 139–167. A. A. Balkema. Rotterdam.

CARLE, K. J. & E. E. RUPPERT (1983). Comparative ultrastructure of the bryozon funiculus: A blood vessel homologue. Z. zool. Syst. Evolut.-forsch. **21**, 181–193.

CARLSON, S. J. (1995). Phylogenetic relationships among extant brachiopods. Cladistics **11**, 131–197.

CARRANZA, S., BAGUÑÀ, J. & M. RIUTORT (1997). Are the Platyhelminthes a monophyletic primitive group? An assessment using 18S rDNA sequences. Mol. Biol. Evol. **14**, 485–497.

CARROLL, R. L. (1993). Paläontologie und Evolution der Wirbeltiere. Thieme Verlag. Stuttgart, New York.

CAVALIER-SMITH, T. (1998). A revised six-kingdom system of life. Biol. Rev. **73**, 203–266.

CHARIG, A. (1983). A new look at the dinosaurs. Heinemann. London. British Museum (Natural History). London.

CHEN, P.-J., DONG, Z.-M. & S.-N. ZHEN (1998). An exceptionally well-preserved theropod dinosaur from the Yixian formation of China. Nature **391**, 147–152.

CHRISTEN, R. & J.-C. BRACONNOT (1998). Molecular phylogeny of tunicates. A preliminary study using 28S ribosomal RNA partial sequences: implications in terms of evolution and ecology. In Q. BONE (Ed.). The biology of pelagic tunicates, 265–271. Oxford University Press. Oxford, New York, Tokyo.

CHUANG, S.-H. (1977). Larval development in Discinisca (Inarticulate brachiopod). Amer. Zool. **17**, 39–53.

CLÉMENT, P. & E. WURDAK (1991). Rotifera. In F. W. HARRISON & E. E. RUPPERT (Eds.). Microscopic Anatomy of Invertebrates **4**, 219–297.

COATES, M. T. & J. A. CLACK (1990). Polydactyly in the earliest known tetrapod limbs. Nature **347**, 66–69.

COHEN, B. L. (2000). Monophyly of brachiopods and phoronids: reconciliation of molecular evidence with Linnean classification (the subphylum Phoroniformea nov.). Proc. Royal Soc. London, B **267**, 225–231.

COHEN, B. L. & A. B. GAWTHROP (1996). Brachiopod molecular phylogeny. In P. COPPER & J. JIN (Eds.). Brachiopoda. 73–80. A. A. Balkema. Rotterdam, Brookfield.

COHEN, B. L. & A. B. GAWTHROP (1997). The brachiopod genome. In WILLIAMS, A., BRUNTON, C. H. C. & S. J. CARLSON. Part H. Brachiopoda. Revised. Vol. I: Introduction, 189–211. Treatise on Invertebrate Palaeontology. The Geological Society of America & The University of Kansas. Boulder, Colorado and Lawrence, Kansas.

COHEN, B. L., STARK, S., GAWTHROP, A. B., BURKE, M. E. & C. W. THAYER (1998). Comparison of articulate brachiopod nuclear and mitochondrial gene trees leads to a clade-based redefinition of protostomes (Protostomozoa) and deuterostomes (Deuterostomozoa). Proc. R. Soc. London B **265**, 475–482.

CRACRAFT, J. (1974). Phylogeny and evolution of the ratite birds. Ibis **116**, 494–521.

CRACRAFT, J. (1988). The major clades of birds. In M. J. BENTON (Ed.). The phylogeny and classification of the tetrapods. Vol. I. Syst. Ass. Spec. Vol. **35 A**, 339–361. Clarendon Press. Oxford.

CROMPTON, D. W. T. (1985). Reproduction. In CROMPTON, D. W. T. & B. B. NICKOL (Eds.). Biology of the Acanthocephala. 213 –271. Cambridge University Press. Cambridge.

CUÉNOT, L. (1948). Anatomie, éthologie et systématique des Échinodermes. Traité de Zoologie **XI**, 3–363. Paris.

CULLEN, D. J. (1973). Bioturbation of superficial marine sediments by interstitial meiobenthos. Nature **242**, 323–324.

CZIHAK, G. (1960). Untersuchungen über die Coelomanlagen und die Metamorphose des Pluteus von Psammechinus miliaris (Gmelin). Zool. Jb. Anat. **78**, 235–256.

DAGET, J. (1958). Sous-classe des Brachioptérygiens (Brachiopterygii). Traité de Zoologie **XIII**, 2501–2521. Paris.

DAUGHERTY, H., CREE, A., HAY, J. M. & M. B. THOMPSON (1990). Neglected taxonomy and continuing extinctions of tuatara (Sphenodon). Nature **347**, 177–179.

DAVID, B. & R. MOOI (1997). Skeletal homology of echinoderms. The Palaeontol. Soc. Papers **3**, 305–335.

DAWYDOFF, C. (1948a). Embryologie des Échinodermes. Traité de Zoologie **XI**, 277–363. Paris.

DAWYDOFF, C. (1948b). Classe des Enteropneustes. Traité de Zoologie **XI**, 369–453. Paris.

DAWYDOFF, C. (1948c). Classe des Pterobranches. Traité de Zoologie **XI**, 454–489. Paris.

DESUI, M. (1991). On the origins of mammals. In H. P. SCHULTZE & L. TRUEB. Origins of the higher groups of tetrapods. 579–597. Cornell University Press. Ithaca, London.

DILLY, P. N. (1972). The structure of the tentacles of Rhabdopleura compacta (Hemichordata) with special reference to neurociliary control. Z. Zellforsch. **129**, 20–39.

DILLY, P. N. (1973). The larva of Rhabdopleura compacta (Hemichordata). Mar. Biol. **18**, 69–86.
DILLY, P. N. (1976). Some features of the ultrastructure of the coenecium of Rhabdopleura compacta. Cell. Tiss. Res. **170**, 253–261.
DILLY, P. N. (1985). The habitat and behavior of Cephalodiscus gracilis (Pterobranchia, Hemichordata) from Bermuda. J. Zool. London **207**, 223–239.
DILLY, P. N., WELSCH, U. & G. REHKÄMPER (1986). Fine structure of tentacles, arms and associated coelomic structures of Cephalodiscus gracilis (Pterobranchia, Hemichordata). Acta Zool. **67**, 181–191.
DRACH, P. (1948). Embranchement des Céphalochordés. Traité de Zoologie **XI**, 931–1035. Paris.
DUNAGAN, T. T. & D. M. MILLER (1991). Acanthocephala. In F. W. HARRISON & E. E. RUPPERT (Eds.). Microscopic Anatomy of Invertebrates **4**, 229–332. Wiley-Liss, Inc. New York.
EERNISSE, D. J. & A. G. KLUGE (1993). Taxonomic congruence versus total evidence, and amniote phylogeny inferred from fossils, molecules and morphology. Molecular Biol. Evolut. **10**, 1170–1195.
EHLERS, U., AHLRICHS, W., LEMBURG, C. & A. SCHMIDT-RHAESA (1996). Phylogenetic systematization of the Nemathelminthes (Aschelminthes). Verh. Dtsch. Zool. Ges. **89**, 1, 8.
EMIG, C. C. (1974). The systematics and evolution of the phylum Phoronida. Z. zool. Syst. Evolut.-forsch. **12**, 128–151.
EMIG, C. C. (1976). Phylogenése des Phoronida. Les Lophophorates et le concept Archimerata. Z. zool. Syst. Evolut. **14**, 10–24.
EMIG, C. C. (1977). Embryology of Phoronida. Amer. Zool. **17**, 21–37.
EMIG, C. C. (1982). The biology of Phoronida. Adv. Mar. Biol. **19**, 1–89.
EMIG, C. C. (1985). Phylogenetic systematics in Phoronida (Lophophorata). Z. zool. Syst. Evolut.-forsch. **23**, 184–193.
EMIG, C. C. (1997). Les lophophorates constituent-ils un embranchement? Bull. Soc. Zool. France **122**, 279–288.
EPP, R. W. & W. M. LEWIS (1979). Sexual dimorphism in Brachionus plicatilis (Rotifera): evolutionary and adaptive significance. Evolution **33**, 919–928.
ERBER, W. (1983). Der Steinkanal der Holothurien: Eine morphologische Studie zum Problem der Protocoelampulle. Z. zool. Syst. Evolut.-forsch. **21**, 217–234.
ESTES, R., DE QUEIROZ, K. & J. GAUTHIER (1988). Phylogenetic relationships within Squamata. In R. ESTES & G. PREGILL (Eds). Phylogenetic relationships of the lizard families. 119–281. Stanford University Press. Stanford, CA.
FEDUCCIA, A. (1980). The age of birds. Harvard University Press. Cambridge, Massachusetts, and London.
FELL, H. B. (1962). Evidence for the validity of Matsumoto's classification of the Ophiuroidea. Publ. Seto Marine Biol. Lab. **10**, 145–152.
FELL, H. B. (1963). The phylogeny of sea-stars. Phil. Trans. Roy. Soc. London (B) **246**, 381–435.
FELL, H. B. (1982). Echinodermata. In S. P. PARKER (Ed.). Synopsis and classification of living organisms **2**, 785–818. Mc Graw-Hill. New York.
FENAUD, R. (1986). The house of Oikopleura dioica (Tunicata, Appendicularia). Zoomorphology **106**, 224–231.
FENAUD, R. (1998a). Anatomy and functional morphology of the Appendicularia. In Q. BONE (Ed.). The biology of pelagic tunicates, 25–34. Oxford University Press. Oxford, New York, Tokyo.
FENAUD, R. (1998b). Life history of the Appendicularia. In Q. BONE (Ed.). The biology of pelagic tunicates, 151–159. Oxford University Press. Oxford, New York, Tokyo.
FENAUD, R. (1998c). The classification of Appendicularia. In Q. BONE (Ed.). The biology of pelagic tunicates, 295–306. Oxford University Press. Oxford, New York, Tokyo.
FIORONI, P. (1987). Allgemeine und vergleichende Embryologie der Tiere. Springer-Verlag. Berlin, Heidelberg.
FLOOD, P. R. (1991). Architectúre of, and water circulation and flow rate in, the house of the planktonic tunicate Oikopleura labradoriensis. Mar. Biol. **111**, 95–111.
FLOOD, P. R. & D. DEIBEL (1998). The appendicularian house. In Q. BONE (Ed.). The biology of pelagic tunicates, 105–124. Oxford University Press. Oxford, New York, Tokyo.

FRANZ, V. (1927). Morphologie der Akranier. Ergebn. Anat. Entwicklungsgesch. **27**, 464–692.
FRANZÉN, A. (1960). Monobryozoon limicola n. sp., a ctenostomatous bryozoan from the detritus layer of soft sediment. Zool. Bidr. Uppsala **33**, 135–148.
FRITSCH, B. & M. H. WAKE (1988). The inner ear of gymnophione amphibians and its nerve supply: a comparative study of regressive events in a complex sensory system (Amphibia, Gymnophiona). Zoomorphology **108**, 201–217.
FUSENIG, N. E. (1986). Mammalian epidermal cells in culture. In J.BEREITER-HAHN, A. G. MATOLTSY & K. SYLVIA RICHARDS (Eds.). Biology of the integument **2** Vertebrates, 409–442. Springer-Verlag. Berlin, Heidelberg.
GADOW, H. (1896). On the evolution of the vertebral column of Amphibia and Amniota. Phil. Trans. Royal Soc. London, Ser. B. **187**, 1–57.
GAFFNEY, E. S. (1980). Phylogenetic relationships of the major groups of amniotes. In A. L. PANCHEN (Ed.). The terrestrial environment and the origin of land vertebrates. Syst. Ass. Spec. Vol. **15**, 593–610. Academic Press. London, New York.
GAFFNEY, E. S. & P. A. MEYLAN (1988). A phylogeny of the turtles. In M. J. BENTON (Ed.). The phylogeny and classification of the tetrapods. Vol. I. Syst. Ass. Spec. Vol. **35** A, 157–219. Clarendon Press, Oxford.
GALE, A. S. (1987). Phylogeny and classification of the Asteroidea (Echinodermata). Zool. J. Linn. Soc. **89**, 107–132.
GANS, C. (1989). Stages in the origin of vertebrates: analysis by means of scenarios. Biol. Rev. **64**, 221–268.
GARDINER, B. G. (1982). Tetrapod classification. Zool. J. Linn. Soc. **74**, 207–232.
GARDINER, B. G. (1993). Haematothermia: warm blooded amniotes. Cladistics **9**, 369–395.
GARDINER, B. G., MAISEY, J. G. & D. T. J. LITTLEWOOD (1996). Interrelationships of basal neopterygians. in M. L. J. STIASSNY, L. R. PARENTI & G. D. JOHNSON (Eds). Interrelationships of fishes, 117–146. Academic Press. San Diego, California.
GAREY, J. R. & A. SCHMIDT-RHAESA (1998). The essential role of „minor" phyla in molecular studies of animal evolution. Amer. Zool. **38**, 907–917.
GARSTANG, W. (1928). The morphology of the Tunicata, and its bearing on the phylogeny of the Chordata. Quart. J. Microsc. Sci. **72**, 51–187.
GAUTHIER, J. (1986). Saurischian monophyly and the origin of birds. Mem. California Acad. Science **8**, 1–56.
GAUTHIER, J., ESTES, R. & K. DE QUEIROZ (1988). A phylogenetic analysis of the Lepidosauromorpha. In R. ESTES & G. PREGILL (Eds.). Phylogenetic relationships of the lizard families, 15–118. Stanford University Press. Stanford, CA.
GAUTHIER, J., KLUGE, A. G. & T. ROWE (1988a). Amniote phylogeny and the importance of fossils. Cladistics **4**, 105–209.
GAUTHIER, J., KLUGE, A. G. & T. ROWE (1988b). The early evolution of the Amniota. In M. J. BENTON (Ed.). The phylogeny and classification of the tetrapods. Vol. I. Syst. Ass. Spec. Vol. **35 A**, 103–155. Clarendon Press. Oxford.
GEE, H. (1996). Before the backbone. Views on the origin of vertebrates. Chapman & Hall. London.
GEE, H. (2000). Deep time. Cladistics, the revolution in evolution. Fourth Estate London.
GERLACH, S. A. (1954). Die Nematodenbesiedlung des Sandstrandes und des Küstengrundwassers an der italienischen Küste. II. Ökologischer Teil. Arch. Zool. Ital. **39**, 311–359.
GERLACH, S. A. (1956). Über einen aberranten Vertreter der Kinorhynchen aus dem Küstengrundwasser. Kieler Meeresforsch. **12**, 120–124.
GERLACH, S. A. (1969). Cateria submersa sp. n., ein cryptorhager Kinorhynch aus dem sublitoralen Mesopsammal der Nordsee. Veröffentl. Inst. Meeresforsch. Bremerhaven **12**, 161–168.
GERLACH, S. A. (2000). Checkliste der Fauna der Kieler Bucht und eine Bibliographie zur Biologie und Ökologie der Kieler Bucht. In: Bundesanstalt für Gewässerkunde (Hrsg.). Die Biodiversität in der deutschen Nord- und Ostsee. Band 1. Bericht BfG-1247, Koblenz.
GHIRARDELLI, R. (1995). Chaetognaths: two unsolved problems: the coelom and their affinities. In G. LANCAVECCHIA, R. VALVASSORI & M. D. CANDIA CARNEVALI (Eds.). Body cavities: function and phylogeny. Selected Symposia and Monographs U. Z. I., **8**, 167–185. Mucchi, Modena.

GILBERT, S. F. & A. M. RAUNIO (1997). Embryology. Constructing the Organism. Sinauer Associates, Inc. Publishers. Sunderland.

GILMOUR, T. H. J. (1978). Ciliation and function of the food-collecting and waste-rejecting organs of lophophorates. Can. J. Zool. **56**, 2142–2155.

GILMOUR, T. H. J. (1979). Feeding in pterobranch hemichordates and the evolution of gill slits. Can. J. Zool. **57**: 1136–1142.

GILMOUR, T. H. J. (1982). Feeding in tornaria larvae and the development of gill slits in enteropneust hemichordates. Can. J. Zool. **60**, 3010–3020.

GIRIBET, G., DISTEL, D. L., POLZ, M., STERRER, W. & W. C. WHEELER (2000). Triploblastic relationships with emphasis on the acoelomates and the position of Gnathostomulida, Cycliophora, Plathelminthes and Chaetognatha: a combined approach of 18S rDNA sequences and morphology. Syst. Biol. **49**, 539–562.

GODEAUX, J. (1998). The relationships and systematics of the Thaliacea, with keys for identification. In Q. BONE (Ed.). The biology of pelagic tunicates, 273–294. Oxford University Press. Oxford, New York, Tokyo.

GODEAUX, J., BONE, Q. & J. C. BRACONNOT (1998). Anatomy of Thaliacea. In Q. BONE (Ed.). The biology of pelagic tunicates, 1–24. Oxford University Press. Oxford, New York, Tokyo.

GOLDSCHMID, A. (1996a). Hemichordata (Branchiotremata). In W. WESTHEIDE & R. RIEGER (Hrsg.). Spezielle Zoologie. Teil 1, 763–777. G. Fischer. Stuttgart, Jena, New York.

GOLDSCHMID, A. (1996b). Echinodermata, Stachelhäuter. In W. WESTHEIDE & R. RIEGER (Hrsg.). Spezielle Zoologie. Teil 1, 778–834. G. Fischer. Stuttgart, Jena, New York.

GOLDSCHMID, A. (1996c). Chordata, Chordatiere. In W. WESTHEIDE & R. RIEGER (Hrsg.). Spezielle Zoologie. Teil 1, 835–862. G. Fischer. Stuttgart, Jena, New York.

GOODRICH, E. S. (1916). On the classification of the Reptilia. Proc. Royal Soc. London, Ser. B **89**, 261–276.

GRASSÉ, P.-P. (1955). Ordre des Monotrèmes. Traité de Zoologie **XVII** (1), 47–92. Paris.

GRUNER, H.-E. (1994). Stamm Branchiotremata oder Hemichordata. In: Urania Tierreich. Wirbellose 2, 616–630. Urania-Verlag. Leipzig, Jena, Berlin.

HAECKEL, E. (1866). Generelle Morphologie der Organismen. Band I und II. G. Reiner, Berlin.

HAFFNER, K. v. (1950). Organisation und systematische Stellung der Acanthocephala. Neue Erg. u. Probl. Zool. (Klatt-Festschrift). 243–274. Leipzig.

HALANYCH, K. M. (1993). Suspension feeding by the lophophore-like arms of the pterobranch hemichordate Rhabdopleura normani. Biol. Bull. **185**, 417–427.

HALANYCH, K. M. (1995). The phylogenetic position of the pterobranch hemichordates based on 18S rDNA sequence data. Mol. Phylo. Evol. **4**, 72–76.

HALANYCH, K. M. (1996a). Testing hypothesis of chaetognath origins: long branches revealed by 18S ribosomal DNA. Syst. Biol. **45**, 223–246.

HALANYCH, K. M. (1996b). Convergence in the feeding apparatuses of lophophorates and pterobranch hemichordates revealed by 18S rDNA: an interpretation. Biol. Bull. **190**, 1–5.

HALANYCH, K. M., BACHELLER, J. D., AGUINALDO, A. M. A., LIVAM, S. M., HILLIS, D. M. & J. A. LAKE (1995). Evidence from 18S ribosomal DNA that the lophophorates are protostome animals. Science **267**, 1641–1643.

HALLERMANN, J. (1998). The ethmoidal region of Dibamus taylori (Squamata: Dibamidae), with a phylogenetic hypothesis on dibamid relationships within Squamata. Zool. J. Linn. Soc. **122**, 385–426.

HASZPRUNAR, G. (1996a). The Mollusca: Coelomate turbellarians or mesenchymate annelids. In J. TAYLOR (Ed.). Origin and evolutionary radiation of the Mollusca. 1–28. Oxford University Press. Oxford, New York, Tokyo.

HASZPRUNAR, G. (1996b). Plathelminthes and Plathelminthomorpha – paraphyletic taxa. J. Zool. Syst. Evol. Research **34**, 41–48.

HAUDE, R. (1994). Fossil holothurians: Constructional morphology of the sea cucumber, and the origin of the calcareous ring. In B. DAVID, A. GUILLE, J.-P. FÉRAL & M. ROUX (Eds.). Echinoderms through time. A. A. Balkema. Rotterdam, Brookfield.

HAUDE, R. & F. LANGENSTRASSEN (1976). Rotasaccus dentifer n. g., n. sp., ein devonischer Ophiocistioide (Echinodermata) mit „holothurioiden“ Wandskleriten und „echinoidem“ Kauapparat. Paläont. Z. **50**, 130–150.

HEALY, J. M., ROWE, F. W. E. & D. T. ANDERSON (1988). Spermatozoa and spermiogenesis in Xyloplax (Class Concentricycloidea); a new type of spermatozoon in the Echinodermata. Zool. Scripta **17**, 297–310.

HECHT, M. K., OSTROM, J. H., VIOHL, G. & P. WELLNHOFER (1985) (Eds.). The beginnings of birds. Proceedings of the international Archaeopteryx conference Eichstätt 1984. Eichstätt.

HEINZELLER, T. & U. WELSCH (1994). Crinoida. In F. W. HARRISON & F.-S. CHIA (Eds.). Microscopic Anatomy of Invertebrates **14**, 9–148. Wiley-Liss, Inc. New York.

HENNIG, W. (1966). Phylogenetic systematics. University of Illinois Press. Urbana, Chicago, London.

HENNIG, W. (1983). Stammesgeschichte der Chordaten. Fortschr. zool. Syst. Evolut.-forsch. **2**. P. Parey. Hamburg, Berlin.

HERLYN, H. (1996). Ultrastruktur des Verdauungssystems von Gnathostomula paradoxa (Ax 1956) mit ausführlicher Betrachtung von Kiefer und Basalplatte. Phylogenetische Bewertung der Merkmale. Diplomarbeit. Universität Göttingen.

HERLYN, H. (2000). Zur Ultrastruktur, Morphologie und Phylogenie der Acanthocephala. Logos Verlag. Berlin. 131 Seiten.

HERLYN, H. & U. EHLERS (1997). Ultrastructure and function of the pharynx of Gnathostomula paradoxa (Gnathostomulida). Zoomorphology **117**, 135–145.

HERMANS, C. O. (1983).The duo-gland adhesive system. Oceanogr. Mar. Biol. Ann. Rev. **21**, 283–339.

HIGGINS, R. P. (1968). Taxonomy and postembryonic development of the Cryptorhagae, a new suborder for the mesopsammic kinorhynch genus Cateria. Trans. Amer. Microsc. Soc. **87**, 21–39.

HIGGINS, R. P. (1990). Zelinkaderidae, a new family of cyclorhagid Kinorhyncha. Smithson. Contr. Zool. **500**, 1–26.

HIGGINS, R. P. & R. M. KRISTENSEN (1986). New Loricifera from southeastern United States coastal waters. Smith. Contr. Zool. **438**, 1–70.

HIGGINS, R. P. & R. M. KRISTENSEN (1988). Loricifera. In R. P. HIGGINS & H. THIEL (Eds.). Introduction to the study of Meiofauna. Smithsonian Institution Press, 319–321. Washington D. C., London.

HIGGINS, R. P. & V. STORCH (1991): Evidence for the direct development in Meiopriapulus fijiensis. Trans. Amer. Microsc. Soc. **110**, 37–46.

HOLMER, L. E., POPOV, L. E., BASETT, M. G. & J. LAURIE (1995). Phylogenetic analysis and ordinal classification of the Brachiopoda. Palaeontology **38**, 713–741.

HONDT, J.-L. d' (1971). Gastrotricha. Oceanogr. Mar. Biol. Ann. Rev. **9**, 141–192.

HOPSON, J. A. (1991). Systematics of the nonmammalian Synapsida and implications for patterns of evolution in synapsids. In H. P. SCHULTZE & L. TRUEB. Origins of the higher groups of tetrapods. 635–693. Cornell University Press. Ithaca, London.

HORST, C. J. van der (1939). Hemichordata. Bronns Klassen und Ordnungen des Tierreichs 4, IV, 2, 2, 1–737.

HUMMON, W. D. (1974). Gastrotricha from Beaufort, North Carolina, USA. Cah. Biol. Mar. **15**, 431–446.

HUMMON, W. D. (1982). Gastrotricha. In S. P. PARKER (Ed.). Synopsis and classification of living organisms. **1**, 857–863. McGraw-Hill Book Company. New York.

HYMAN, L. H. (1955). The Invertebrates: Echinodermata. Vol. IV. McGraw-Hill Book Company, Inc. New York, Toronto, London.

HYMAN, L. H. (1959). The Invertebrates: Smaller Coelomate Groups. Vol. V. McGraw-Hill Book Company. New York.

JÄGERSTEN, G. (1972). Evolution of the metazoan life cycle. A comprehensive theory. Academic Press. London, New York.

JANVIER, P. (1998). Early vertebrates. Clarendon Press. Oxford.

JARVIK, E. (1980). Basic structure and evolution of vertebrates. Vol. 1 & 2. Academic Press. Inc. London.

JEBRAM, D. (1973). Ecological aspects of the phylogeny of the Bryozoa. Z. zool. Syst. Evolut.-forsch. **11**, 275–283.

JEFFERIES, R. P. S. (1986).The ancestry of the vertebrates. British Museum (Natural History). London.

JEFFERIES, R. P. S. (1997). A defence of the calchicordates. Lethaia **30**, 1–10.

JEFFERIES, R. P. S., LEWIS, M. & K. DONOVAN (1987). Protocystites menevensis – a stem-group chordate (Cornuta) from the Middle Cambrium of South Wales. Palaeontology **30**, 429–484.

JENSEN, M. (1981). Morphology and classification of Euechinoidea Bronn, 1860 – a cladistic analysis. Vidensk. Medd. Dansk. Naturh. Foren. Kobenhavn **143**, 7–99.

KAESTNER, A. (1963). Lehrbuch der Speziellen Zoologie. Band I. Wirbellose. 5. Lieferung. G. Fischer, Jena.

KAPP, H. (1991). Morphology and anatomy. In Q. BONE, H. KAPP & A. C. PIERROT-BULTS (Eds.). The biology of chaetognaths. 5–17. Oxford University Press. Oxford, New York, Tokyo.

KAPP, H. (1996). Zum Ursprung der Chaetognatha – der aktuelle Stand von DNA-Analysen und morphologisch-anatomischer Forschung. Verh. Dtsch. Zool. Ges. **89**.1, 13.

KAPP, H. (1997). Die einzigartige Embryonalentwicklung der Chaetognathen. Verh. Dtsch. Zool. Ges. **90.** 1, 77.

KAPP, H. (2001). The unique embryology of Chaetognatha. Zool. Anz. **239**, 263–266.

KARDONG, K. V. (1998). Vertebrates. Comparative anatomy, function, evolution. WCB/McGraw-Hill. Boston, Massachusetts.

KATAYAMA, T., YAMAMOTO, M., WADA, H. & N. SATOH (1993). Phylogenetic position of acoel turbellarians inferred from partial 18S DNA sequences. Zool. Scr. **10**, 529–536.

KEMP, T. S. (1982). Mammal-like reptiles and the origin of mammals. Academic Press. London, New York.

KESLING, R. V. & D. LE VASEUR (1971). Strataster ohioensis, a new early Mississipian brittle star, and the paleoecology of its community. Contribut. Mus. Palaeont. Univ. Michigan **23**, 305–341.

KISIELEWSKI, J. (1987a). Two new interesting genera of Gastrotricha (Macrodasyida and Chaetonotida) from the Brazilian freshwater psammon. Hydrobiologia **153**, 23–30.

KISIELEWSKI, J. (1987b). New records of marine Gastrotricha from the French coasts of Manche and Atlantic. I. Macrodasyida, with description of seven new species. Bull. Mus. natn. Hist. Nat. Paris, sér 4, **9**, A, 837–877.

KLAUSEWITZ, W. (1963). Echte Fische. Handbuch der Biologie **6**, 515–628

KOMATSU, M., MURASE, M. & C. OGURO (1988). Morphology of the barrel-shaped larva of the seastar, Astropecten latespinosus. In R.D. BURKE, P.V. MLADENOV, P. LAMBERT & R. L. PARSLEY (Eds.). Echinoderm Biology, 267–272. A. A. Balkema. Rotterdam, Brookfield.

KOZLOFF, E. N. (1990). Invertebrates. Saunders College Publishing. Philadelphia.

KRISTENSEN, R. M. (1983). Loricifera, a new phylum with Aschelminthes characters from the meiobenthos. Z. zool. Syst. Evolut.-forsch. **21**, 163–180.

KRISTENSEN, R. M. (1991a). Loricifera – a general biological and phylogenetic overview. Verh. Dtsch. Zool. Ges. **84**, 213–246.

KRISTENSEN, R. M. (1991b). Loricifera. In F.W. HARRISON & E. E. RUPPERT (Eds.). Microscopic Anatomy of Invertebrates **4**, 377–404. Wiley-Liss, Inc. New York.

KRISTENSEN, R. M. & P. FUNCH (2000). Micrognathozoa: a new class with complicated jaws like those of Rotifera and Gnathostomulida. J. Morph. **246**, 1–49.

KUHN, H.-J. (1971). Die Entwicklung und Morphologie des Schädels von Tachyglossus aculeatus. Abh. senckenb. naturf. Ges. **528**, 1–192.

KÜHNE, W. G. (1973). The systematic position of monotremes reconsidered (Mammalia). Z. Morph. Tiere **75**, 59–64.

KÜHNE, W. G. (1977). On the Marsupionta, a reply to Dr. Parrington. J. Nat. Hist. **11**, 225–228.

KÜMMEL, G. & J. BRANDENBURG (1961). Die Reusengeißelzellen (Cyrtocyten). Z. Naturforsch. **16 b**, 692–697.

LACALLI,, T. C. (1988). Ciliary band patterns and pattern rearrangements in the development of the doliolaria larva. In R. D. BURKE, P. V. MLADENOV, P. LAMBERT & R. L. PARSLEY (Eds.). Echinoderm Biology, 273–275. A. A. Balkema. Rotterdam, Brookfield.

LACALLI, T. C. & J. E. WEST (1986). Ciliary band formation in the doliolaria larva of Florometra I. The development of normal epithelial pattern. J. Embryol. exp. Morph. **96**, 303–323.

LACALLI, T. C. & J. E. WEST (1987). Ciliary band formation in the doliolaria larva of Florometra II. Development of anterior and posterior half-embryos and the role of the mesentoderm. Development **99**, 273–284.

LAFAY, B., SMITH, A. B. & R. CHRISTEN (1995). A combined morphological and molecular approach to the phylogeny of Asteroids (Asteroidea; Echinodermata). Syst. Biol. **44**, 190–208.

LAND, J. VAN DER (1975). Priapulida. In A.C. GIESE & J. S. PEARSE (Eds.) Reproduction of Marine Invertebrates II, 55–65.

LANDMANN, L. (1986). The skin of reptiles. Epidermis and dermis. In J. BEREITER-HAHN, A. G. MATOLTSY & K. SYLVIA RICHARDS. Biology of the Integument. 2. Vertebrates, 150–187. Springer-Verlag. Berlin, Heidelberg, New York, Tokyo.

LANG, K. (1953). Die Entwicklung des Eies von Priapulus caudatus LAM. und die systematische Stellung der Priapuliden. Ark. f. Zool. (2), **5**, 321–348.

LAUDER, G. V. & K. L. LIEM (1983). The evolution and interrelationships of the actinopterygian fishes. Bull. Mus. Comp. Zool. **150**, 95–197.

LAURANT, R. F. (1986). Sous class des lissamphibiens (Lissamphibia). Systématique. Traité de Zoologie **XIV**, 594–797. Paris.

LAURIEN, M. & R. R. REISZ (1995). A reevaluation of early amniote phylogeny. Zool. J. Linn. Soc. **113**, 165–223.

LEE, K., FEINSTEIN, J. & J. CRACRAFT (1997). The phylogeny of ratite birds: resolving conflicts between molecular and morphological data sets. In D. P. MINDELL (Ed.). Avian molecular evolution and systematics. Academic Press. San Diego, London.

LEE, M. S. Y. (1995). Historical burden in systematics and the interrelationships of „parareptiles". Biol. Rev. **70**, 459–547.

LEE, M. S. Y. (1997). Paraiasaur phylogeny and the origin of turtles. Zool. J. Linn. Soc. **120**, 197–280.

LEE, M. S. Y. (1998). Convergent evolution and character correlation in burrowing reptils: towards a resolution of squamate relationships. Biol. J. Linn. Soc. **65**, 369–453.

LEMBURG, C. (1995a). Ultrastructure of the introvert and associated structures of the larvae of Halicryptus spinulosus (Priapulida). Zoomorphology **115**, 11–29.

LEMBURG, C. (1995b). Ultrastructure of sense organs and receptor cells of neck and lorica of the Halicryptus spinulosus larva (Priapulida). Microfauna Marina **10**, 7–30.

LEMBURG, C. (1998). Electron microscopic localization of chitin in the cuticle of Halicryptus spinulosus and Priapulus caudatus (Priapulida) using gold-labelled wheat germ agglutinin: phylogenetic implications for the evolution of the cuticle within the Nemathelminthes. Zoomorphology **118**, 137–158.

LEMBURG, C. (1999). Ultrastrukturelle Untersuchungen an den Larven von Halicryptus spinulosus und Priapulus caudatus. Hypothesen zur Phylogenie der Priapulida und deren Bedeutung für die Evolution der Nemathelminthes. Cuvillier Verlag. Göttingen.

LEMBURG, C. & A. SCHMIDT-RHAESA (1999). Priapulida. Encyclopedia of Reproduction **3**, 1053–1058.

LESCURE, J., RENOUS, S. & J.-P. GASC (1986). Proposition d'une nouvelle classification des amphibiens gymnophiones. Mém. Soc. Zool. France **43**, 145–177.

LESTER, S. M. (1985). Cephalodiscus spec. (Hemichordata: Pterobranchia): Observations of functional morphology, behaviour and occurence in shallow water around Bermuda. Mar. Biol. **85**, 263–268.

LESTER, S. M. (1988a). Ultrastructure of adult gonads and development and structure of the larva of Rhabdopleura normani (Hemichordata: Pterobranchia). Acta Zool. **69**, 95–109.

LESTER, S. M. (1988b). Settlement and metamorphosis of Rhabdopleura normani (Hemichordata: Pterobranchia). Acta Zool. **69**, 111–120.

LEWIS, D. N. & P. C. ENSOM (1982). Archaeocidaris whatleyensis sp. nov. (Echinoida) from the carboniferous limestone of Somerset, and notes on echinoid phylogeny. Bull. Br. Mus. nat. Hist. (Geol.) **36**, 77–104.

LITTLEWOOD, D. T. J., ROHDE, K., BRAY, R. A. & E. A. HERNIOU (1999). Phylogeny of the Plathelminthes and the evolution of parasitism. Biol. J. Linn. Soc. **68**, 257–287.

LITTLEWOOD, D.T. J., ROHDE, K. & K. A. CLOUGH (1999). The interrelationships of all major groups of Plathelminthes: phylogenetic evidence from morphology and molecules. Biol. J. Linn. Soc. **66**, 75–114.

LITTLEWOOD, D. T. J. & A. B. SMITH (1995). A combined morphological and molecular phylogeny for echinoids. Philos. Trans. R. Soc. London, B **347**, 213–234.

LITTLEWOOD, D. T. J., SMITH, A. B., CLOUGH, K. A. & R. H. EMSON (1997). The interrelationships of the echinoderm classes: morphological and molecular evidence. Biol. J. Linn. Soc. **61**, 409–438.

LITTLEWOOD, D. T. J., SMITH, A. B., CLOUGH, K. A. & R. H. EMSON (1998a). Five class of echinoderm and one school of thought. In R. MOOI & M. TELFORD (Eds.). Echinoderms: San Francisco, 47–50. A. A. Balkema. Rotterdam.

LITTLEWOOD, D. T. J., TELFORD, M. J., CLOUGH, K. A. & K. ROHDE (1998b). Gnathostomulida - an enigmatic metazoan phylum from both morphological and molecular perspectives. Mol. Phyl. Evol. **9**, 72–79.

LÖNNBERG, E., FAVORO, G., MOZEJKO, B. & M. RAUTHER (1924). Pisces (Fische). Bronns Klassen und Ordnungen des Tierreichs. **VI**, 1, 1, 1–710.

LORENZEN, S. (1981). Entwurf eines phylogenetischen Systems der freilebenden Nematoden. Veröffentl. Inst. Meeresforsch. Bremerhaven, Suppl. **7**, 1–472.

LORENZEN, S. (1985). Phylogenetic aspects of pseudocoelomate evolution. In S. CONWAY MORRIS, J. B. GEORGE, R. GIBSON & H. M. PLATT (Eds.). The origins and relationships of lower invertebrates. The Systematics Ass. Special Vol. **28**, 210–223.

LORENZEN, S. (1986). Odontobius (Nematoda, Monhysteridae) from the baleen plates of whales and its relationship to Gammarinema living on crustaceans. Zool. Scripta **150**, 101–106.

LORENZEN, S. (1996). Nemathelminthes (Aschelminthes). In W. WESTHEIDE & R. RIEGER (Eds.). Spezielle Zoologie. Teil 1. 682–732. G. Fischer. Stuttgart, Jena, New York.

LØVTRUP, S. (1977). The phylogeny of Vertebrata. Wiley & Sons. London, New York, Sidney, Toronto.

LØVTRUP, S. (1985). On the classification of the taxon Tetrapoda. Syst. Zool. **34**, 463–470.

LUCKETT, W. P. & U. ZELLER (1989). Developmental evidence for dental homologies in the monotreme Ornithorhynchus and its systematic implications. Z. Säugetierkunde **54**, 193–204.

LUDWIG, H. & O. HAMANN (1907). Die Seelilien. Bronn‘s Klassen und Ordnungen des Tierreiches. 2. Band, 3. Abt. Echinodermen (Stachelhäuter). V. Buch, 1415–1602.

LÜTER, C. (1995). Ultrastructure of the metanephridia of Terebratulina retusa and Crania anomala (Brachiopoda). Zoomorphology **115**, 99–107.

LÜTER, C. (1996). The median tentacle of the larva of Lingula anatina (Brachiopoda) from Queensland, Australia. Aust. J. Zool. **44**, 355–366.

LÜTER, C. (1997). Neue Befunde zur Ontogenese der Brachiopoda und deren phylogenetische Deutung. Verh. Dtsch. Zool. Ges. **90**. 1, 180.

LÜTER, C. (1998a). Zur Ultrastruktur, Ontogenese und Phylogenie der Brachiopoda. Cuvillier Verlag. Göttingen. 184 p.

LÜTER, C. (1998b). Note: Embryonic and larval development of Calloria inconspicua (Brachiopoda, Terebratellidae). J. Roy. Soc. New Zealand **28**, 165–167.

LÜTER, C. (2000a). Ultrastructure of larval and adult setae of Brachiopoda. Zool. Anz. **239**, 75–90.

LÜTER, C. (2000b). The origin of the coelom in Brachiopoda and its phylogenetic significance. Zoomorphology **120**, 15–28.

LÜTER, C. & T. BARTOLOMAEUS (1997). The phylogenetic position of Brachiopoda - a comparison of morphological and molecular data. Zool. Scripta **26**, 245–253.

MADIN, L. P. & D. DEIBEL (1998). Feeding and energetics of Thaliacea. In Q. BONE (Ed.). The biology of pelagic tunicates. Oxford University Press. Oxford, New York, Tokyo.

MAGNUS, D. B. E. (1964). Gezeitenströmung und Nahrungsfiltration bei Ophiuren und Crinoiden. Helgol. Wiss. Meeresunters. **10**, 104–117.

MAGNUS, D. B. E. (1965). Wasserströmung und Nahrungserwerb bei Stachelhäutern des Roten Meeres (Untersuchungen an Schlangensternen und Federsternen). Ber. Phys.-Med. Ges. Würzburg, N. F. **71**, 128–141 (1962–1964).

MAISEY, J. G. (1986). Heads and tails: a chordate phylogeny. Cladistics **2**, 201–256.

MALAKHOV, V. V. (1994). Nematodes. Structure, Development, Classification and Phylogeny. Smithsonian Institution Press. Washington, London.

MALAKHOV, V. V. (1998). Embryological and histological peculiarities of the order Enoplida, a primitive group of nematodes. Russ. J. Nematodology **6**, 41–46.

MALLAT, J. (1981). The suspension feeding mechanism of the larval lamprey Petromyzon marinus. J. Zool. London **194**, 103–142.

MALLAT, J. (1984a). Early vertebrate evolution: pharyngeal structure and the origin of gnathostomes J. Zool. London **204**, 169–183.

MALLAT, J. (1984b). Feeding ecology of the earliest vertebrates. Zool. J. Linn. Soc. **82**, 261–272.

MALLAT, J. & J. SULLIVAN (1998). 28S and 18S rDNA sequences support the monophyly of lampreys and hagfishes. Mol. Biol. Evol. 15, 1706–1718.

MARCUS, E. (1926). Beobachtungen und Versuche an lebenden Süßwasserbryozoen. Zool. Jb. Abt. Syst. **52**, 279–349.

MARCUS, E. (1938). Bryozoarios marinhos brasileiros II. Bol. Fac. Fil. Ciênc. Letr. Univ. Sao Paulo, Zoologia **2**, 1–196.

MARINELLI, W. (1960). Deuterostomia. Handbuch der Biologie **6**, 311–408.

MARINELLI, W. & A. STRENGER (1954–1973). Vergleichende Anatomie und Morphologie der Wirbeltiere. F. Deuticke, Wien.

MARKEVICH, G. I. & L. A. KUTIKOVA (1989). Mastax morphology under SEM and ist usefulness in reconstructing rotifer phylogeny and systematics. Hydrobiologia **186/187**, 285–289.

MÄRKEL, K. (1976). Das Wachstum der „Laterne des Aristoteles" und seine Anpassung an die Funktion der Laterne (Echinodermata, Echinoidea). Zoomorphologie **86**, 25–40.

MÄRKEL, K. (1978). On the teeth of the recent cassiduloid Echinolampas depressa Gray, and on some liassic fossil teeth nearly identical in structure (Echinodermata, Echinoida). Zoomorphologie **89**, 125–144.

MÄRKEL, K. (1979). Structure and growth of the cidaroid socket-joint lantern of Aristotle compared to the hinge-joint lantern of non-cidaroid regular echinoids (Echinodermata, Echinoidea). Zoomorphologie **94**, 1–32.

MARSHALL, C. R. (1994). Molecular approaches to echinoderm phylogeny. In B. DAVID, A. GUILLE, J.-P. FÉRAL & M. ROUX (Eds.). Echinoderms through time. A. A. Balkema. Rotterdam, Brookfield.

MARSHALL, L. G. (1979). Evolution of the metatherian and eutherian (mammalian) characters: a review based on cladistic methodology. Zool. J. Linn. Soc. **66**, 369–410.

MARSHALL, L. G. (1984). Monotremes and Marsupials. In S. ANDERSON & J. K. JONES (Eds.). Orders and families of recent mammals of the world. J. Wiley & Sons. New York.

MEGLITSCH, P. A. & F. R. SCHRAM (1991). Invertebrate Zoology. Oxford University Press. Oxford, New York.

MENKER, D. (1970). Lebenszyklus, Jugendentwicklung und Geschlechtsorgane von Rhabdomolgus ruber (Holothurioida: Apoda). Marine Biology **6**, 167–186.

MENKER, D. & P. AX (1970). Zur Morphologie von Arenadiplosoma migrans n. g. n. sp., einer vagilen Ascidien-Kolonie aus dem Mesopsammal der Nordsee (Tunicata, Ascidiacea). Z. Morph. Tiere **66**, 323–336.

MEVES, A. (1973). Elektronenmikroskopische Untersuchungen über die Zytoarchitektur des Gehirns von Branchiostoma lanceolatum. Z. Zellforsch. **139**, 511–532.

MILLOT, J. & J. ANTHONY (1958). Crossoptérygiens actuels. *Latimeria chalumnae.* Traité de Zoologie **XIII**, 2553–2597. Paris.

MILNER, A. R. (1988). The relationships and origin of living amphibians. In M. J. BENTON (Ed.). The phylogeny and classification of the tetrapods. Vol. I. Syst. Ass. Spec. Vol. **35 A**, 59–102. Clarendon Press, Oxford.

MINDELL, D. P., SORENSON, M. D., DIMCHEFF, D. E., HASEWAGA, M., AST, J. C. & Y. TAMAKI (1999). Interordinal relationships of birds and other reptiles based on whole mitochondrial genomes. Syst. Biol. **48**, 138–152.

MOCK, H. (1979). Chaetonotoida (Gastrotricha) der Nordseeinsel Sylt. Mikrofauna Meeresboden **78**, 1–107.

MONNIOT, F. (1965). Ascidies interstitielles des côtes d' Europe. Mém. Mus. natl. Hist. Nat. Paris. Sér. A **35**, 1–154.

MORTENSEN, T. (1933). Papers from Dr. Th. Mortensen's Pacific expedition 1914–1916. LX. On an extrordinary Ophiurid, Ophiocanops fugiens. Vidensk. Medd. Naturhist. Foren. Kobenhavn **93**, 1–22.

MUKAI, H., TERAKADO, K. & C. R. REED (1997). Bryozoa. In F. W. HARRISON & R. M. WOOLLACOTT (Eds.). Microscopic Anatomy of Invertebrates **13**, 45–206. Wiley-Liss, Inc. New York.

NEBELSICK, M. (1993). Introvert, mouth cone, and nervous system of Echinoderes capitatus (Kinorhyncha, Cyclorhagida) and implications for the phylogenetic relationships of Kinorhyncha. Zoomorphology **113**, 211–232.

NEUHAUS, B. (1987). Ultrastructure of the protonephridia in Dactylopodola baltica and Mesodasys laticaudatus (Macrodasyida): Implications for the ground pattern of the Gastrotricha. Microfauna Marina **3**, 419–438.

NEUHAUS, B. (1994). Ultrastructure of alimentary canal and body cavity, ground pattern, and phylogenetic relationships of the Kinorhyncha. Microfauna Marina **9**, 61–156.

NEUHAUS, B. (1995). Postembryonic development of Paracentrophyes praedicta (Homalorhagida): neoteny questionable among the Kinorhyncha. Zool. Scripta **24**, 179–192.

NEUHAUS, B. (1999). Kinorhyncha. Encyclopedia of Reproduction **2**, 933–937.

NEUHAUS, B., BRESCIANI, J. & W. PETERS (1977). Ultrastructure of the pharyngeal cuticle and lectin labelling with wheat germ agglutinin-gold conjugate indicating chitin in the pharyngeal cuticle of Oesophagostomum dentatum (Strongylida, Nematoda). Acta Zool. **78**, 205–213.

NEUHAUS, B., KRISTENSEN, R. M. & C. LEMBURG (1996).Ultrastructure of the cuticle of the Nemathelminthes and electronmicroscopical localisation of chitin. Verh. Dtsch. Zool. Ges. **89**, 1, 221.

NEUMANN, G. (1933). Zweite Klasse der Tunicata. Acopa = Caducichordata. In W. KÜKENTHAL & T. KRUMBACH. Handb. d. Zool. **5**, 2, 203–532. W. de Gruyter. Berlin u. Leipzig.

NICHOLS, D. (1967). Echinoderms. Hutchinson & Co, London.

NIELSEN, C. (1970). On metamorphosis and ancestrula formation in cyclostomatous bryozoans. Ophelia **7**, 221–256.

NIELSEN, C. (1971). Entoproct life-cycles and the entoproct/ectoproct relationship. Ophelia **9**, 209–341.

NIELSEN, C. (1991). The development of the brachiopod Crania (Neocrania) anomala (O. F. Müller) and its phylogenetic significance. Acta Zool. **72**, 7–28.

NIELSEN, C. (1995). Animal evolution. Interrelationships of the living phyla. Oxford University Press. Oxford, New York, Tokyo.

NIELSEN, C. (1998a). Origin and evolution of animal life cycles. Biol. Rev. **73**, 125–155.

NIELSEN, C. (1998b). Morphological approaches to phylogeny. Amer. Zool. **38**, 942–952.

NIELSEN, C. (1999). Origin of the chordate central nervous system – and the origin of chordates. Dev. Genes. Evol. **209**, 198–205.

NIELSEN, C. & A. JESPERSEN (1997). Entoprocta. In F. W. HARRISON & R. M. WOOLLACOTT (Eds.). Microscopic Anatomy of Invertebrates **13**, 13–43. Wiley-Liss, Inc. New York.

NIELSEN, C. & K. J. PEDERSEN (1979). Cystid structure and protrusions of the polypide in Crisia (Bryozoa, Cyclostomata). Acta Zool. **60**, 65–88.

NIELSEN, C. & H. U. RIISGARD (1998). Tentacle structure and filter-feeding in Crisia eburnea and other cyclostomatous bryozoans, with a review of upstream-collecting mechanisms. Mar. Ecol. Progr.Ser. **168**, 163–186.

NIELSEN, C. & J. ROSTGAARD (1976). Structure and function of an entoproct tentacle with a discussion of ciliary feeding types. Ophelia **15**, 115–140.

NIELSEN, C., SCHARFF, N. & D. EIBYE-JACOBSON (1996). Cladistic analysis of the animal kingdom. Biol. J. Linn. Soc. **57**, 385–410.

NOGRADY, T., WALLACE, R. L. & T. W. SNELL (1993). Rotifera. Vol. 1: Biology, Ecology and Systematics. Guides to the identification of the microinvertebrates of the continental waters of the world 4. SPB Academic Publishing. The Hague.

NOVACEK, M. J. (1993). Reflections on higher mammalian phylogenetics. J. Mammal. Evolution **1**, 3–30.

NOVACEK, M. J. & A. R. WYSS (1986). Higher-level relationships of the recent eutherian orders: morphological evidence. Cladistics **2**, 257–287.

NÜBLER-JUNG, K. & D. ARENDT (1999). Dorsoventral axis inversion: Enteropneust anatomy links invertebrates to chordates turned upside down. J. Zool. Syst. Evol. Research **37**, 93–100.

PADIAN, K. & L. M. CHIAPPE (1998). The origin and early evolution of birds. Biol. Rev. **73**, 1–42.

PARKES, K. C. & G. A. CLARK (1966). An additional character linking ratites and tinamous and an interpretation of their monophyly. Condor **68**, 459–471.

PARSONS, T. S. & E. E. WILLIAMS (1962). The teeth of Amphibia and their relation to amphibian phylogeny. J. Morph. **110**, 375–389.

PARSONS, T. S. & E. E. WILLIAMS (1963). The relationships of the modern Amphibia: a re-examination. Quart. Rev. Biol. **38**, 26–53.

PATTERSON, C. (1982). Morphology and interrelationships of primitive actinopterygian fishes. Amer. Zool. **22**, 241–259.

PAUL, C. R. C. & A. B. SMITH (1984). The early radiation and phylogeny of echinoderms. Biol. Rev. **59**, 443–481.

PEARSE, V. B. & J. S. PEARSE (1994). Echinoderm phylogeny and the place of concentricycloids. In B. DAVID, A. GUILLE, J.-P. FÉRAL & M. ROUX. Echinoderms through time. A. A. Balkema. Rotterdam, Brookfield.

PETERSON, K. J. (1994). The origin and early evolution of the Craniata. In D. R. PROTHERO & R. M. SCHOCH (Eds.). Major features of vertebrate evolution, 14–37. Short courses in Palaeontology 7. The Palaeontological Society and the University of Tennessee, Knoxville, Tenn.

PETERSON, K. J. (1995). A phylogenetic test of the calcichordate scenario. Lethaia **28**, 25–38.

PINNA, M. C. C. de (1996). Teleostean monophyly. In M. L. J. STIASSNY, L. R. PARENTI & G. D. JOHNSON (Eds.). Interrelationships of fishes, 147–162. Academic Press. San Diego, California.

POPOV, L. E., BASSETT, G., HOLMER, L. E. & J. LAURIE (1993). Phylogenetic analysis of higher taxa of Brachiopoda. Lethaia **26**, 1–5.

POR, F. D. & H. J. BROMLEY (1974). Morphology and anatomy of Maccabeus tentaculatus (Priapulida: Seticoronaria). J. Zool. London **173**, 173–197.

POUGH, F. H., ANDREWS, R. M., CADLE, J. E., CRUMP, M. L., SAVITZKY. A. H. & K. D. WELLS (1998). Herpetology. Prentice Hall. Upper Saddle River, NJ.

PREUSCHOFT, H., REIF, W. E., LOITSCH, C. & E. TEPE (1991). The function of labyrinthodont teeth: big teeth in small jaws. In N. SCHMIDT-KITTLER & K. VOGEL (Eds.). Constructional morphology and evolution. 151–171. Springer. Berlin, Heidelberg, New York.

PURASJOKI, K. J. (1944). Beiträge zur Kenntnis der Entwicklung und Ökologie der Halicryptus spinulosus-Larve (Priapulida). Ann. Zool. Soc. Zool.-Bot. Fenn. Vanamo **9**, Nr. 6, 1–14.

QIANG, J., CURRIE, P. J., NORELL, M. A. & J. SHU-AN (1998). Two feathered dinosaurs from north eastern China. Nature **393**, 753–761.

RÄHR, H. (1979). The circulatory system of Amphioxus (Branchiostoma lanceolatum (Pallas)). A light-microscopic investigation based on intravascular injection technique. Acta Zool. **60**, 1–18.

RÄHR, H. (1981). The ultrastructure of the blood vessels of Branchiostoma lanceolatum (Pallas) (Cephalochordata). Zoomorphology **97**, 53–74.

RAFF, R. A., FIELD, K. G., GHISELIN, M. T., LANE, D. L., OLSEN, G. J., PACE, N. R., PARKS, A. L., PARR, B. A. & E. C. RAFF (1988). Molecular analysis of distant phylogenetic relationships of echinoderms. In C. R. C. PAUL & A. B. SMITH (Eds.). Echinoderm phylogeny and evolutionary biology. 29–41. Clarendon Press. Oxford.

RAO, K. S. (1973). Studies on freshwater Bryozoa – III. The Bryozoa of the Narmada River System. In G. P. LARWOOD (Ed.). Living and fossil Bryozoa, 529–537. Academic Press. London, New York.

RASMUSSEN, H. W. (1978). Articulata. Crinoida. In C. MOORE & C. TEICHERT (Eds.). Treatise on Invertebrate Palaeontology. Part T, Echinodermata 2, Vol. **3**, 814–928.

REMANE, A. (1924). Neue aberrante Gastrotrichen. I: Macrodasys buddenbrocki nov. gen. nov. spec. Zool. Anz. **61**, 289–297.

REMANE, A. (1929–33). Rotatoria. Bronns Klassen und Ordnungen des Tierreichs. 4. Bd., II. Abt., 1. Teil. Akademische Verlagsgesellschaft. Leipzig.

REMANE, A. (1936). Gastrotricha und Kinorhyncha. Bronns Klassen und Ordnungen des Tierreichs. 4. Bd., II. Abt., 1. Buch, 2. Teil. 1–385. Akademische Verlagsgesellschaft. Leipzig.

REMANE, A. (1950). Die morphologische Ableitung des Räderorgans der Rotatoria Bdelloidea. Neue Erg. u. Probl. Zool. (Klatt-Festschrift). 805–812. Leipzig.

REMANE, A. (1952). Zwei neue Turbanella-Arten aus dem marinen Küstengrundwasser. Kieler Meeresforsch. **9**, 62–65.

REMANE, A. (1961). Neodasys uchidai, eine zweite Neodasys-Art (Gastrotricha Chaetonotoidea). Kieler Meeresforsch. **17**, 85–88.

RENFREE, M. B. (1993). Ontogeny, genetic control and phylogeny of female reproduction in monotreme and therian mammals. In F. S. SZALAY, M. J. NOVACEK & M. C. McKENNA. Mammal phylogeny. Mesozoic differentiation, multituberculates, monotremes, early therians, and marsupials. 4–20. Springer-Verlag. New York, Berlin, Heidelberg.

RICCI, C. (1983). Rotifera or Rotatoria? Hydrobiologia **104**, 1–2.

RICCI, C. (1998). Are lemnisci and proboscis present in the Bdelloidea? Hydrobiologia **387/388**, 93–96.

RIEGER, G. E. & R. M. RIEGER (1977). Comparative fine structure study of the gastrotrich cuticle and aspects of cuticle evolution within the Aschelminthes. Z. zool. Syst. Evolut.-forsch. **15**, 81–124.

RIEGER, R. M. (1984). Evolution of the cuticle in the lower Eumetazoa. In J. BEREITER-HAHN, A. G. MATOLTSY & K. SYLVIA RICHARDS. Biology of the integument. 1 Invertebrates. 389–399. Springer-Verlag. Berlin.

RIEGER, R. M. & S. TYLER (1995). Sister-group relationship of Gnathostomulida and Rotifera-Acanthocephala. Invertebrate Biol. **114**, 186–188.

RIEMANN, F. (1972). Corpus gelatum und ciliäre Strukturen als lichtmikroskopisch sichtbare Bauelemente des Seitenorgans freilebender Nematoden. Z. Morph. Tiere **72**, 46–76.

RIEMANN, F. & M. SCHRAGE (1978). The mucus-trap hypothesis of feeding of aquatic nematodes and implications for biodegradation and sediment texture. Oecologia **34**, 75–88.

RIEPPEL, O. (1988). The classification of the Squamata. In M. J. BENTON (Ed.). The phylogeny and classification of tetrapods. Vol. I. Syst. Ass. Spec. Vol. **35 A**, 261–293. Clarendon Press, Oxford.

RIEPPEL, O. (1994). The Lepidosauromorpha: an overview with special emphasis on the Squamata. In N. C. FRAZER & H. D. SUES (Eds.). In the shadow of the dinosaurs. Early Mesozoic tetrapods. 23–37. Cambridge University Press.

RIEPPEL, O. (2000). The braincases of mosasaurs and Varanus and the relationships of snakes. Zool. J. Linn. Soc. **129**, 489–514.

RIISGARD, H. U. & P. MANRÍQUES (1997). Filter-feeding in fifteen marine ectoprocts (Bryozoa): particle capturing and water pumping. Mar. Ecol. Prog. Ser. **154**, 223–239.

RIUTORT, M., FIELD, K. G., RAFF, R. A. & J. BAGUÑÀ (1993). 18S RNA sequences and phylogeny of Plathelminthes. Biochem. Syst. Ecol. **21**, 71–77.

ROMER, A. S. & T. S. PARSONS (1977). The vertebrate body. Saunders College. Philadelphia.

ROSEN, D. E., FOREY, P. L. , GARDINER, B. G. & C. PATTERSON (1981). Lungfishes, tetrapods, palaeontology and plesiomorphy. Bull. Am. Mus. Nat. Hist. **167**, 159–276.

ROWE, F. W. E. (1988). Review of the extant class Concentricycloidea and reinterpretation of the fossil class Cyclocystoidea. In R. D. BURKE, P. V. MLADENOV, P. LAMBERT & R. L. PARSLEY (Eds.). Echinoderm Biology, 3–15. A. A. Balkema. Rotterdam, Brookfield.

ROWE, F. W. E. (1989). A review of the family Caymanostellidae (Echinodermata: Asteroida) with the description of a new species of Caymanostella Belyaev and a new genus. Proc. Linn. Soc. N. S. W. **111**, 293–307.

ROWE, F. W. E., BAKER, A. N. & H. E. S. CLARK (1988). The morphology, development and taxonomic status of Xyloplax Baker, Rowe and Clark (1986) (Echinodermata: Concentricycloidea), with the description of a new species. Proc. R. Soc. London B **233**, 431–459.

ROWE, F. W. E., HEALY, J. M. & D. T. ANDERSON (1994). Concentricycloidea. In F. W. HARRISON & F.-S. CHIA (Eds.). Microscopic Anatomy of Invertebrates **14**, 149–167 Wiley-Liss. Inc. New York.

ROWE, T. (1993). Phylogenetic systematics and the early history of mammals. In F. S. SZALAY, M. J. NOVACEK & M. C. McKENNA. Mammal phylogeny. Mesozoic differentiation, multituberculates, monotremes, early therians, and marsupials. 129–145. Springer-Verlag. New York, Berlin, Heidelberg.

RUPPERT, E. E. (1978a). The reproductive system of gastrotrichs. II. Insemination in Macrodasys. A unique mode of sperm transfer in Metazoa. Zoomorphology **89**, 207–228.

RUPPERT, E. E. (1978b). The reproductive system of gastrotrichs. III. Genital organs of Thaumastodermatinae subfam. n. and Diplodasyinae subfam. n. with discussion of reproduction in the Macrodasyida. Zool. Scripta **7**, 93–114.

RUPPERT, E. E. (1982). Comparative ultrastructure of the gastrotrich pharynx and the evolution of myoepithelial foreguts in Aschelminthes. Zoomorphology **99**, 181–220.

RUPPERT, E. E. (1991). Gastrotricha. In F. W. HARRISON & E. E. RUPPERT (Eds.). Microscopic Anatomy of Invertebrates **4**, 41–109. Wiley-Liss. Inc. New York.

RUPPERT, E. E. (1997a). Introduction. Microscopic anatomy of the notochord, heterochrony and chordate evolution. In F.W. HARRISON & E. E. RUPPERT (Eds.). Microscopic Anatomy of Invertebrates **15**, 1–13. Wiley-Liss, Inc. New York.

RUPPERT, E. E. (1997b). Cephalochordata (Acrania). In F. W. HARRISON & E. E. RUPPERT (Eds.). Microscopic Anatomy of Invertebrates **15**, 349–504. Wiley-Liss, Inc. New York.

RUPPERT, E. E. & E. J. BALSER (1986). Nephridia in the larvae of hemichordates and echinoderms. Biol. Bull. **171**, 188–196.

RUPPERT, E. E. & R. D. BARNES (1991). Invertebrate Zoology. Saunders College Publishing. Fort Worth.

RUPPERT, E. E. & P. R. SMITH (1988). The functional organization of filtration nephridia. Biol. Rev. **63**, 231–258.

RYBCZYNSKI, N. (2000). Cranial anatomy and phylogenetic position of Suminia getmanovi, a basal anomodont (Amniota: Therapsida) from the Late Permian of Eastern Europe. Zool. J. Linn. Soc. **130**, 329–373.

RYLAND, J. S. (1970). Bryozoans. Hutchinson & Co (Publishers) LTD. London.

SALVINI-PLAWEN, L. v. (1974). Zur Morphologie und Systematik der Priapulida: Chaetostephanus praeposteriens, der Vertreter einer neuen Ordnung Seticoronaria. Z. zool. Syst. Evolut.-forsch. **12**, 31–54.

SALVINI-PLAWEN, L. v. (1988). The epineural (vs. gastroneural) cerebral-complex of Chaetognatha. Z. zool. Syst. Evolut.-forsch. **26**, 425–429.

SALVINI-PLAWEN, L. v. (1989). Mesoderm heterochrony and metamery in Chordata. Fortschr. Zool. **35**, 213–219.

SALVINI-PLAWEN, L. v. (1998).The urochordate larva and archichordate organization: chordate origin and anagenesis revisited. J. Zool. Syst. Evol. Research **36**, 129–145.

SALVINI-PLAWEN, L. v. & E. MAYR (1977). On the evolution of photoreceptors and eyes. Evolut. Biol. **10**, 207–263.

SCHAEFFER,B. (1941).The morphological and functional evolution of the tarsus in amphibians and reptiles. Bull. Am. Mus. Nat. Hist. **78**, 395–472.

SCHAEFFER, B. (1987). Deuterostome monophyly and phylogeny. Evol. Biol. **21**, 179–235.

SCHIERENBERG, E., WIEGNER, O., BOSSINGER, O., SKIBA, F. & M. KUTZOWITZ (1997/98). Pattern formation and cell specification in nematode embryos: A theme with considerable variations. Zoology **100**, 320–327.

SCHMIDT, G. D. (1985). Development and life cycles. In CROMPTON, D. W. T. & B. B. NICKOL (Eds.). Biology of the Acanthocephala. 273–305. Cambridge University Press, Cambridge.

SCHMIDT-RHAESA, A. (1996a). Zur Morphologie, Biologie und Phylogenie der Nematomorpha. Untersuchungen an Nectonema munidae und Gordius aquaticus. Cuvillier Verlag. Göttingen.

SCHMIDT-RHAESA, A. (1996b). Ultrastructure of the anterior end in three ontogenetic stages of Nectonema munidae (Nematomorpha). Acta Zool. **77**, 267–278.

SCHMIDT-RHAESA, A. (1996c). The nervous system of Nectonema munidae and Gordius aquaticus, with implications for the ground pattern of the Nematomorpha. Zoomorphology **116**, 133–142.

SCHMIDT-RHAESA, A. (1996d). Die Biologie der Saitenwürmer (Nematomorpha). Rätselhaftes Leben wenig bekannter Würmer. Mikrokosmos **85**, 279–283.

SCHMIDT-RHAESA, A. (1997a). Ultrastructural features of the female reproductive system and female gametes of Nectonema munidae Brinkmann 1930 (Nematomorpha). Parasitol. Res. **83**, 77–81.
SCHMIDT-RHAESA, A. (1997b). Ultrastructural observations of the male reproductive system and spermatozoa of Gordius aquaticus L., 1758. Invertebr. Reprod. Dev. **32**, 31–40.
SCHMIDT-RHAESA, A. (1997c). Nematomorpha. Süßwasserfauna von Mitteleuropa. Vol. **4**, 4.
SCHMIDT-RHAESA, A.(1997/98). Phylogenetic relationships of the Nematomorpha – a discussion of current hypothesis. Zool. Anz. **236**, 203–216.
SCHMIDT-RHAESA, A. (1998). Muscular ultrastructure in Nectonema munidae and Gordius aquaticus (Nematomorpha). Invertebr. Biol. **117**, 37–44.
SCHMIDT-RHAESA, A. (1999). Nematomorpha. Encyclopedia of Reproduction **3**, 333–341.
SCHMIDT-RHAESA, A., BARTOLOMAEUS, T., LEMBURG, C., EHLERS, U. & J. R. GAREY (1998). The position of the Arthropoda in the phylogenetic system. J. Morph. **238**, 263–285.
SCHULTZE, H.-P. (1969). Die Faltenzähne der rhipidistiiden Crossopterygier, der Tetrapoden und der Actinopterygier-Gattung Lepisosteus; nebst einer Beschreibung der Zahnstruktur von Onychodus (struniiformer Crossopterygier). Palaeontographia Italica **65**, 63–137.
SCHULTZE, H.-P. (1977). Ausgangsform und Entwicklung der rhombischen Schuppen der Osteichthyes (Pisces). Paläont. Z. **51**, 152–168.
SCHULTZE, H.-P. (1986). Dipnoans as sarcopterygians. J. Morph. Suppl. 1, 39–74.
SCHULTZE, H.-P. (1991). Der Ursprung der Tetrapoda – ein lebhaft diskutiertes altes Problem. Verh. Dtsch. Zool. Ges. **84**, 135–151.
SCHULTZE, H.-P. (1994). Comparison of hypothesis on the relationships of sarcopterygians. Syst. Biol. **43**, 155–173.
SCHULTZE, H.-P. & G. ARRATIA (1989). The composition of the caudal skeleton of teleosts (Actinopterygii: Osteichthyes). Zool. J. Linn. Soc. **97**, 189–231.
SCHWENK, K. (1998). Comparative morphology of the lepidosauria tongue and its relevance to squamate phylogeny. In R. ESTES & G. PREGILL (Eds.). Phylogenetic relationships of the lizard families, 569–598. Stanford University Press. Stanford, Ca.
SEGERS, H. & G. MELONE (1998). A comparative study of trophi morphology in Seisonidea (Rotifera). J. Zool. Lond. **244**, 201–207.
SHINN, G. L. (1997). Chaetognatha. In F. W. HARRISON & E. E. RUPPERT (Eds.). Microscopic Anatomy of Invertebrates **15**, 103–220. Wiley-Liss, Inc. New York.
SHINN, G. L. & M. E. ROBERTS (1994). Ultrastructure of hatching chaetognaths (Ferosagitta hispida): Epithelial arrangement of the mesoderm and its phylogenetic implications. J. Morphol. **219**, 143–163.
SIBLEY, C. G. & J. E. AHLQUIST (1990). Phylogeny and classification of birds. A study in molecular evolution. Yale University Press. New Haven & London.
SIEWING, R. (1969). Lehrbuch der vergleichenden Entwicklungsgeschichte der Tiere. P. Parey. Hamburg, Berlin.
SILÉN, L. (1942). Origin and development of the cheilo-ctenostomatous stem of Bryozoa. Zool. Bidr. Uppsala **22**, 1–59.
SILÉN, L. (1944). The anatomy of Labiostomella gisleni Silén (Bryozoa Protocheilostomata). Kungl. Svenska Vetensk., Akad. Handl. Ser. 3, Vol. **21**, No. 6, 1–111.
SKALING, B. & B. M. MACKINNON (1988). The absorptive surfaces of Nectonema sp. (Nematomorpha: Nectonematoidea) from Pandalus montagui: histology, ultrastructure, and absorptive capabilities of the body wall and intestine. Can. J. Zool. **66**, 289–295.
SMILEY, S. (1988). The phylogenetic relationships of holothurians: a cladistic analysis of the extant echinoderm classes. In C. R. C. PAUL & A. B. SMITH (Eds.). Echinoderm phylogeny and evolutionary biology, 69–84. Clarendon Press. Oxford.
SMILEY, S. (1994). Holothuroidea. In F. W. HARRISON & F.-S. CHIA (Eds.). Microscopic Anatomy of Invertebrates **14**, 401–471. Wiley-Liss. Inc. New York.
SMITH, A. B. (1981). Implications of lantern morphology for the phylogeny of post-Palaeozoic echinoids. Palaeontology **24**, 779–801.
SMITH, A. B. (1984a). Echinoid palaeobiology. George Allen and Unwin. London.

SMITH, A. B. (1984b). Classification of Echinodermata. Palaeontology **27**, 431–459.

SMITH, A. B. (1988a). Fossil evidence for the relationships of extant echinoderm classes and their time of divergence. In C. R. C. PAUL & A. B. SMITH (Eds.). Echinoderm phylogeny and evolutionary biology, 85–97. Clarendon Press. Oxford.

SMITH, A. B. (1988b). To group or not to group: the taxonomic position of Xyloplax. In R. D. BURKE, P. V. MLADENOV, P. LAMBERT & R. L. PARSLEY (Eds.). Echinoderm Biology, 17–23. A. A. Balkema. Rotterdam, Brookfield.

SMITH, A. B. (1990). Evolutionary diversification of echinoderms during the early Palaeozoic. In P. D. TAYLOR & G. P. LARWOOD (Eds.). Major evolutionary radiations. Systematics Association Special Volume **42**, 265–286.

SMITH, A. B. (1994). Systematics and the fossil record: documenting evolutionary patterns. Oxford. Blackwell Scientific.

SMITH, A. B., LAFAY, B. & R. CHRISTEN (1992). Comparative variation of morphological and molecular evolution through geological time: 28S ribosomal RNA versus morphology in echinoids. Philos. Trans. R. Soc. London. B **338**, 365–382.

SMITH, A. B., LITTLEWOOD, D. T. J. & G. A. WRAY (1995). Comparing patterns of evolution: Larval and adult life history stages and small subunit ribosomal RNA of post-Palaeozoic echinoids. Philos. Trans. R. Soc. Lond. B **349**, 11–18.

SMITH, A. B. & G. L. J. PATERSON (1995). Ophiuroid phylogeny and higher taxonomy: morphological, molecular and palaeontological perspectives. Zool. J. Linn. Soc. **114**, 213–243.

SMITH, M. J., ARNDT, A., GORSKI, S. & E. FAJBER (1993). The phylogeny of echinoderm classes based on mitochondrial gene arrangements. J. Mol. Evol. **36**, 545–554.

STARCK, D. (1978). Vergleichende Anatomie der Wirbeltiere 1. Springer. Berlin, Heidelberg, New York.

STARCK, D. (1979). Vergleichende Anatomie der Wirbeltiere 2. Springer. Berlin, Heidelberg, New York.

STARCK, D. (1982). Vergleichende Anatomie der Wirbeltiere 3. Springer. Berlin, Heidelberg, New York.

STARCK, D. (1995). Säugetiere. In A. KAESTNER. Lehrbuch der Speziellen Zoologie. Wirbeltiere. 5. Teil: Säugetiere. 1–1241. G. Fischer. Jena, Stuttgart, New York.

STEBBING, A. R. D. (1970). Aspects of reproduction and life cycles of Rhabdopleura compacta (Hemichordata). Mar. Biol. **5**, 205–212.

STIASNY-WINHOFF, G. & G. STIASNY (1931). Die Tornarien. Kritik der Beschreibungen und Vergleich sämtlicher bekannter Enteropneustenlarven. Ergebn. Fortschr. Zoologie **7**, 38–208.

STIASSNY, M. L. J., R. L. PARENTI & G. D. JOHNSON (1996). Interrelationships of fishes. Academic Press. London.

STORCH, V. & R. P. HIGGINS (1989). Ultrastructure of developing and mature spermatozoa of Tubiluchus corallica (Priapulida). Trans. Am. Microsc. Soc. **108**, 45–50.

STORCH, V. HIGGINS, R. P. & M. P. MORSE (1989). Internal anatomy of Meiopriapulus fijiensis (Priapulida). Trans. Am. Microsc. Soc. **108**, 245–261.

STORCH, V. & F. RIEMANN (1973). Zur Ultrastruktur der Seitenorgane (Amphiden) des limnischen Nematoden Tobrilus aberrans (W. Schneider, 1925) (Nematoda, Enoplida). Z. Morph. Tiere **74**, 163–170.

STORCH, V. & U. WELSCH (1997). Systematische Zoologie. G. Fischer. Stuttgart, Jena, Lübeck, Ulm.

STRATHMANN, R. R. (1973). Function of lateral cilia in suspension feeding of lophophorates (Brachiopoda, Phoronida, Ectoprocta). Mar. Biol. **23**, 129–136.

STRATHMANN, R. R. (1988a). Larvae, phylogeny and von Baer's law. In C. R. C. PAUL & A. B. SMITH (Eds.). Echinoderm phylogeny and evolutionary biology, 53–68. Clarendon Press. Oxford.

STRATHMANN, R. R. (1988b). Functional requirements and the evolution of developmental patterns. In R. D. BURKE, P. V. MLADENOV, P. LAMBERT & R. L. PARSLEY. Echinoderm Biology, 55–69. A. A. Balkema. Rotterdam, Brookfield.

STRATHMANN, R. R. & L. M. Mc EDWARD (1986). Cyphonautes ciliary sieve breaks a biological rule of inference. Biol. Bull. **171**, 694–700.

STRENGER, A. (1963). Cyclostomata. Handbuch der Biologie **6**, 499–514.

STRENGER, A. & W. ERBER (1983). Zur Larvalentwicklung bei Asteroidea und Kritik am systematischen Begriffspaar Pelmatozoa – Eleutherozoa bei Echinodermen. Z. zool. Syst. Evolut.-forsch. **21**, 235–239.

STRICKER, S. A., REED, C. G. & R. L. ZIMMER (1988a). The cyphonautes larva of the marine bryozoan Membranipora membranacea. I. General morphology, body wall and gut. Can. J. Zool. **66**, 368–383.

STRICKER, S. A., REED, C. G. & R. L. ZIMMER (1988b). The cyphonautes larva of the marine bryozoan Membranipora membranacea. II. Internal sac, musculature, and pyriform organ. Can. J. Zool. **66**, 384–398.

SUMIDA, S. S. & K. L. M. MARTIN (Eds.) (1997). Amniote origins. Completing the transition to land. Academic Press, Inc. San Diego, California.

SWALLA, B. J., CAMERON, C. B., CORLEY, L. S. & J. R. GAREY (2000). Urochordates are monophyletic within the deuterostomes. Syst. Biol. **49**, 52–64.

SZALAY, F. S. (1993). Metatherian taxon phylogeny: Evidence and interpretation from the cranioskeletal system. In F. S. SZALAY, M. J. NOVACEK & M. C. McKENNA (Eds.). Mammal phylogeny. Mesozoic differentiation, multituberculates, monotremes, early therians, and marsupials. 216–242. Springer-Verlag. New York, Berlin, Heidelberg.

SZALAY, F. S. (1994). Evolutionary history of the marsupials and an analysis of osteological characters. Cambridge University Press. Cambridge.

TESCHE, M. & H. GREVEN (1989). Primary teeth in Anura are nonpedicellate and bladed. Z. zool. Syst. Evolut.-forsch. **27**, 326–329.

TEUCHERT, G. (1968). Zur Fortpflanzung und Entwicklung der Macrodasyoidea (Gastrotricha). Z. Morph. Tiere **63**, 343–418.

TEUCHERT, G. (1974). Aufbau und Feinstruktur der Muskelsysteme von Turbanella cornuta Remane (Gastrotricha, Macrodasyoidea). Mikrofauna Meeresboden **39**, 1–26.

TEUCHERT, G. (1977). The ultrastructure of the marine gastrotrich Turbanella cornuta Remane (Macrodasyoidea) and its functional and phylogenetical importance. Zoomorphologie, **88**, 189–246,

THULBORN, R. A. (1984). The avian relationships of Archaeopteryx, and the origin of birds. Zool. J. Linn. Soc. **82**, 119–158.

TRAVIS, P. B. (1983). Ultrastructural study of body wall organization and Y-cell composition in the Gastrotricha. Z. zool. Syst. Evolut.-forsch. **21**, 52–68.

TRUEB, L. & R. CLOUTIER (1991). A phylogenetic investigation of the inter- and intrarelationships of the Lissamphibia (Amphibia: Temnospondyli). In H.-P. SCHULTZE & L. TRUEB. Origins of the higher groups of tetrapods. Controversy and consensus. 223–313. Cornell University Press, Ithaca.

TURBEVILLE, J. M., SCHULTZ, J. R. & R. A. RAFF (1994). Deuterostome phylogeny and the sister group of the chordates: Evidence from molecules and morphology. Mol. Biol. Evol. **11**, 648–655.

TZSCHASCHEL, G. (1979). Marine Rotatoria aus dem Interstitial der Nordseeinsel Sylt. Mikrofauna Meeresboden **71**, 1–64.

UNWIN, D. M. (1998). Feathers, filaments and theropod dinosaurs. Nature **391**, 119–120.

VAN CLEAVE, H. J. (1949). Morphological and phylogenetic interpretations of the cement glands in the Acanthocephala. J. Morphol. **84**, 427–457.

VOGEL, W. O. P. (1985). Systemic vascular anastomoses, primary and secondary vessels in fish, and the phylogeny of lymphatics. In K. JOHANSEN and W. W. BURGGREN (Eds.). Cardiovascular shunts. Alfred Benzon Symposium **21**, 143–159.

VOGEL, W. O. P., G. M. HUGHES & U. MATTHEUS (1998). Non-respiratory blood vessels in Latimeria gill filaments. Phil. Trans. R. Soc. London B, **353**, 465–475.

VOGEL, W. O. P. & U. MATTHEUS (1998). Lymphatic vessels in lungfishes. I. The lymphatic vessels system in Lepidosireniformes. Zoomorphology **117**, 199–212.

VORONOV, D. A. & Y. R. PANCHIN (1998). Cell lineage in marine nematode Enoplus brevis. Development **125**, 143–150.

WADA, H., KOMATSU, M. & N. SATOH (1996). Mitochondrial rDNA phylogeny of the Asteroidea suggests the primitiveness of the Paxillosida. Molec. Phyl. Evol. **6**, 97–106.

WADDELL, P. J., OKADA, N. & M. HASEGAWA (1999). Towards resolving the interordinal relationships of placental mammals. Syst. Biol. **48**, 1–5.

WÄGELE, J. W., ERIKSON, T., LOCKHART, P. & B. MISOF (1999). The Ecdysozoa: Artifact or monophylum? J. Zool. Syst. Evol. Research **37**, 211–223.

WAKE, M. H. (Ed.). (1979). Hyman's comparative vertebrate anatomy. The University of Chicago Press. Chicago and London.

WAKE, M. H. (1985). The comparative morphology and evolution of the eyes of caecilians (Amphibia, Gymnophiona). Zoomorphology **105**, 277–295.

WEBB, J. E. & M. B. HILL (1958). The ecology of Lagos Lagoon. **IV**. On the reactions of Branchiostoma nigeriense to its environment. Phil. Trans.Roy. Soc. B, **241**, 355–391.

WERNER, B. (1959). Das Prinzip des endlosen Schleimfilters beim Nahrungserwerb wirbelloser Meerestiere. Int. Rev. ges. Hydrobiol. **44**, 181–216.

WERNER, B. & E. WERNER (1954). Über den Mechanismus des Nahrungserwerbs der Tunicaten, speziell der Ascidien. Helgoländer wiss. Meeresunters. **5**, 57–92.

WHETSTONE, K. N. (1983). Braincase of mesozoic birds: I. New preparation of the „London" Archaeopteryx. J. Vertebr. Palaeont. **2**, 439–452.

WIEDERMANN, A. (1995). Zur Ultrastruktur des Nervensystems bei Cephalodasys maximus (Macrodasyida, Gastrotricha). Microfauna Marina **10**, 173–233.

WILEY, E. O. (1976). The phylogeny of fossil and recent gars (Actinopterygii: Lepisosteidae). Univ. Kansas Mus. Nat. Hist. Misc. Publ. **64**, 1–11.

WILKIE, I. C. (1988). Design for disaster: the ophiuroid intervertebral ligament as a typical mutable collagenous structure. In R. D. BURKE, P. V. MLADENOV, P. LAMBERT & R. L. PARSLEY (Eds.). Echinoderm Biology, 25–38. A. A. Balkema. Rotterdam, Brookfield.

WILKIE, I. C. & R. H. EMSON (1988). Mutable collagenous tissues and their significance for echinoderm palaeontology and phylogeny. In C. R. C. PAUL & A. B. SMITH (Eds.). Echinoderm phylogeny and evolutionary biology, 311–330. Clarendon Press. Oxford.

WILLIAMS, A., CARLSON, S. J., BRUNTON, C. H. C., HOLMER, L. E. & L. POPOV (1996). A supra-ordinal classification of the Brachiopoda. Phil. Trans. R. Soc. Lond. B **351**, 1171–1193.

WILLIAMS, A., JAMES, M. A., EMIG, C. C., MACHAY, S. & M. C. RHODES (1997). Anatomy. In WILLIAMS, A., BRUNTON, C. H. C. & S. J. CARLSON. Part H. Brachiopoda. Revised. Vol. I: Introduction, 7–188. Treatise on Invertebrate Palaeontology. The Geological Society of America & The University of Kansas. Boulder, Colorado and Lawrence, Kansas.

WOOLLACOTT, R. M. & F. W. HARRISON (1997). Introduction. In F. W. HARRISON & R. M. WOOLLACOTT (Eds.). Microscopic Anatomy of Invertebrates **13**, 1–11. Wiley-Liss, Inc. New York.

XU, X., WANG, X.-L. & X.-C. WU (1999). A dromaeosaurid with a filamentous integument from the Yixian formation of China. Nature **401**, 262–266.

YALDEN, D. W. (1985). Feeding mechanisms as evidence for cyclostome monophyly. Zool. J. Linn. Soc. **84**, 291–300.

YOUNG, J. Z. (1995). The life of vertebrates. Clarendon Press. Oxford.

ZELLER, U. (1989). Die Entwicklung und Morphologie des Schädels von Ornithorhynchus anatinus (Mammalia: Prototheria: Monotremata). Abh. senckenb. naturf. Ges. **545**, 1–188.

ZRZAVÝ, J., MIHULKA, S., KEPKA, P., BEZDĚK, A. & D. TIETZ (1998). Phylogeny of the Metazoa based on morphological and 18S ribosomal DNA evidence. Cladistics **14**, 249–285.

Index

- Standard print of page numbers is leading to statements in the continual text.
- Page numbers printed in bold refer to autapomorphies of the mentioned taxa.
- Page numbers printed in italics are pointing to names in a diagram of phylogenetic kinship as well as in the illustrations.
- Names of non-monophyletic taxa are in quotation marks.
- An obelisk marks fossil taxa.

Printing (Computer to Plate): Saladruck Berlin
Binding: Stürtz AG, Würzburg